ADVANCES IN TRIAZOLE CHEMISTRY

ADVANCES IN TRIAZOLE CHEMISTRY

TAHIR FAROOQ
Assistant Professor,
Department of Applied Chemistry,
Government College University,
Faisalabad, Pakistan

ELSEVIER

Elsevier
Radarweg 29, PO Box 211, 1000 AE Amsterdam, Netherlands
The Boulevard, Langford Lane, Kidlington, Oxford OX5 1GB, United Kingdom
50 Hampshire Street, 5th Floor, Cambridge, MA 02139, United States

Library of Congress Cataloging-in-Publication Data
A catalog record for this book is available from the Library of Congress

British Library Cataloguing-in-Publication Data
A catalogue record for this book is available from the British Library

ISBN: 978-0-12-817113-4

Publisher: Susan Dennis
Acquisitions Editor: Emily McCloskey
Editorial Project Manager: Ruby Smith
Production Project Manager: Sruthi Satheesh
Cover Designer: Miles Hitchen

Typeset by SPi Global, India

Dedication

I dedicate this piece of work to my lovely kids (Zoraiz and Areshva) and late father who didn't live long enough to see me prosper.

Contents

Contributors

Tahir Farooq
Department of Applied Chemistry, Government College University, Faisalabad, Pakistan

Amjad Hameed
Nuclear Institute for Agriculture and Biology (NIAB), Faisalabad, Pakistan

Arruje Hameed
Department of Biochemistry, Government College University, Faisalabad, Pakistan

Ali Raza
Department of Applied Chemistry, Government College University, Faisalabad, Pakistan

Preface

Triazoles are five-membered, π-excessive, heteroaromatic ring structures containing two pyridine-type and one pyrrole-type ring nitrogen atoms. Both 1,2,3- and 1,2,4-isomers are widespread in biologically active compounds and functional materials, and as a consequence their synthesis, reactivity, and properties are of high interest.

Prior to the pioneering work of Huisgen developing 1,3-dipolar cycloaddition reactions in the 1960s, synthetic routes to 1,2,3-triazoles were limited and the products themselves largely unremarked in the chemical literature save for their high nitrogen content and consequent potential for exothermic decomposition. That changed with the Munich group's discovery that 1,2,3-triazoles can be readily accessed via the thermal, uncatalyzed [3+2]-cycloaddition reaction between an azide and an alkyne. This reaction, and 1,3-dipolar cycloadditions in general, enjoyed high visibility in the organic chemistry community in part due to interest in understanding the mechanism of the reactions. This mechanistic interest endured through the 1970s with the advent of frontier molecular orbital (FMO), and particularly from the 1980s triazoles began to find applications as constituents of new materials and as synthetic intermediates such as the acyl benzotriazoles popularized by Katritsky. Then, in 2002, Meldal and Sharpless independently reported the copper-catalyzed variant of the azide/terminal alkyne [3+2]-cycloaddition—a reaction that became known as the click reaction. Whereas the thermal azide/alkyne cycloaddition generally furnishes mixtures of regioisomers, the copper-catalyzed reaction provides only 1,4-disubstituted 1,2,3-triazoles in high yields and under mild conditions. Suddenly, interest in, and application of, triazoles in almost all areas of chemistry exploded! Notably, as azides and terminal alkynes are largely absent in living systems, click reactions with unnatural azide and alkyne-tagged biomolecules proved to be hugely versatile for performing biorthogonal "ligation" or "coupling" processes in biological fluids, with the 1,2,3-triazole linkage being considered a "bioisostere" for a secondary amide bond. In 2004, developments in 1,2,3-triazole synthesis came full-circle as Bertozzi introduced strain-promoted, copper-free click reactions of cyclooctyne derivatives with azides: biorthogonal Huisgen couplings!

Methods for the synthesis of 1,2,4-triazoles are generally via classical condensation reactions in which one reactant already contains a hydrazine (i.e., N–N bond-containing) moiety. The evolution of methods that are increasingly efficient, versatile, and mild and furthermore allow for regioselective introduction of substituents has been fueled by the ubiquity of these heterocycles in compounds of interest in the pharmaceutical and agrochemical sectors. Medicinal interest in 1,2,4-triazoles has centered on their antibacterial and antifungal properties with the discovery by Pfizer of the antifungal blockbuster fluconazole in 1981 being a notable milestone. The agrochemical sector's fascination with these "symmetric" triazoles as herbicides and fungicides was also intense in the 1970s and 80s following Bayer's discovery of the cereal protective triazole fungicide triadimefon.

In *Advances in Triazole Chemistry*, Tahir Farooq guides the reader expertly through a myriad of methods for the synthesis of triazoles. He explores the use of 1,2,3- and 1,2,4-triazoles as bioisosteres in medicinal chemistry and as novel bases in (oligo)nucleotide chemistry. He also reviews the use of triazoles as components of smart polymers, glycoconjugates, and functionalized nanomaterials as well as highlighting the role of the triazole motif in plant growth regulators, in peptidomimetics, in coordination complexes, and in myriad areas of materials science. It is a feast of chemistry that beautifully illustrates the incredible versatility and power of synthesis to create new structures with designed properties of utility across a broad swath of science.

The humble triazole has come a long way in the last 60 years, and this book provides a great overview of this journey and the plethora of applications for which this five-membered aromatic heterocycle has proven to be a critical structural component. It should prove of value to chemists both wanting to familiarize themselves with this fascinating area of chemistry and wanting to appraise themselves of latest developments.

Alan C. Spivey, FRSC, FRSB, SFHEA
Professor of Synthetic Chemistry
Assistant Provost (Teaching and Learning)
Office 501C, Molecular Sciences Research Hub (MSRH)
80 Wood Lane, London W12 0BZ
White City Campus, Imperial College London

CHAPTER 1

Introduction, Classification, and Synthesis of Triazoles

Tahir Farooq*
Department of Applied Chemistry, Government College University, Faisalabad, Pakistan
*Corresponding author. E-mail: tahirfarooqfsd@gmail.com

Introduction

In recent times, heterocyclic chemistry has been recognized as a most challenging field, but it is still a significantly rewarding forefront because of heterocycles as the major class in organic chemistry.[1] Predominantly, heterocycles are found in biologically active pharmaceuticals, agrochemicals, and most additives and modifiers commonly employed in various industrial applications. Owing to their incredible capacity to accommodate a diverse range of substituents around a core structure, in addition to their characteristic structural features, they have always been under keen consideration of medicinal chemists for the construction of new bioactive compounds. In connection to this, the design and development of nitrogen-rich heterocycles has attracted much interest over the recent past years. Triazoles are the most promising heterocycles exhibiting a broad spectrum of chemical, agrochemical, and biological properties.[2] Ever-increasing worth of this privileged motif has quickened the development of many facile synthetic strategies during the last few years.

Triazole is a five-membered heterocycle containing three nitrogen atoms at 1, 2, and 3 or 1, 2, and 4 positions. Triazole exists in the following forms (Figure 1):

1,2,3-Triazoles

Over the past few years, 1,2,3-triazoles have received much attention by the scientific community as exhibited by its scope of applications in various disciplines including material science, organometallic, combinatorial, and synthetic medicinal chemistry as well as in agrochemicals.[2–4] Furthermore, 1,2,3-triazolic compounds are extensively used as dyes and related materials, corrosion inhibitors, optical brightening agents, and also as photostabilizers.[5]

https://doi.org/10.1016/B978-0-12-817113-4.00005-6

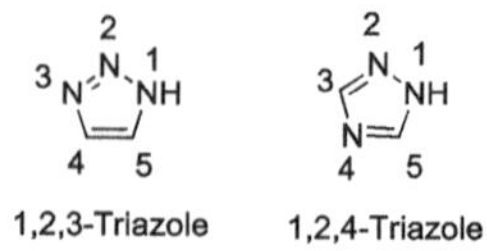

Figure 1 Major types of triazoles.

Indeed, their notable stability across a range of severe conditions has activated their usefulness in pure and applied chemical sciences.[6]

Some of the structural features advantageously required in the context of developing drug delivery systems and for nanomedicine have also been represented by triazole moiety. For example, as a consequence of high aromatic stabilization, they can withstand acid or basic hydrolysis, oxidizing and or reducing conditions. Similarly, this splendid moiety survives over a wide scale of pH in various solvents. This has also been exhibited by its remarkable resistance to various metabolic degradation processes in living systems (Figure 2).[7,8]

In fact, the 1,4-disubstituted 1,2,3-triazoles are not a newly introduced category for medicinal chemists. Before the recent notable developments in strategies for triazole synthesis, more than 7000 1,4-disubstituted 1,2,3-triazolic compounds were known with excellent bioactivities.[9] They were known to display a broad spectrum of activities including selective β_3 adrenergic receptor inhibition,[10] antibacterial,[11] anti-histamine activity,[12] anti-HIV,[13] and potent anti-viral[14] properties. Desired improvements in the pharmacokinetic profiles of antibiotics were achieved by incorporating a triazole motif. The β-lactamase inhibitory potential of tazobactam was found to be dependent on the presence of triazole ring.[15]

In Figure 3 are some of the triazolic drugs that are well-known compounds with their bioactive potential before the advent of modern click chemistry.[9,17,18]

Synthesis of 1,2,3-triazoles

The broad utility of 1,2,3-triaoles across a range of scientific disciplines for the construction of novel molecular architectures has resulted in procedural modifications and development of new synthetic methods

Figure 2 Types of 1,2,3-triazoles.

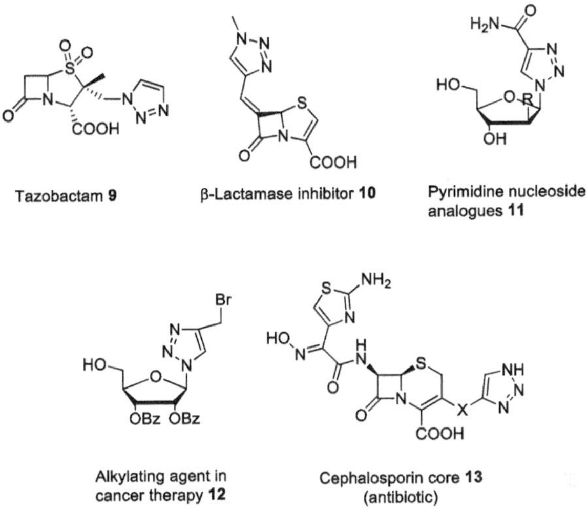

Figure 3 Triazolic-compounds known before advent of CuAAC.[16]

for their construction.[21] Until recently, thermal 1,3-dipolar [3+2]-cycloaddition of alkyne and azide was considered a premier method for the synthesis of 1,2,3-triazoles originally explored by Huisgen during the years 1950 to 1970.[22,23] He thoroughly explored the potential of 1,3-dipolar [3+2]-cycloaddition by using alkynes and azides with a diverse variety of substituents.[22] However, these reactions are non-selective and usually produce a mixture of 1,4- and 1,5-regioisomers, because differently substituted azides and alkynes produce no directing effect (Scheme 1).[22] Furthermore, as demonstrated by the following examples, these reactions require higher temperatures and longer reaction times.

Regioselective approach for 1,4-disubstituted 1,2,3-triazoles

In 2002, both Sharpless[24] and Meldal[25] groups observed a remarkable acceleration of [3+2]-cycloadditions of terminal alkyne and organic azides by up to 10^7 times by introducing Cu(I) salts as catalysts without higher temperature requirements (Scheme 2).[26,27] More importantly, copper-(I)-catalyzed cycloadditions became regioselective and afford only 1,4-regioisomer with minimum or no work-up involvement.[24] These high-yielding regioselective cycloadditions with a capacity to tolerate a range of substituents are regarded as the "cream of the crop".[28] However, Cu(I)-catalysis failed to favor the [3+2]-cycloadditions of internal alkynes with organic azides.

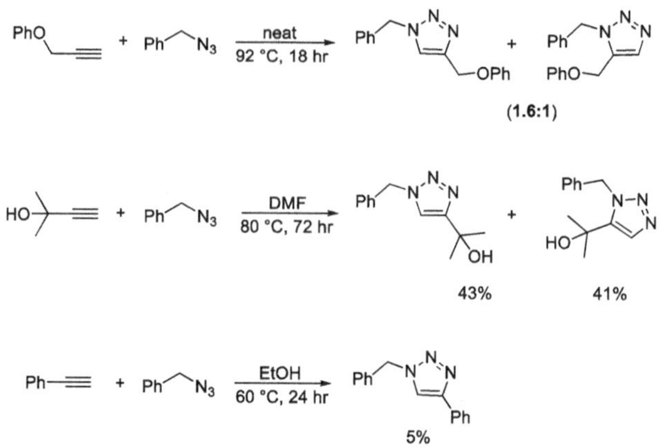

Scheme 1 Non-selective thermal 1,3-dipolar cycloaddition reactions.[19,20]

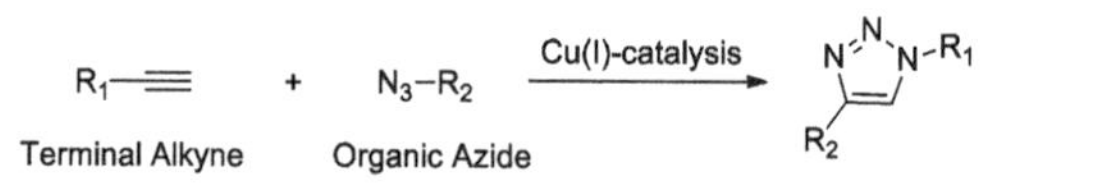

Scheme 2 Cu(I)-catalyzed alkyne azide cycloaddition reactions.

One can perceive observable differences in yields and regioselectivity by comparing the following copper-catalyzed reactions with their aforementioned thermal version (Scheme 3).

Following the distinguished efforts of Sharpless and Meldal, various reports described the synthesis of 1,4-disubstituted 1,2,3-triazoles with different modes to generate active Cu(I) species, as Cu(II) failed to catalyze the cycloaddition reactions. In this connection, different copper(I) salts,[25] in-situ reduction of copper(II),[24] and comproportionation of Cu(II) and Cu(0) were commonly practiced approaches.[31] Such efforts have always been adopted to achieve perfection of copper-based catalysis systems (Scheme 4).

Regioselective synthesis of 1,5-disubstituted 1,2,3-triazoles

The researchers focused the regioselective synthesis of 1,5-disubstituted 1,2,3-triazole after the successful development of CuAAC regioselective approach for the synthesis of 1,4-disubstituted 1,2,3-triazoles. Although according to one existing procedure, the 1,5-regioisomer could be synthesized as a major product by the reaction of bromomagnesium acetylide with organic azides (Scheme 5).[32] However, this method was found limited in scope and lost its versatility.

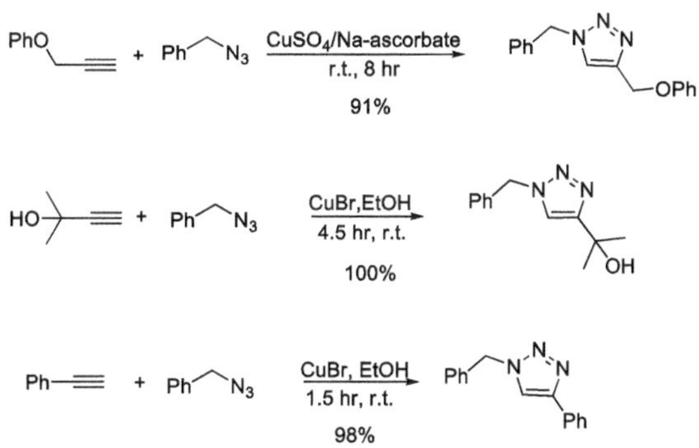

Scheme 3 Copper(I)-catalyzed [3 + 2]-cycloaddition.[29]

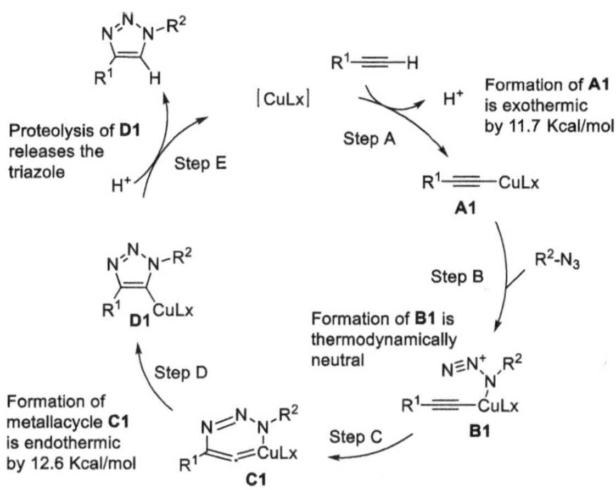

Scheme 4 Mechanistic explanation proposed for Cu(I)-catalyzed cycloaddition.[30]

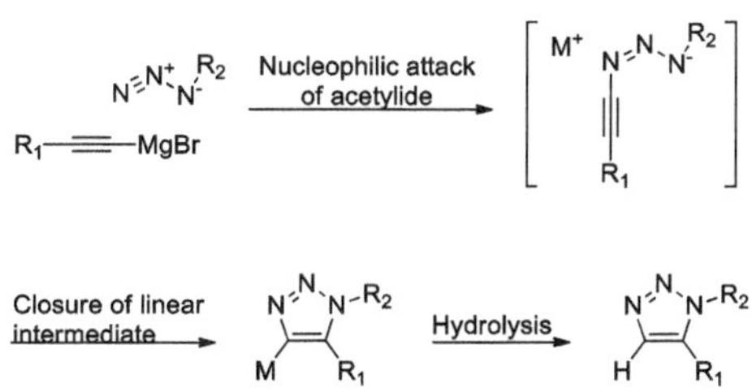

Scheme 5 Reaction of bromomagnesium acetylide with organic azides.

In 2005, the Sharpless group[33] again came up with excellent development regarding the regioselective synthesis of 1,5-disubstituted 1,2,3-triazoles. According to this new methodology, Ru(II)-catalysts direct 1,3-dipolar cycloaddition of azides and alkynes (RuAAC) to produce a complimentary regioisomer, the 1,5-disubstituted 1,2,3-triazole as an exclusive product (Schemes 6 and 7). Furthermore, Ru(II) complexes catalyzed the cycloadditions of both terminal and internal alkynes with organic azides while Cu-catalysis supported cycloadditions of terminal alkynes only.[33]

According to the given mechanism, initial coordination of alkyne and azide gives intermediate **A**, oxidative coupling gives ruthenacycle **B** or **C**, and reductive elimination leads to 1,5-disubstituted 1,2,3-triazole as an exclusive product (Scheme 8).[33,35]

In general, the copper(I)-catalyzed cycloaddtions are not influenced by steric or electronic properties of substituents attached to alkynes or azides. So, alkynes with a variety of substituents react well with azides also carrying different substituents including electron-deficient or electron-rich groups; aliphatic, aromatic, or heterocyclic substituents; and primary, secondary, or tertiary group.[30]

While structural features of azides could significantly affect the outcome of Ru-catalyzed 1,3-dipolar cycloaddtions in terms of regioselectivity and catalytic efficiency, the nature of alkynes never influence such reactions.[33]

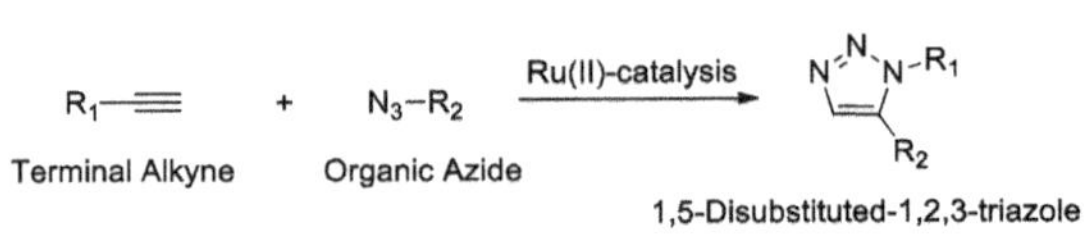

Scheme 6 Ru(II)-catalyzed alkyne azide cycloaddition reactions.

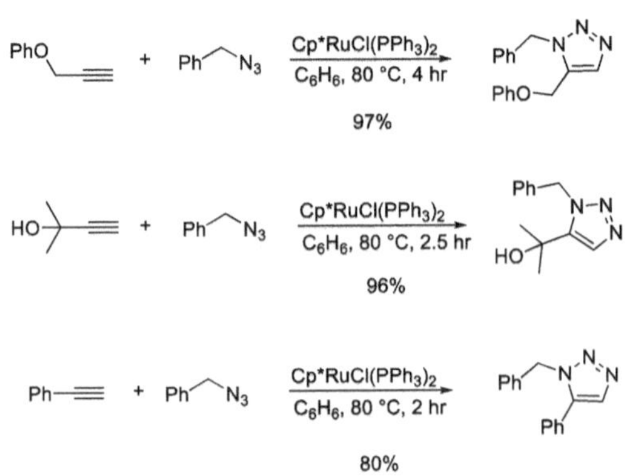

Scheme 7 Ru(II)-catalyzed [3 + 2]-cycloaddition.[33,34]

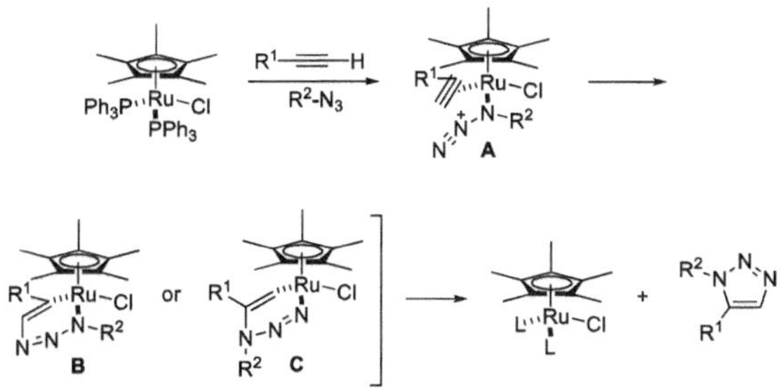

Scheme 8 Proposed mechanism for Ru(II)-catalyzed cycloaddition.

1,4,5-Trisubstituted 1,2,3-triazoles

Ru-catalyzed synthesis of 1,4,5-trisubstituted 1,2,3-triazoles

In addition to earlier well-explained, regioselective, Ru-catalyzed methodology for the synthesis of 1,5-disubstituted 1,2,3-triazoles, Sarpless group also reported the synthesis of 1,4,5-trisubstituted 1,2,3-triazoles. Initially, they presented the following single case with symmetrical alkyne (Scheme 9).

In 2006, Weinreb[35] and his team further explored the scope and generality of the ruthenium-catalyzed cycloadditions of unsymmetrical alkynes and alkyl azides (Scheme 10).

As indicated by this outcome, it was observed that during the synthesis of 1,4,5-trisubstituted 1,2,3-triazoles, the substitution patterns of electronically or sterically biased unsymmetrical alkynes principally control the regioselectivity of the product. However, no rational explanation was given explaining the regioisomeric ratios.

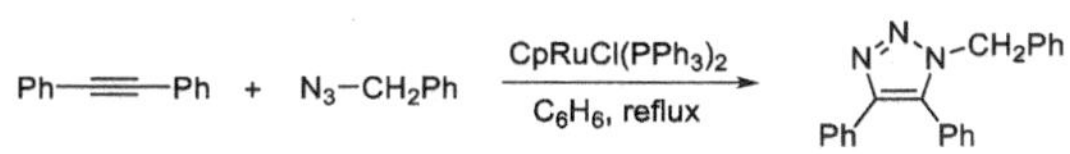

Scheme 9 1,4,5-Trisubstituted 1,2,3-triazoles from symmetrical alkynes.

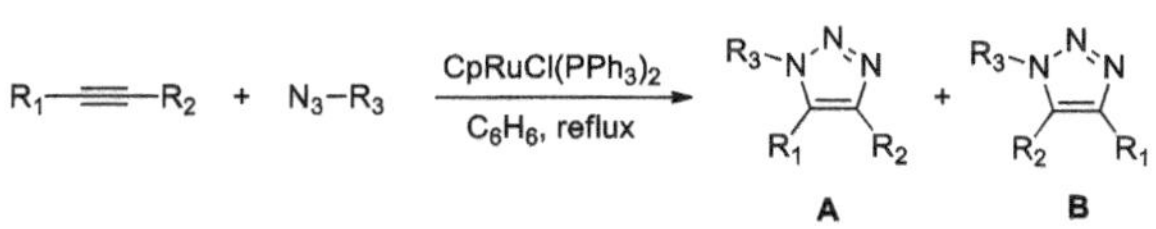

Sr. No	R_1	R_2	R_3	Yield	Ratio (A:B)
1	Ph	Me	CH_2Ph	95	38:62
2	Ph	Pr	CH_2Ph	80	13:87
3	Ph	CO_2Et	CH_2Ph	85	100:0
4	Ph	CO_2Me	CH_2Ph	100	100:0
5	Ph	CH_2OH	CH_2Ph	70	0:100
6	Ph	$CH(OEt)_2$	CH_2Ph	75	50.50
7	Et	CMe_2OH	CH_2Ph	80	0:100
8	Ph	Me	$CHPh_2$	65	33:67
9	Ph	Me	1-ademantyl	10	0:100

Scheme 10 1,4,5-Trisubstituted 1,2,3-triazoles from unsymmetrical alkynes.

Cu-catalyzed synthesis of 1,4,5-trisubstituted 1,2,3-triazoles

The Cu(I)-catalyzed one-pot regiospecific synthesis of 1,4,5-trisubstituted 1,2,3-triazoles using terminal alkynes was reported by Wu and his co-workers.[36] Initially, this rational method explained the synthesis of 5-iodo-1,4-disubstituted 1,2,3-triazoles, which subsequently could be proceeded to a variety of 1,4,5-trisubstituted 1,2,3-triazoles (Scheme 11).

In these cases, the presence of amine ligand was found as a key factor for the successful completion of reaction. Furthermore, the structural features of amine ligands seriously influence the rate and chemoselectivity of the reactions.

Later a practical, controlled, and operationally facile approach for the regio- and o-selective synthesis of 5-iodo-1,4-disubstituted 1,2,3-triazoles and their transformation to 1,4,5-trisubstituted 1,2,3-triazoles at significant high rates was well explained by Sharpless and his co-researchers.[37] In their comprehensive study, they proved the generality of the reaction with notable solvent and functional group compatibility and broad substrate scope.

This method effortlessly provides 5-iodo-1,2,3-triazoles, which serve as useful intermediates and are functionalized into a variety of useful products. The 5-iodotriazoles could be transformed exclusively into 1,4,5-triaryl-1,2,3-triazoles when they react with arylbronic acid under Pd^0-catalyzed cross-coupling reactions. The intermediates are not isolated or purified and are directly converted to corresponding substituted 1,2,3-triazoles. It is regarded as a perfect regiocontrolled reaction as the Ru-catalyzed cycloaddtion reactions fail to meet such high regioselectivity requirements.

Different sets of reaction conditions with a variety of Cu(I) and Cu(II) salts, amine ligands, and solvent systems were employed to optimize reaction condition for the synthesis of corresponding 5-iodo-1,4,5-trisubstituted 1,2,3-triazoles from iodoalkynes and azides. The presence of amine ligands

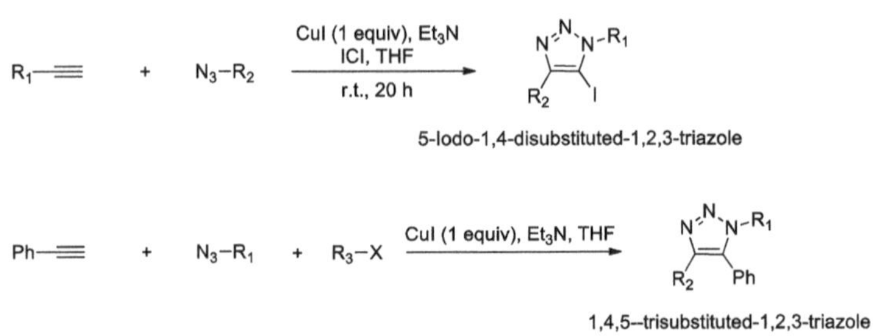

Scheme 11 Cu-catalyzed preparation of 1,4,5-trisubstituted 1,2,3-triazoles.

was detected as a key factor for smooth and faster product formation. This was confirmed by no product formation in preliminary experiments run without TEA. Maximum product formation was observed when TEA was used in excess (2 equiv.). Furthermore, the chemical nature of ligand was also found influential in controlling the reaction rate and chemoselectivity. This trend was observed with low triazole production when 1,2-diamines were employed instead of TEA (Scheme 12).

However, 5-iodotriazoles were produced in excellent yields when tris((1,2,3-triazolyl)-methyl)-amine efficiently promoted the cycloadditions. The tris ligands like TBA and TTTA proficiently accelerated the product formation, and the reaction completed in only 45 min instead of 6 h. The ligands, in fact, accelerate the rate of triazole-forming route and consequently results in high chemoselectivity. The facile, chemoselective, and high production of 5-iodo-1,2,3-triazoles with TTTA recognized it as the ideal ligand for such chemical transformations (Figure 4). The cycloaddition catalyst systems CuI-TEA and CuI-TTTA showed considerable compatibility with a range of solvents.

This Cu(I)-catalyzed cycloaddition seems to follow the CuAAC pathway, however, the copper activates the terminal alkynes and iodoalkynes in an entirely different pattern (Scheme 13). The known possible mechanistic pathway is outlined as follows.

Moses and his team[38] further explored the regiospecific preparation of 5-iodo-1,2,3-triazoles after the remarkable presentation by Sharpless and his research team. They also found CuI-TTTA system th best suitable for the aforementioned cycloadditions (Scheme 14). The plausible mechanism was outlined as follows,

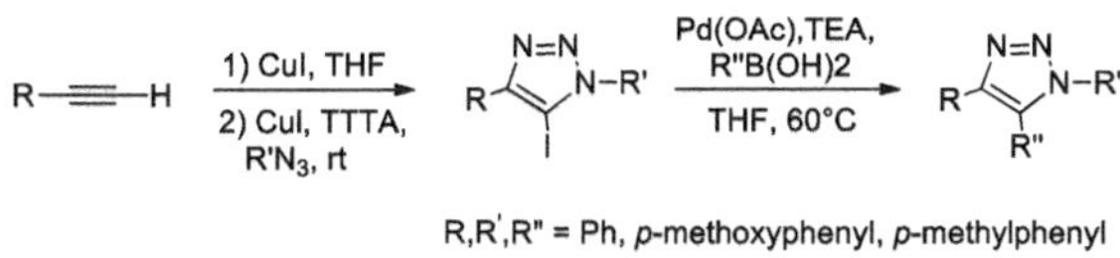

Scheme 12 One-pot three step synthesis of 1,4,5-trisubstituted 1,2,3-triazoles.

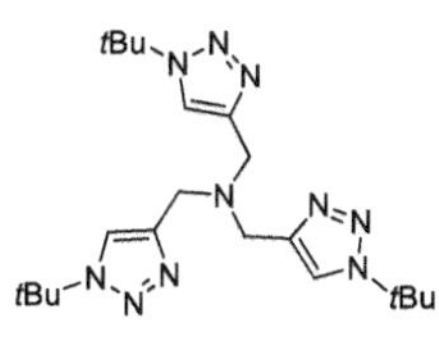

Figure 4 Chemical structure of TTTA.

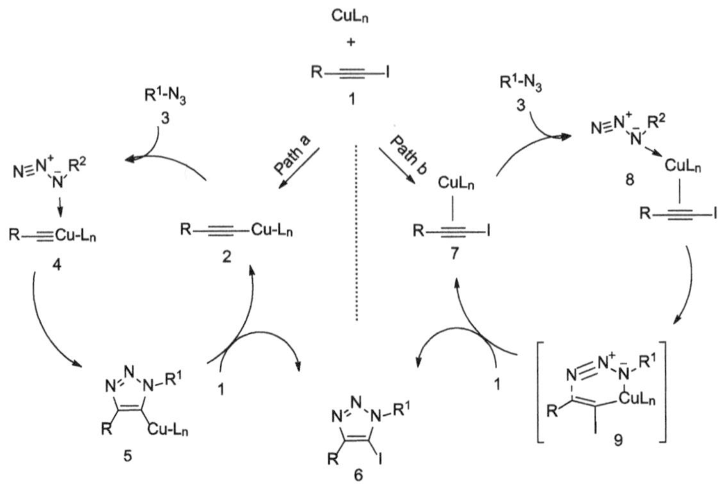

Scheme 13 Suggested mechanism for Cu[I]-catalyzed azide-iodoalkyne cycloadditions.

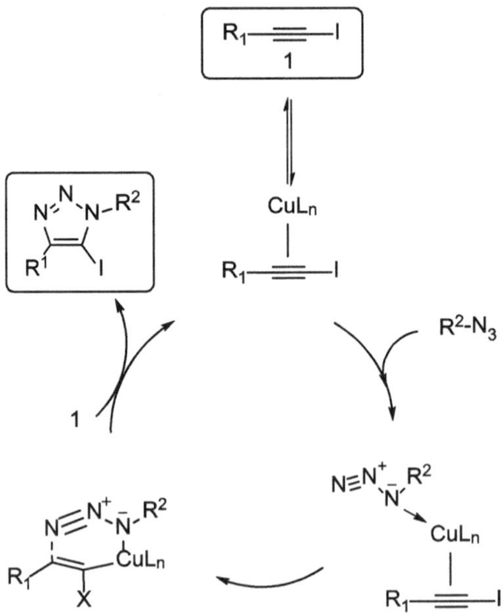

Scheme 14 Proposed mechanism for synthesis of 5-iodo-1,2,3-triazole.

In addition to the catalyst system, the presence of iodonating agent directs the Cu(I)-catalyzed cycloaddition between terminal alkyne and azide to produce iodotriazole as a major product. Whereas, the protiotriazole 5-protio-1,2,3-triazole appears as a side product.

It has been reported that reactions initiating with terminal alkyne accepts the electrophilic iodination of a Cu^1-triazolide intermediate as a compulsory step. The selective approaches for exclusive synthesis of 5-phosphonate, 5-chloro-, and 5-amino-1,2,3-triazoles have been well explained by taking into account the electrophilic additions to Cu^1-triazolide intermediates (Scheme 15).

The Barsoum's group[39] very recently argued that iodotriazoles are obtained through the iodination of Cu^1-triazolide intermediates. They explained that the protiotriazole formation starts after the whole consumption of iodoalkyne (Scheme 16).

The following exemplary reaction was followed by ^1H NMR after specific intervals of time. The speedy development of iodoalkyne further directs the selective synthesis of iodotriazoles with fractions of protiotriazoles. In fact, at the start of reaction, the controlled formation of iodoalkyne could lead to complete conversion and exclusive production of iodotriazole. Such observations were recorded for the following reaction (Scheme 17).

A commonly suggested mechanistic pathway for CuAAC proposes Cu^1-acetylide as the first and triazolide as the late intermediate. The Cu^1-triazolide is then intercepted by iodinating agent to form iodotriazole instead of protiotriazole (Scheme 18).

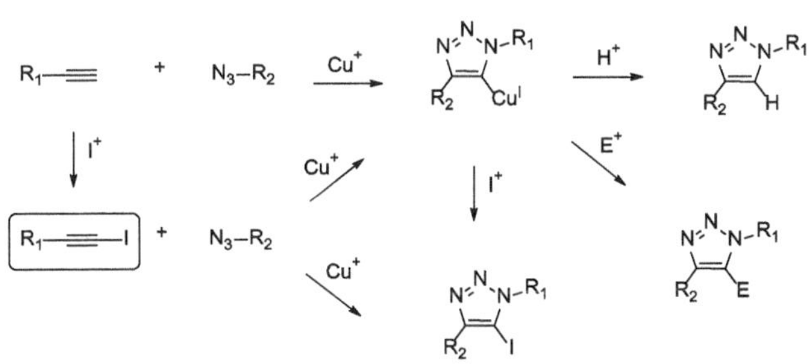

Scheme 15 Postulated pathways to iodotriazoles and related products.

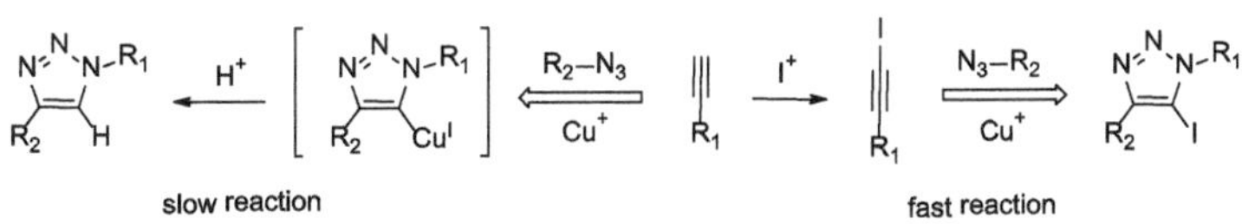

Scheme 16 Mechanistic picture for iodo/protio-triazole formations.

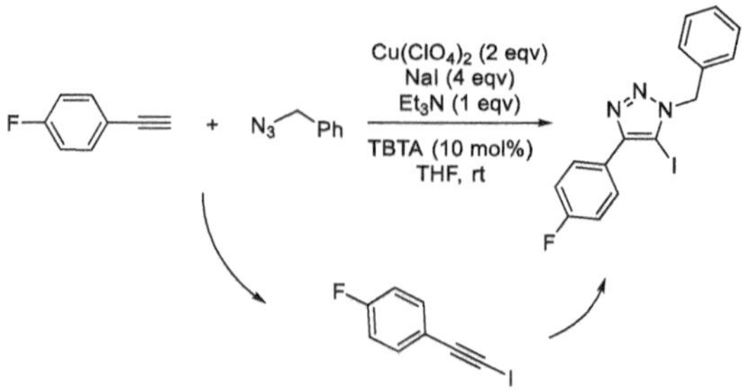

Scheme 17 Synthesis of iodotriazole.

Monosubstituted 1,2,3-triazoles

The monosubstituted 1,2,3-triazoles are also known as N-unsubstituted 1,2,3-triazoles. For their facile synthesis, various synthetic methodologies have been devised due to ever-increasing attraction for compounds encompassing monosubstituted 1,2,3-triazole as functional moiety.

The Cu^1-mediated cycloadditions of sodium azide with non-activated terminal alkynes proceed efficiently to generate reasonable yields of N-unsubstituted 1,2,3-triazolic compounds (Scheme 19).[40]

During the proceedings of the previous reaction at lower temperature conditions, a Cu^1-azide complex is produced. When the temperature of the reaction exceeds 70°C, Cu-alkyne complex is constructed. Subsequently, azide ion in dissociated form or coordinated with copper reacts with Cu^1-alkyne complex to produce only monosubstituted 1,2,3-triazole.

A straightforward preparation of N-unsubstituted 1,2,3-triazoles was achieved by reacting trimethylsilyl azides with nonactivated terminal alkynes under copper-(I)-catalyzed cycloadditions (Schemes 20 and 22).

This methodology helped to tackle the issues related to the practicality of existing methods (Scheme 21).[41]

In Pd-catalyzed cross-coupling reaction, the alkenyl halide and azide reacts to produce monosubstituted 1*H*-1,2,3-triazole.[42] The synthesis of monosubstituted-triazoles instead of expected vinyl azides established a new dimension of Pd-catalyzed cross-coupling reactions (Scheme 23).

In preliminary experiments, *β*-bromostyrene was made to react with sodium azide under the following conditions. Furthermore, the influential role of under given different supporting ligands was studied. In most of the cases, only the starting material was recovered. In the same reaction, quantitative amounts of 1*H*-1,2,3-triazole was produced when xantphos, a bidentate ligand, was used as a supporting ligand.

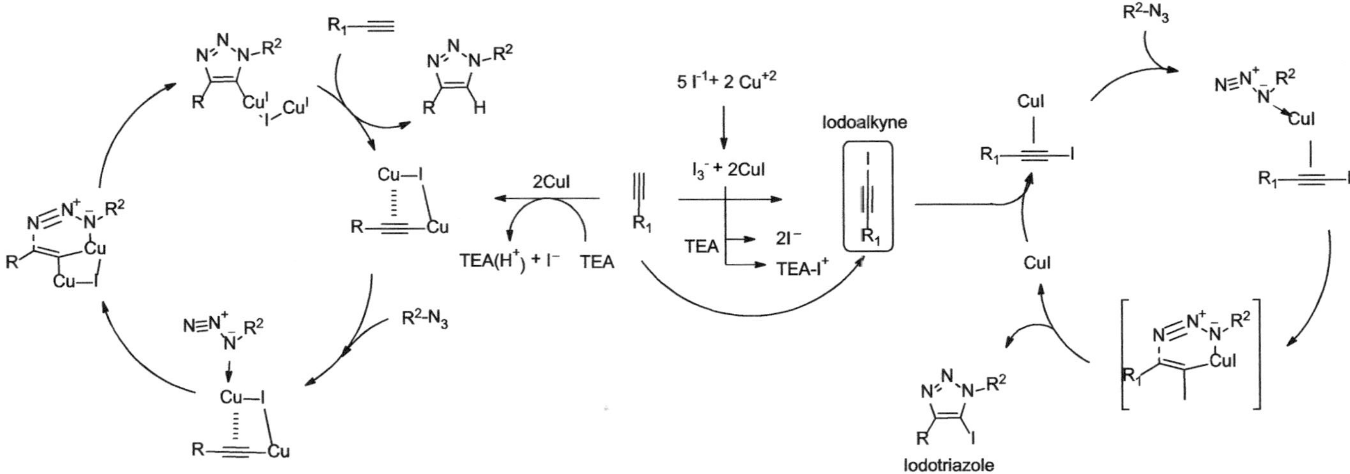

Scheme 18 Mechanistic pathway for iodotriazole and protiotriazole.

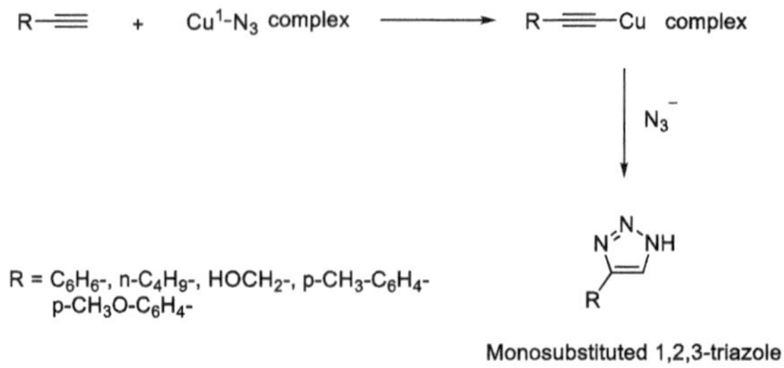

Scheme 19 Cycloadditions of sodium azide with non-activated terminal alkyne.

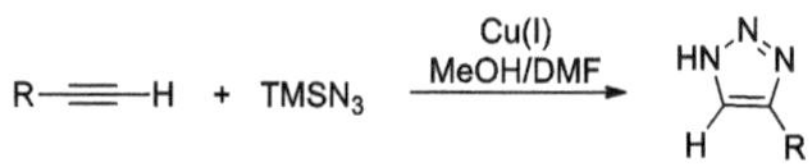

Scheme 20 Preparation of N-unsubstituted 1,2,3-triazoles.

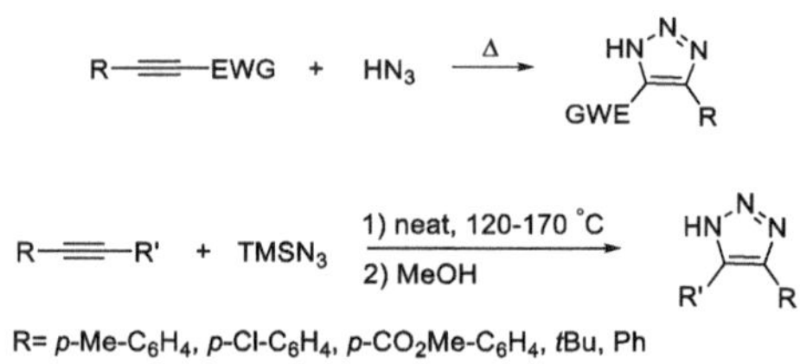

Scheme 21 Monosubstituted 1,2,3-triazole formation by [3 + 2]-cycloadditions.

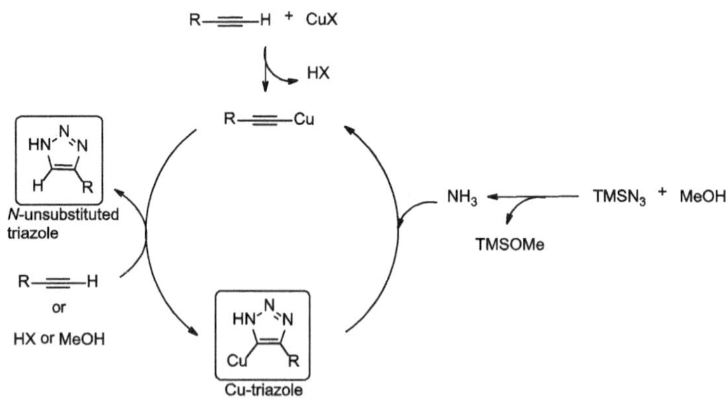

Scheme 22 Mechanistic explanation for monosubstituted 1,2,3-triazole synthesis.

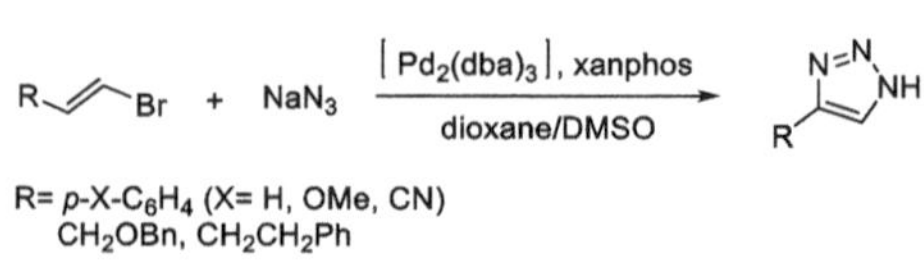

Scheme 23 Preparation of 1*H*-1,2,3-triazoles from β-bromostyrene.

The reaction showed a great dependency on ligand structural specifications. A significant promotion of reaction was observed when diphosphines xantphos and DPEphos with large bite angle were used as supporting ligands, while tri-phenylphosphine showed little influence on reaction rate and product formation. No conversion of reactants was observed when binap, a bidentate phosphine with relatively small bite angle, was tested as a supporting ligand. Similar results were recorded when monophosphines like johnphos, davephos, and xphos were employed (Figure 5).

Again, no product formation was achieved in controlled experiments run without ligands or without metal. Hence the key role of a catalytic system was established for further explorations. The optimized conditions were used to study the scope of reaction. The substituted β-bromostyrenes produced triazoles in near quantitative yields. Differently substituted aromatic ring provided similar results as electron-releasing, electron-withdrawing, and neutral groups did not establish any perceptible difference in the outcome of the reaction. Different functional groups such as nitriles or methyl esters, heteroaromatic rings such as the 2-furan, and orthosubstituents on the ring are well tolerated by the reaction under the aforementioned optimized conditions. The reaction was found to be very chemoselective in nature. The alkyl-substituted bromoethylenes furnish triazolic products under different conditions.

The reaction dynamic was initially explained by a plausible mechanistic pathway (Scheme 24). It takes into account the formation of complex **II** by the oxidative addition of bromoalkene to I. Subsequently, nucleophilic substitution constitutes complex III. Then alkenyl azide IV is produced with release of Pd^0 catalyst as a result of reductive elimination step. In the final

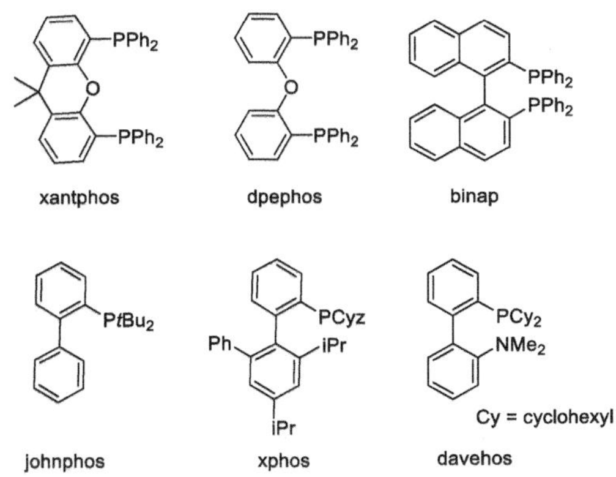

Figure 5 Ligands used for synthesis of 1*H*-1,2,3-triazoles.

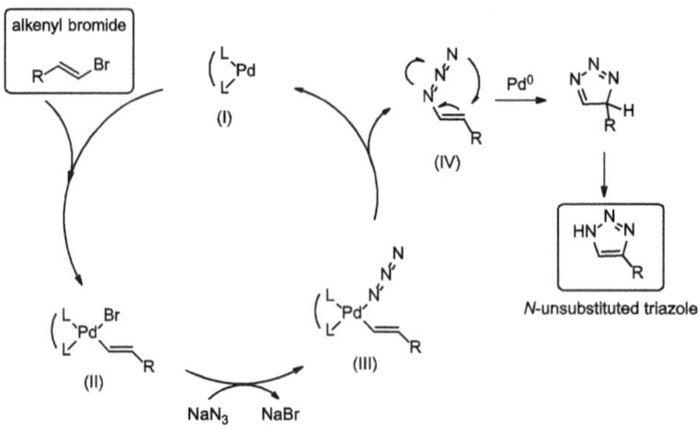

Scheme 24 Mechanistic explanation for synthesis of 1*H*-1,2,3-triazoles.

step, the Pd^0 accelerated 1,5-electrocyclization, and then tautomerization provides 1*H*-1,2,3-triazole as the exclusive product.

Previously, it has been observed that vinyl azides in general do not cyclize on heating to corresponding triazolic product; however, nitriles and 2*H*-azirines are produced along with nitrogen gas. So, if the given mechanism (Scheme 24) is supposed to be followed by the reaction, then vinyl azide should electrocyclize under the catalytic influence of Pd^0. By applying computational inputs, some special experiments were performed for the validation of this hypothetical view.

The occurrence of oxidative addition was evidently confirmed. An easily practicable method disclosed by Buchwald group was adopted to construct a xantphos-legated alkenylpalladium bromide complex II (Scheme 24). Furthermore, the reaction with sodium azide provided the expected triazolic compound. These results confirmed the formation of intermediate II, hence the catalytic oxidative addition was indeed validated.

However, the reductive elimination step has not been supported by experimental studies. By using Pd-xantphos catalyst, attempts were made to produce triazole from styryl azide. Under different sets of conditions, the vinyl azide decomposed and no triazole formation was confirmed. Therefore, it was concluded that reductive elimination might not be responsible for vinyl azide generation; accordingly, the suggestive mechanism (Scheme 25) was discounted under the given circumstances.

In the second mechanistic approach (Scheme 25), vinylpalladium complex II was considered as the intermediate that underwent a [3 + 2] cycloaddition with N_3 through a stepwise or concerted manner to furnish triazole. The triazolide VII and hydridopalladium complex VIII were detected when

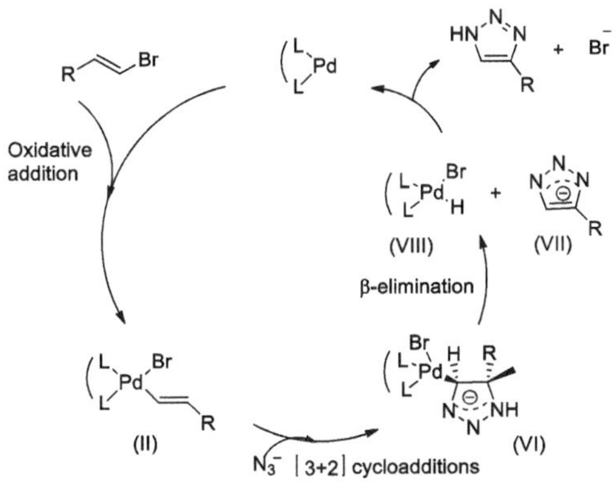

Scheme 25 Triazole formation from alkenyl bromides and sodium azides.

dihydrotriazolylpalladium complex VI underwent β-elimination process. In the final step, the HBr produced from reductive elimination accelerates the triazolyl anion and liberated the Pd^0 complex.

Amantini and his research fellows[43] published a new strategy for the 1*H*-1,2,3-triazoles using electron poor olefin and $TMSN_3$. This TBAF-catalyzed widely applicable approach was considered as an alternative to highly efficient CuAAC reactions. Following the suggested procedure, a number of exemplary 4-aryl-1*H*-1,2,3-triazolic compounds were received in good yield under solvent-free conditions. For these chemically efficient and ecofriendly reactions, the demands of inert atmosphere and dried glassware are not binding at all (Scheme 26).

The reaction conditions were optimized by obtaining the following target molecule under different conditions (Scheme 27). The following table represents the details of such attempts.

The TBAF was found to be an efficient catalyst for [3 + 2]-cycloaddition reactions of electron-poor olefin and $TMSN_3$.

Zhang and his co-workers reported a simplistic one-pot procedure involving *anti*-3-aryl-2,3-dibromopropanoic acids and sodium salt for the synthesis of 4-aryl-1*H*-1,2,3-triazoles. The $Pd_2(dba)_3$ and Xantphos

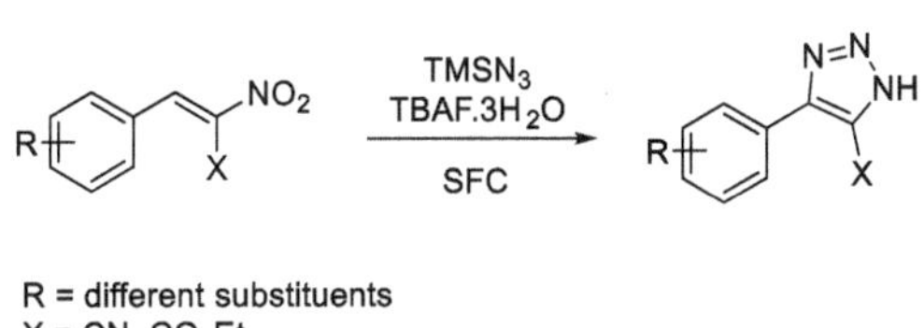

Scheme 26 Synthesis of 4-aryl-1*H*-1,2,3-triazolic compounds.

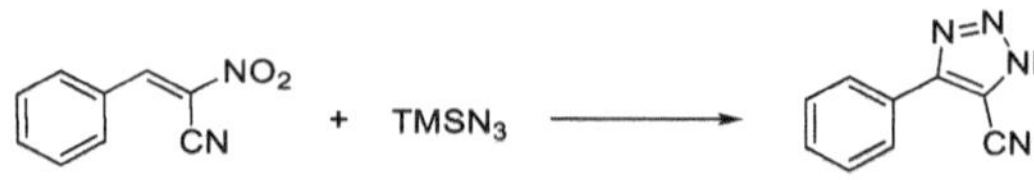

Scheme 27 Reaction optimization for 4-aryl-1*H*-1,2,3-triazolic compound.

catalyzed these reactions to produce triazoles in excellent yield. A number of *anti*-3-aryl-2,3-dibromopropanoic acids were reacted to generate corresponding 4-aryl-1*H*-1,2,3-triazoles with electron-withdrawing or electron-donating substituents (Scheme 28).

The following proposed mechanism represents debromination, decarboxylation, and cycloaddition steps.

The (*Z*)-β-arylvinyl bromide A is generated when 1 undergoes a debrominative decarboxylation step. Then vinylpalladium complex (*Z*)-**B** is instantly constructed from **A**. Subsequently, (*Z*)-**B** isomerizes to more stable (*E*)-**B**, which then undergoes [3+2] cycloadditions with HN_3 to afford dihydrotriazolylpalladium complex **C**. As a result of β-elimination, this complex generates 4-aryl-1*H*-1,2,3-triazole with the release of hydridopalladium complex **D**. In the final step, **D** undergoes reductive elimination with the releasing of HBr to regenerate Pd^0-complex (Scheme 29).

Kamijo and his team[44] managed to prepare triazoles from nonactivated alkynes that were not achievable from other known procedures. In this new strategy, the [3+2] cycloadditions are promoted by a bimetallic Pd^0-Cu^1 catalyst system. The following given three-component-couplings of alkyne, $TMSN_3$ =, and allyl methyl carbonate gives allyltriazoles, which subsequently undergo deallylation to produce corresponding triazoles (Schemes 30–32).

Methods have also been developed to achieve 1*H*-1,2,3-triazoles in high yields by removing substituents on N(1) of triazole. The C4- and C5-monosubstituted-N*H*-1,2,3-triazole were prepared from corresponding

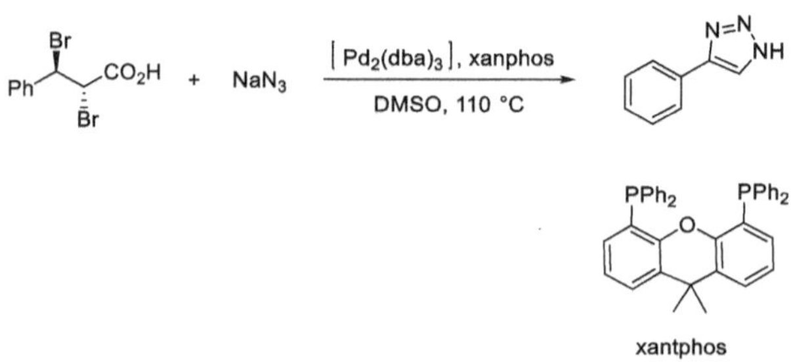

Scheme 28 Synthesis of 4-aryl-1*H*-1,2,3-triazoles from *anti*-3-aryl-2,3-dibromopropanoic acids.

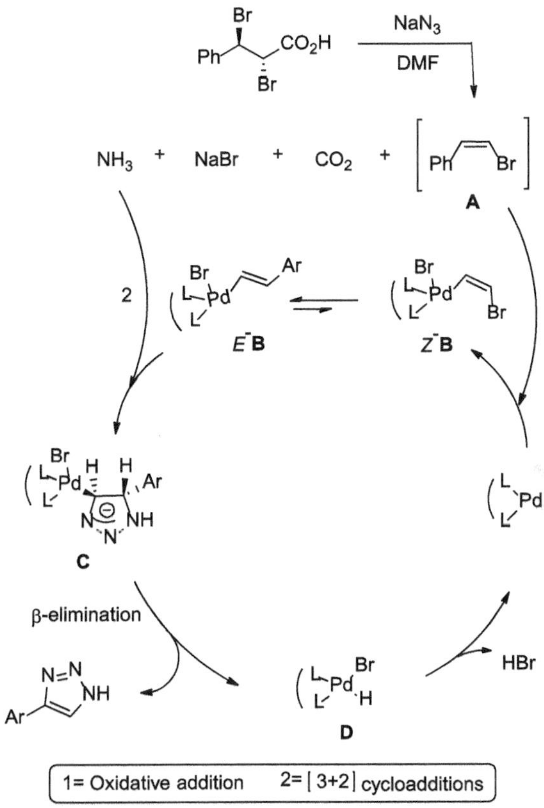

Scheme 29 Proposed path for 4-aryl-1*H*-1,2,3-triazoles from *anti*-3-aryl-2,3-dibromopropanoic acids.

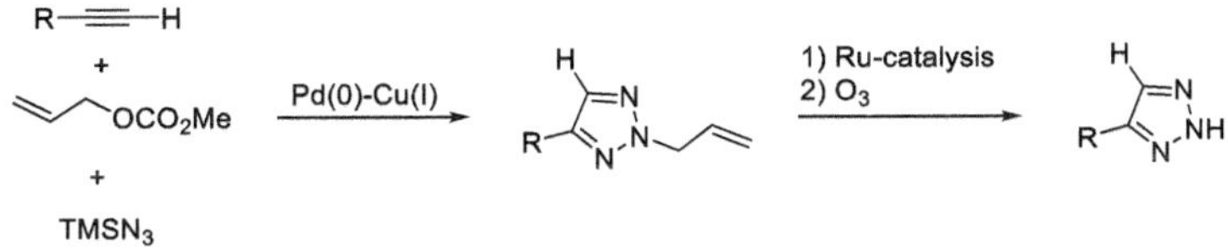

Scheme 30 Synthesis of triazoles using Pd^0-Cu^1 catalyst system.

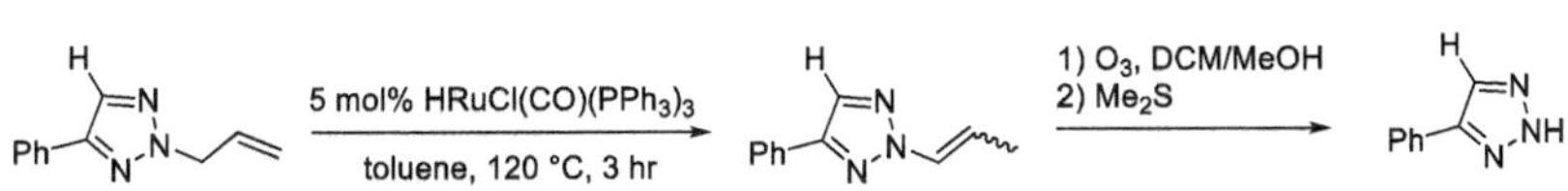

Scheme 31 Conversion of allyltriazole to 1*H*-1,2,3-triazole.

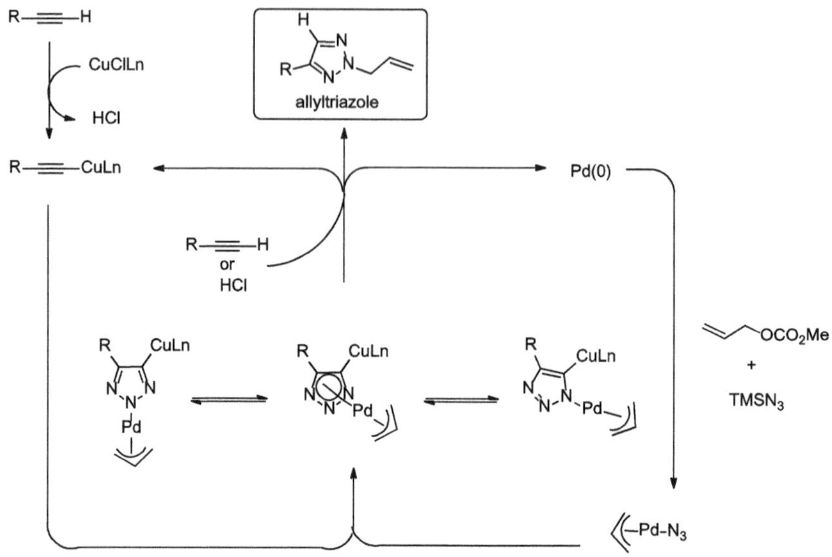

Scheme 32 Schematic proposal for synthesis of triazoles by bimetallic Pd^0-Cu^1 catalyst system.

TSE-protected 1,4- and 1,5-disubstituted 1,2,3-triazoles under slightly basic conditions (Scheme 33).[45]

The *N*-benzyl substituted 1,2,3-triazoles undergo debenzylation by hydrogenolysis to furnish high yields of C4-substituted-N*H*-1,2,3-triazoles (Scheme 34).[46]

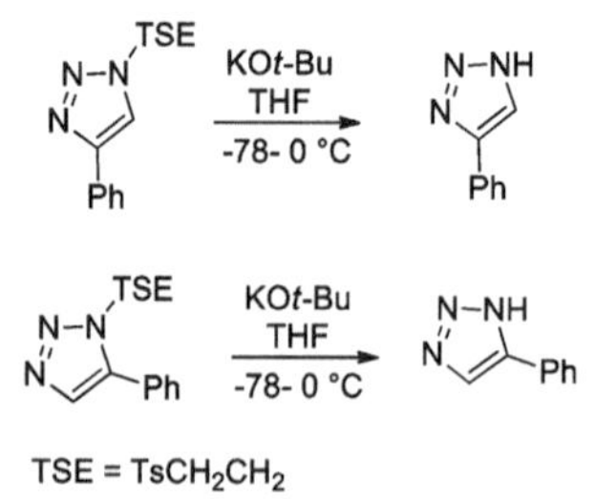

Scheme 33 Removal of TSE for N*H*-1,2,3-triazoles.

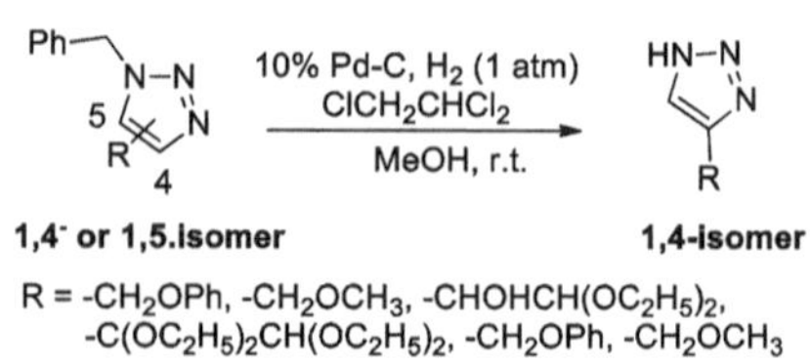

Scheme 34 *N*-Debenzylation of 1,4- and 1,5-disubstituted 1,2,3-triazoles.[46]

The successful debenzylation of 4- and 5-substituted 1-benzyl-1,2,3-triazoles was found to be dependent on alkoxy substituents at C4 or C5. However, triazoles containing alkoxy groups at N-atom did not undergo hydrogenolysis under the same reaction conditions. The alkoxy substituents at C4 or C5 produce only C4-alkoxy substituted-N*H*-1,2,3-triazole after the debenzylation step.

An efficient procedure for the synthesis of 2-substituted 1,2,3-triazoles was reported by Kalisiak *et al.*[47] as a welcome advance in triazole chemistry. The excellent yields of 2-hydroxymethyl-2*H*-1,2,3-triazoles were received when terminal alkynes, formaldehyde, and sodium azide underwent Cu(I)-catalyzed cycloadditions. These triazoles are further convertible to a variety of functional molecules. The N*H*-1,2,3-triazoles are received when hydroxymethyl group is removed under oxidative and reductive conditions (Scheme 35).

Bis-triazoles could be further converted to diarylated molecules according to the procedure presented by Gevogyan *et al.* (Scheme 36).

So the 2*H*-hydroxymethyl-1,2,3-triazoles are easy to generate by employing this effortless one-pot strategy. As a second option, one can also produce them in better yields when N*H*-triazoles undergo hydroxymethylation with HCHO. A range of 2*H*-substituted 1,2,3-triazoles were obtained from hydroxymethyl-triazoles, which in fact are valuable precursors for vital heterocyclic compounds widely used in materials and medicinal fields.

Synthesis of 1,2,4-triazoles

In ethanolic potassium hydroxide, carbohydrazides react with carbon disulfide to generate dithiocarbazate, which further reacts with hydrazine hydrates to furnish 1,2,4-triazoles in excellent yields (Scheme 37).[48]

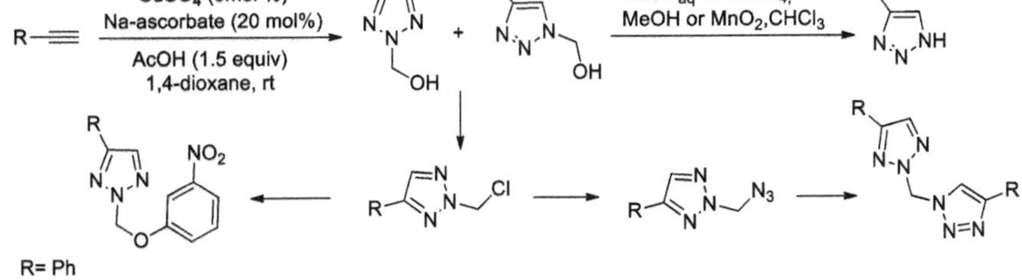

Scheme 35 Synthesis of 2-substituted-2*H*-1,2,3-triazole.

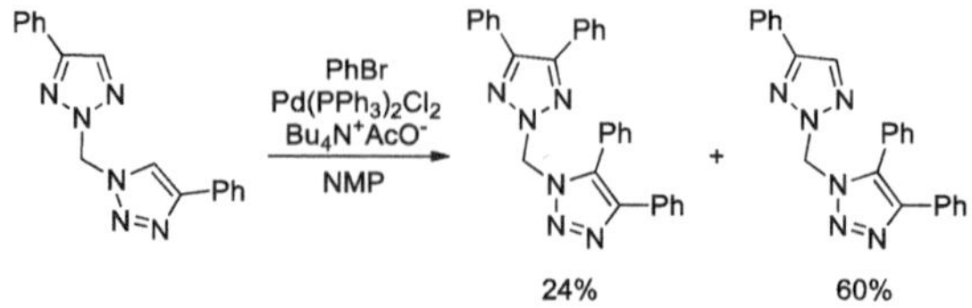

Scheme 36 Pd-catalyzed arylation of bistriazole.

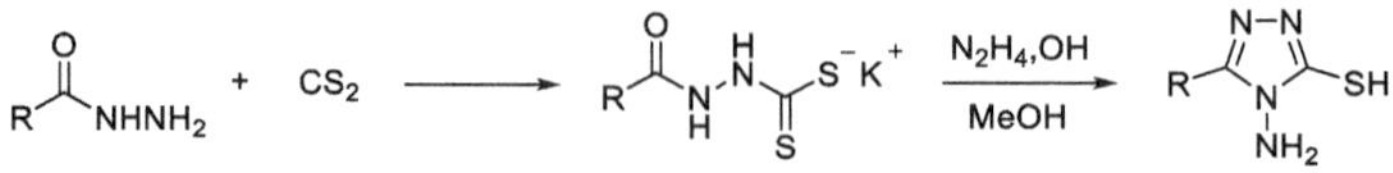

Scheme 37 Synthesis of 3-mercapto-1,2,4-triazoles.

The thiosemicarbazides undergo cyclodehydration in a basic medium to give 1,2,4-triazoles in notable yields (Scheme 38).

Microwave-assisted, solvent-free synthesis of thiadiazolyl-substituted 1,2,4-triazoles were well explored and appropriately described by Kidwai and his co-researchers.[49] On solid support of alumina, 5-substituted-2-amino-1,3,4-thiadiazoles undergo demand-free but superfast insertion reaction with 5-alkyl-2-mercapto-1,3,4-oxadiazoles to produce thiadiazolyl-substituted triazoles in quantitative yields (Scheme 39).

Yeung and his research team[50] came up with an efficiently practicable, facile, and greener route for 3,5-disubstituted 1,2,4-triazoles as exclusive product. Keeping in view the wide-ranging reaction tolerance for different heterocycles and functional groups under proposed conditions, the under given procedure was considered as fairly general in its applications. This base-catalyzed single step procedure involves microwave irradiations of

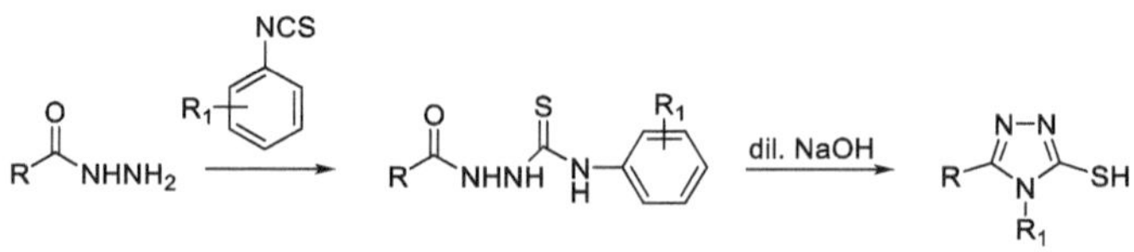

Scheme 38 Cyclodehydration of thiosemicarbazides to 1,2,4-triazoles.

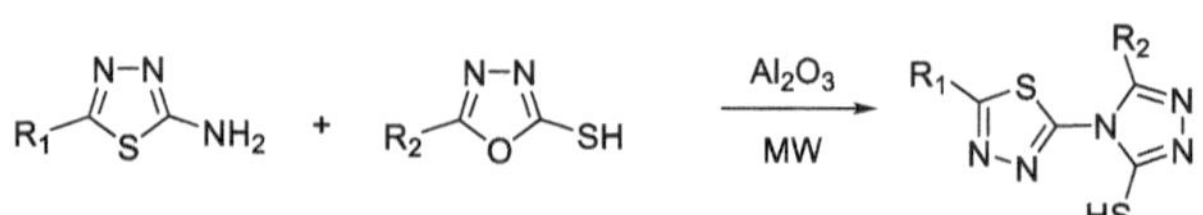

Scheme 39 Efficient synthesis of thiadiazolyl-substituted-triazoles.

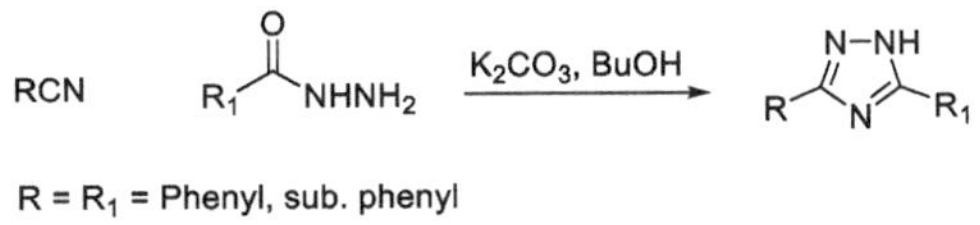

Scheme 40 Synthesis of 3,5-disubstituted 1,2,4-triazole.

nitrile and hydrazide which consequently undergo condensation to produce 3,5-disubstituted 1,2,4-triazole (Scheme 40).

The greener aspect of the reaction was considered for further exploration to achieve 1,2,4-triazoles through an environment friendly processes. In this connection, the use of grinding techniques and microwave irradiation has been focused widely in recent times as they hold a special place in the area of green chemistry. Green chemistry concept helps to achieve the requirements of environmental friendly reactions.

Reddy and his fellows[51] presented the synthesis of 3-mercapto-1,2,4-triazoles from differently substituted coumarins by microwave-accelerated and grinding-facilitated reaction procedure. Consequently, the reaction became effortlessly operational, less demanding, non-damaging, and high yielding with minimum work-up requirements and shorter reaction times. All these specialties painted these reactions as green organic transformations (Scheme 41).

High selectivity, operational simplicity, reusability, and environmental compatibility of solid acidic catalysts such as silica gel, alumina, clays, and zeolites have increased their applications in various areas of synthesis and product isolation in last few decades. Such reactions have the advantages of shorter reaction time, purity of products, and high yields with minimum work-up (Scheme 42).[52]

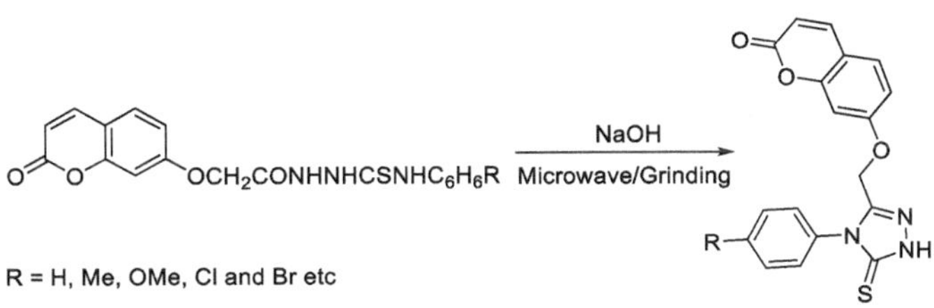

Scheme 41 Synthesis of 3-mercanpto-1,2,4-triazoles from different coumarins.

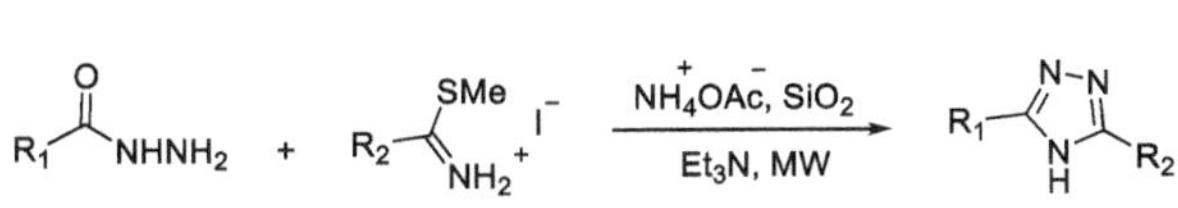

Scheme 42 Silica-supported synthesis of 3,5-disubstituted 1,2,4-triazoles.

A quick and highly regioselective access to a diversity of 1,3,5-trisubstituted 1,2,4-triazoles was presented by Castanedo and his research fellows.[53] This one-pot method involves the reaction of carboxylic acids, primary amidines, and monosubstituted hydrazines. In these reactions, HATU plays a vital role in the couplings of peptide reagents to furnish acylamide intermediate in the presence of diisopropylethylamine (DIPEA) as a base in DMF. A range of substituents could be incorporated at position 5 in the resulting 1,2,4-triazole molecule (Scheme 43).

An effectual approach for direct preparation of 1,2,4-triazole based on the capacity of transition metals to activate the nitriles was well explained by Ueda and his co-workers.[54] This Cu-catalyzed reaction of 2-aminopyridines and nitriles results in N−C bond construction and oxidative coupling of N−N. The molecular oxygen acted as oxidant in catalytic process, and water was generated theoretically as the only by-product This method permits the simplistic synthesis of various bioactive 1,2,4-triazoles (Scheme 44).

For reaction optimization, the 2-aminopyridine and benzonitrile were selected for study (Scheme 44). The reaction was run in different solvents to screen the catalytic activity of different transition metals.

This suggested mechanism takes into account the formation of N−N bond under oxidation conditions (Scheme 45). So, initially amidine might have been formed, which subsequently underwent oxidative N−N bond formation. The Cu through coordinated intermediate manages to promote nucleophilic attack of amino pyridines on nitriles and furnish amidine. Afterward, the triazolopyridine is produced when amidine undergoes

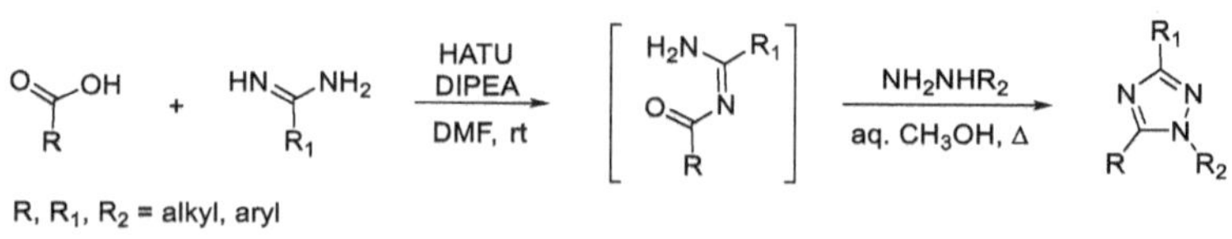

Scheme 43 One-pot synthesis of 1,3,5-trisubstituted 1,2,4-triazoles.

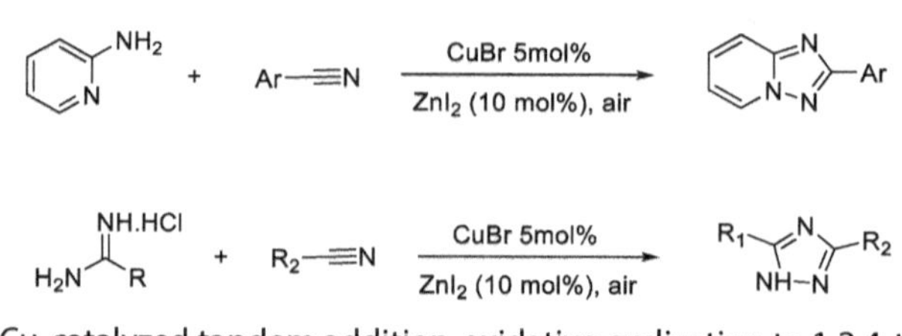

Scheme 44 Cu-catalyzed tandem addition-oxidative cyclization to 1,2,4-triazoles.

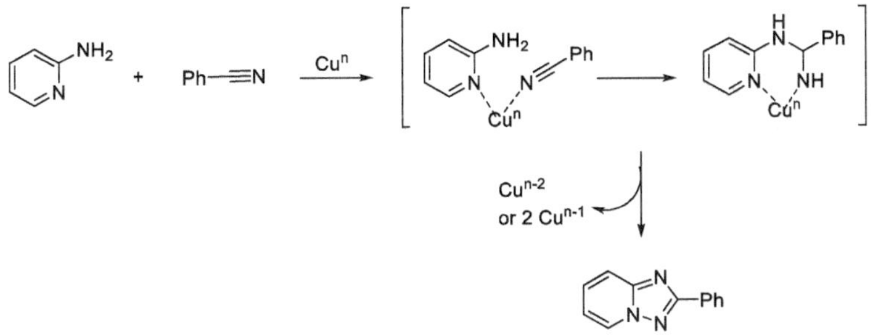

Scheme 45 Suggested mechanism for Cu-catalyzed oxidative synthesis of 1,2,4-triazoles.

oxidative cyclizations promoted by copper. The N–N bond construction under oxidative conditions was also verified by the following independent reaction (Scheme 46).

Yin and his co-researchers[55] developed an advantageous one-pot method for cyanoimidation of aldehydes by employing cyanamide (Scheme 47). The NBS was used as an oxidant, and the reaction proceeded in the absence of a catalyst. In the subsequent step, the 1,2,4-triazoles were achieved in high yields when *N*-cyanobenimidate underwent cyclization reactions.

This methodology was found advantageously economical and high yielding with minimum work-up under milder reaction conditions.

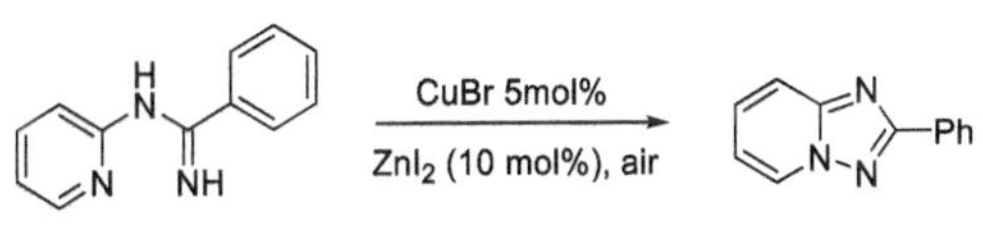

Scheme 46 Conversion of amidine to 1,2,4-triazole.

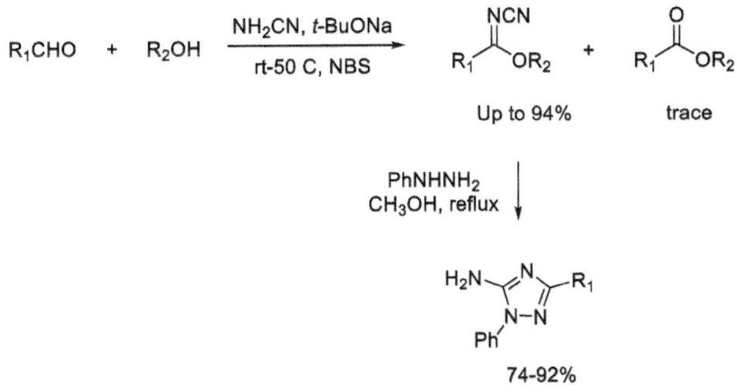

Scheme 47 Preparation of 1,2,4-triazoles from *N*-cyanobenimidates.

A conveniently adoptable and less-demanding method for the synthesis of oxamide-derived amidines in higher yields was introduced by Xu and his fellows.[56] These reagents reacted further with hydrazine hydrochloride salts to furnish 1,5-disubstituted 1,2,4-triazoles with operational simplicity. The generality of the reaction has been well explained as all types of hydrazines were found to quickly react with amidines (Scheme 48).

Taking into account the applications of combinatorial solid-phase synthetic strategies, the synthesis of 3,4,5-trisubstituted 1,2,4-triazoles was achieved by Boegline and his research group from different thioamides and hydrazides in significant higher yields (Scheme 49).[57]

By using different thioamidess and hydrazides, 3,4,5-trisubstituted 1,2,4-triazoles were also prepared in solution form. They were further used for peptidomemtic scaffold preparations (Scheme 50).

A regioselective solid-phase approach to obtain trisubstituted 3-alkylamino-1,2,4-triazolic compounds in high purity was described by

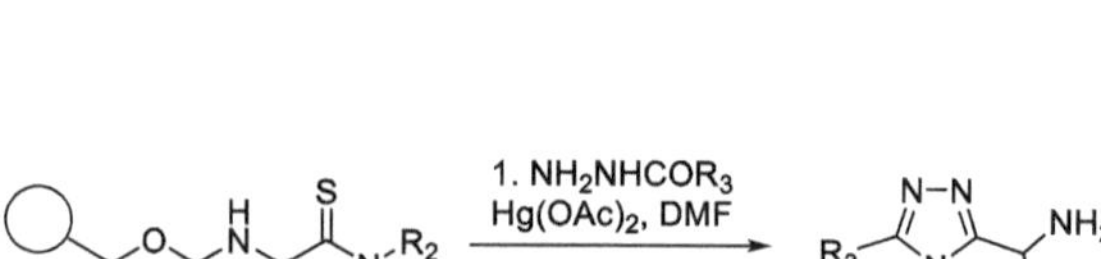

Scheme 48 Synthesis of 1,5-disubstituted 1,2,4-triazoles from hydrazine hydrochlorides.

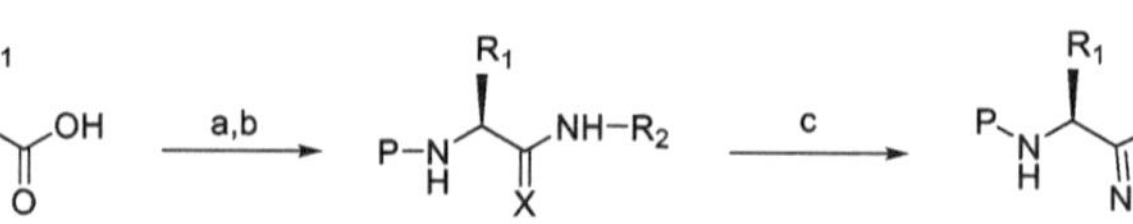

Scheme 49 Solid phase perparation of 3,4,5-trisubstituted 1,2,4-triazoles.

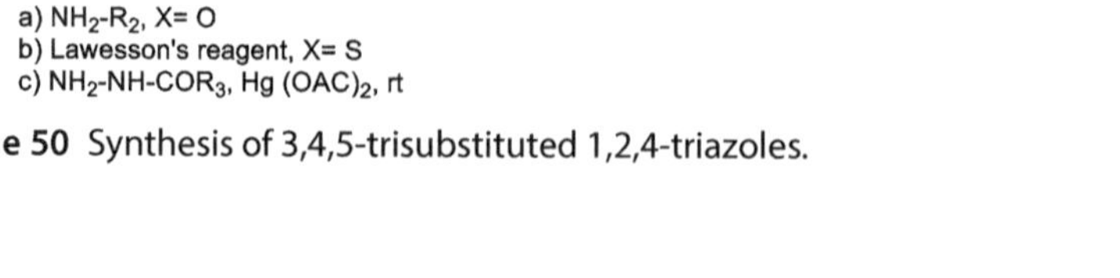

Scheme 50 Synthesis of 3,4,5-trisubstituted 1,2,4-triazoles.

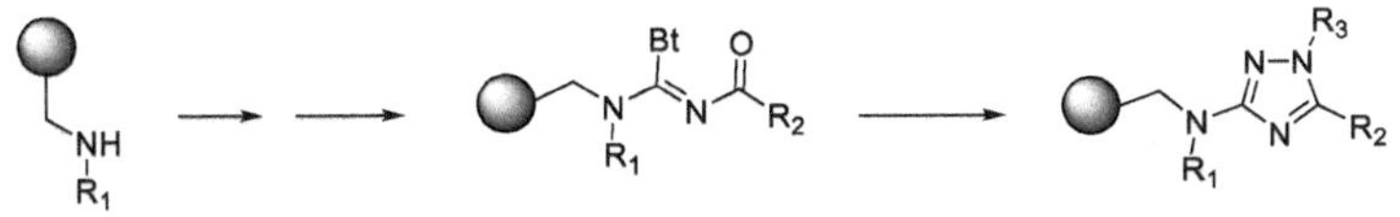

Scheme 51 Solid-phase synthesis of 3-alkylamino-1,2,4-triazoles.

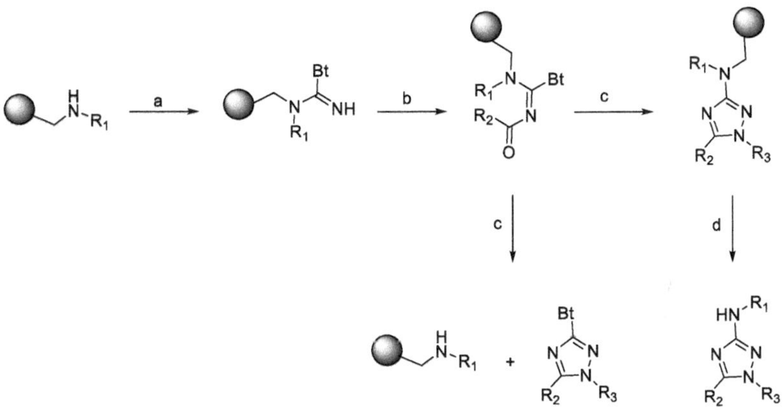

a) 3 eqv. $Bt_1C(NH)Bt_1$+$Bt_1C(NH)Bt_2$/THF/rt
b) 5 eqv. R_2COCl/10 eqv. DIEA/DCM/rt
c) 5 eqv. R_3NHNH_2/12 eqv. DBU/THF/50 °C/argon
d) TFA/DCM (95:5)

Scheme 52 Schematic sequence for solid-phase synthesis of 3-alkylamino-1,2,4-triazoles.

Makara and his research team.[58] This synthetic scheme relies heavily on immobilized *N*-acyl-1*H*-benzotriazole-1-carboximidamides with the role of crucial intermediating molecules. Furthermore, under milder conditions, they undergo cyclizations with hydrazines to generate corresponding 3-alkylamino-1,2,4-triazoles in excellent yield (Schemes 51 and 52).

References

1. Katritzky AR, Ramsden CA, Scriven EFV, Taylor RJK. *Comprehensive heterocyclic chemistry III*. vol. III. Oxford, UK: Pergamon; 2008.
2. Katritzky AR, Zhang Y, Singh SK. 1,2,3-Triazole formation under mild conditions via 1,3-dipolar cycloaddition of acetylenes with azides. *Heterocycles* 2003;**60**(5):1225–39.
3. Helms B, Mynar JL, Hawker JC, Frechet JMJ. *J Am Chem Soc* 2004;**126**:15020.
4. Khanetskyy B, Dallinger D, Kappe CO. *J Comb Chem* 2004;6(6):884.
5. Fan WQ, Katritzky AR. *Comprehensive heterocyclic chemistry*. 2nd ed. vol. 4. Oxford: Elsvier Science; 1996. p. 1–126.
6. Mamidyala SK, Finn MG. *Chem Soc Rev* 2010;**39**(4):1252.
7. Whiting M, Muldoon J, Lin Y-C, Silverman SM, Lindstrom W, Olson AJ, et al. *Angew Chem Int Ed* 2006;**45**(9):1435.

8. Tornøe CW, Sanderson SJ, Mottram JC, Coombs GH, Meldal M. *J Comb Chem* 2004;**6**(3):312.
9. Tron GC, Pirali T, Billington RA, Canonico PL, Sorba G, Genazzani AA. *Med Res Rev* 2008;**28**(2):278.
10. Brockunier LL, Parmee ER, Ok HO, Candelore MR, Cascieri MA, Colwell Jr LF, et al. *Bioorg Med Chem Lett* 2000;**10**(18):2111.
11. Genin MJ, Allwine DA, Anderson DJ, Barbachyn MR, Emmert DE, Garmon SA, et al. *J Med Chem* 2000;**43**(5):953.
12. Buckle DR, Rockell CJM, Smith H, Spicer BA. *J Med Chem* 1986;**29**(11):2262.
13. Velázquez S, Alvarez R, Pérez C, Gago F, De Clercq E, Balzarini J, et al. *Antivir Chem Chemother* 1998;**9**(6):481.
14. Pålhagen S, Canger R, Henriksen O, van Parys JA, Rivière M-E, Karolchyk MA. *Epilepsy Res* 2001;**43**(2):115.
15. Bennet IS, Brooks G, Broom NJP, Calvert SH, Coleman K. *J Antibiot* 1991;**44**:969.
16. Alonso R, Camarasa MJ, Alonso G, De Las Heras FG. *Eur J Med Chem* 1980;**15**:105.
17. Kume M, Kubota T, Kimura U, Nakashimizu K, Motokawa M, Nakano M. *J Antibiot* 1993;**46**:177.
18. Makabe O, Suzuki H, Umezawa S. *Bull Chem Soc Jpn* 1977;**50**:2689.
19. Kitamura Y, Taniguchi K, Maegawa T, Monguchi Y, Kitade Y, Sajiki H. *Heterocycles* 2009;**77**:521.
20. Cabrero-Antonino JR, Garcia T, Rubio-Marques P, Vidal-Moya JA, Leyva-Perez A, Al-Deyab SS, et al. *ACS Catal* 2011;**1**:147.
21. Tome AC. *Science of synthesis*. vol. 13. New York: Thieme; 2004. p. 415–601.
22. Huisgen R. *1,3-Dipolar cycloaddition chemistry*. New York: Wiley; 1984. p. 1.
23. Huisgen R. *J Pure Appl Chem* 1989;**61**.
24. Rostovtsev VV, Green LG, Fokin VV, Sharpless KB. *Angew Chem Int Ed* 2002;**41**:2596.
25. Tornøe CW, Christensen C, Meldal M. *J Organomet Chem* 2002;**67**(9):3057.
26. Appukkuttan P, Dehaen W, Fokin VV, Van der Eycken E. *Org Lett* 2004;**6**(23):4223.
27. Himo F, Lovell T, Hilgraf R, Rostovtsev VV, Noodleman L, Sharpless KB, et al. *J Am Chem Soc* 2004;**127**(1):210.
28. Kolb HC, Finn MG, Sharpless KB. *Angew Chem Int Ed* 2001;**40**(11):2004.
29. Hagiwara H. *Reclaimable azide-alkyne cycloaddition catalysts and preparation of triazole compounds therewith*. JP2009178612A; 2009.
30. Wu P, Fokin VV. *Aldrichimica Acta* 2007;**40**(1):7.
31. Wang Q, Chan TR, Hilgraf R, Fokin VV, Sharpless KB, Finn MG. *J Am Chem Soc* 2003;**125**(11):3192.
32. Krasiński A, Fokin VV, Sharpless KB. *Org Lett* 2004;**6**(8).
33. Zhang L, Chen X, Xue P, Sun HHY, Williams ID, Sharpless KB, et al. *J Am Chem Soc* 2005;**127**(46):15998.
34. Farooq T, Haug BE, Sydnes LK. *Manuscript*; 2012.
35. Majireck MM, Weinreb SM. A study of the scope and regioselectivity of the ruthenium-catalyzed [3 + 2]-cycloaddition of azides with internal alkyness. *J Organomet Chem* 2006;**71**:8680.
36. Wu Y-M, Deng J, Li Y, Chen Q-Y. Regiospecific synthesis of 1,4,5-trisubstituted 1,2,3-triazole via one-pot reaction promoted by copper(I) salt. *Synthesis* 2005;**2005**(EFirst):1314–8.
37. Hein JE, Trip JC, Krasnova LB, Sharpless KB, Fokin VV. Copper(I)-catalyzed cycloadditions of organic azides and 1-iodoalkynes. *Angew Chem Int Ed* 2009;**48**:8018–21.
38. Spiteri C, Moses JE. Copper-catalyzed azide-alkyne cycloaddition: regioselective synthesis of 1,4,5-trisubstituted 1,2,3-triazoles. *Angew Chem Int Ed Engl* 2010;**49**(1):31–3.

39. Barsoum DN, Okashah N, Zhang X, Zhu L. Mechanism of copper(I)-catalyzed 5-Iodo-1,2,3-triazole formation from azide and terminal alkyne. *J Org Chem* 2015;**80**:9542–51.
40. Lu L-H, Wu J-H, Yang C-H. Preparation of 1H-1,2,3-triazoles by cuprous ion mediated cycloaddition of terminal alkyne and sodium azide. *J Chin Chem Soc* 2008;**55**:414–7.
41. Jin T, Kamijo S, Yamamoto Y. *Eur J Org Chem* 2004;**2004**(18):3789.
42. Barluenga J, Vald C, Beltr G, Escribano M, Aznar F. Developments in Pd catalysis: synthesis of 1H1,2,3-triazoles from sodium azide and alkenyl bromides. *Angew Chem Int Ed* 2006;**45**:6893–6.
43. Amantini D, Fringuelli F, Piermatti O, Pizzo F, Zunino E, Vaccaro L. Synthesis of 4-aryl-1H-1,2,3-triazoles through TBAF-catalyzed [3 + 2] cycloaddition of 2-aryl-1-nitroethenes with TMSN3 under solvent-free conditions. *J Organomet Chem* 2005;**70**:6526–9.
44. Kamijo S, Jin T, Huo Z, Yamamoto Y. *J Am Chem Soc* 2003;**125**(26):7786.
45. Yap AH, Weinreb SM. *Tetrahedron Lett* 2006;**47**(18):3035.
46. Farooq T, Sydnes LK, Törnroos KW, Haug BE. Debenzylation of functionalized 4- and 5-substituted 1,2,3-triazoles. *Synthesis* 2012;**44**(EFirst):2070–8.
47. Kalisiak J, Sharpless KB, Fokin VV. Efficient synthesis of 2-substituted 1,2,3-triazoles. *Org Lett* 2008;**10**(15):3171–4.
48. Çoruh I, Rollas S, Turan SÖ, Akbuğa J. Synthesis and evaluation of cytotoxic activities of some 1,4-disubstituted thiosemicarbazides, 2,5-disubstituted 1,3,4thiadiazoles and 1,2,4-triazole-5-thiones derived from benzilic acid hydrazide. *Marmara Pharm J* 2012;**16**:56–63.
49. Kidwai M, Misra P, Bhushan KR, Dave B. A novel route to 1,2,4-triazoles. *Synth Commun* 2000;**30**:3031–40.
50. Yeung KS, Farkas ME, Kadow JF, Meanwell NA. A base catalyzed, direct synthesis of 3,5-disubstituted 1,2,4-triazoles from nitriles and hydrazides. *Tetrahedron Lett* 2005;**46**:3429–32.
51. Reddy KR, Mamatha R, SurendraBabu MS, Shiva Kumar K, Jayaveera KN, Narayanaswamy G. Synthesis and antimicrobial activities of some triazole, thiadiazole, and oxadiazole substituted coumarins. *J Heterocyclic Chem* 2014;**51**:132–7.
52. Rostamizadeh S, Tajik H, Yazdanfarahi S. Solid phase synthesis of 1,2,4-triazoles under microwave-irradiation. *Synth Commun* 2003;**33**:113–7.
53. Castanedo GM, Seng PS, Blaquiere N, Trapp S, Staben ST. Rapid synthesis of 1,3,5-substituted 1,2,4-triazoles from carboxylic acids, amidines, and hydrazines. *J Organomet Chem* 2011;**76**(4):1177–9.
54. Ueda S, Nagasawa H. Facile synthesis of 1,2,4-triazoles via a copper-catalyzed tandem addition-oxidative cyclization. *J Am Chem Soc* 2009;**131**:15080–1.
55. Yin P, Ma W-B, Chen Y, Huang W-C, Deng Y, He L. Highly efficient cyanoimidation of aldehydes. *Org Lett* 2009;**11**(23):5482–5.
56. Xu Y, McLaughlin M, Bolton EN, Reamer RA. Practical synthesis of functionalized 1,5-disubstituted 1,2,4-triazole derivatives. *J Organomet Chem* 2010;**75**:8666–9.
57. Boeglin D, Cantel S, Heitz A, Martinez J, Fehrentz J-A. Solution and solid-supported synthesis of 3,4,5-trisubstituted 1,2,4triazole-based peptidomimetics. *Org Lett* 2003;**5**:4465–8.
58. Makara GM, Ma Y, Margarida L. Solid-phase synthesis of 3-alkylamino-1,2,4-triazoles. *Org Lett* 2002;**4**(10):1751–4.

CHAPTER 2

Triazoles as Bioisosteres in Medicinal Chemistry: A Recent Update

Tahir Farooq*
Department of Applied Chemistry, Government College University, Faisalabad, Pakistan
*Corresponding author. E-mail: tahirfarooqfsd@gmail.com

Introduction

Over the last few decades, the bioisosteric replacement strategy has emerged as a leading approach, widely applied for the improvement of efficacy, targetability, selectivity, bioavailability, and other physiochemical properties of lead molecules. This approach helps to fine-tune the rational structural modifications leading to the therapeutic agent with desired drug-like properties. It also enables medicinal chemists to modulate the issues of drug resistance and toxicity of known drugs though rational applications of more suited isosteres. Over the last two decades, the triazole has emerged as a promising candidate for bioisosteric replacement due to its remarkable stability under oxidative, reductive, and hydrolytic conditions. Furthermore, triazolic functionality represents some advantageous of structural and electronic features like rigidity, planner nature, and polarizability. The triazole does have the potential to act as a hydrogen bond acceptor as well as a donor depending on the surrounding substituents (Figure 1).[1,2]

During the last two decades, the remarkable developments regarding the facile synthesis of triazole including Cu- and Ru-catalyzed Azide Alkyne cycloadditions, metal-free organocatalyzed, and one-pot multicomponent synthesis of triazoles have accelerated their applications in functional molecules, especially in drug synthesis (Figure 2).[3] The regioselective access to triazolic moieties has highlighted their potential role as promising bioisosteres with a wide array of applications. Over the last decade, the 1,2,3-triazole has frequently been used as an amide bond isostere due to its electronic and structural resemblance especially the field of peptidomimetics, which exploited it immensely for the development

Advances in Triazole Chemistry
https://doi.org/10.1016/B978-0-12-817113-4.00010-X

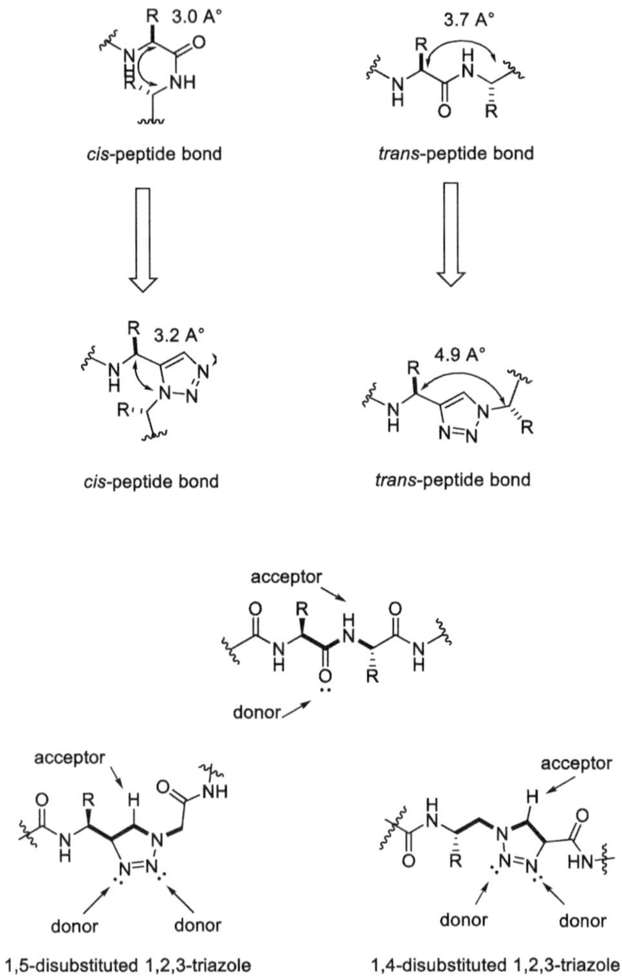

Figure 1 Known structural and electronic features of triazole moiety.

of novel analogues. Triazoles have also been used as ester isosteres in a number of molecules to modulate their enzyme-mediated degradation. Furthermore, they have also been employed as bioisosteres of ester, carboxylic acid, olefins, and heterocycles (Figure 3).[4]

Needless to say that 1,2,4-triazole also shows bioisosteric applications; however, its use is less common compared with 1,2,3-triazoles. Herein, a few recent examples have been presented that highlight the use of 1,2,4-triazole as a promising isostere remodeling the physiochemical properties of important molecules.

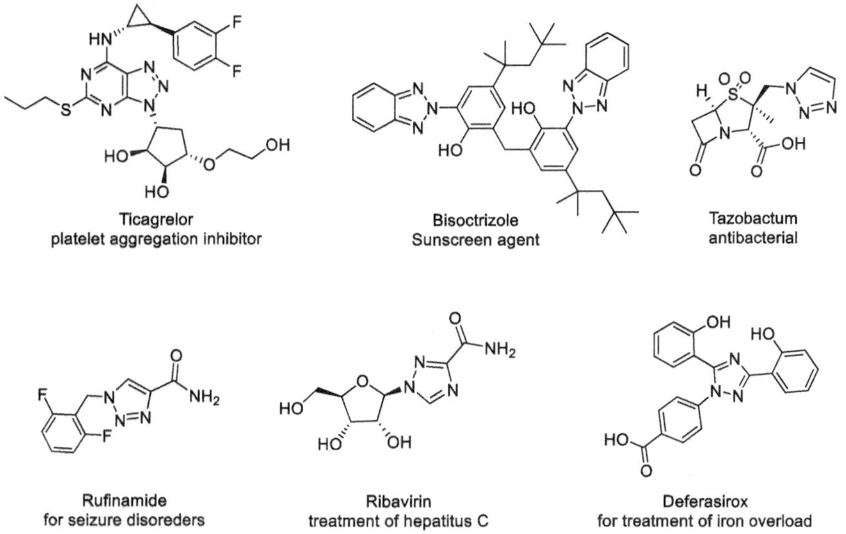

Figure 2 Known drugs with triazolic moiety.

1,2,3-Triazoles as bioisosteres

The sialyltransferases (STs) enzymes add sialic acid (second-most important sugar after glucose) to the cell-surface molecules through a process called sialylation. This vital process ensures the smooth cellular processes including cell adhesion, protein targeting, and cell-cell recognition thus controlling many important functions. The hypersialylation has been found correlated with cancer in many cases.[5] The metastatic potential of tumors increases due to abnormal sialylation, which results in poor patient prognosis. The increased activities of STs reduce the treatment efficacy of cisplatin and paclitaxel in human colorectal and ovarian cancer cell lines. The inhibition of STs could enhance the sensitivity of antimetastatic agents. Thus, efficacy of existing antitumor drugs could be improved using them in combination with STs inhibitors. The cytidine 5′-monophosphate Neu5Ac is the common sugar nucleotide donor for 20 different STs in humans. There has been great interest in the development of specific inhibitors for each subtype of ST as they control synthesis of specific sialylated structures with potent biological roles.[6] Among them, β-galactoside α-2,6-sialyltransferase 1 (ST6Gal 1)-mediated sialylation modulates substrate-oriented receptor internalization, oligomerization, and protein conformations. The upregulation of ST6Gal 1 has been found to induce tumor-initiating potential and chemiresistance, and promotes cancer stem cell phenotypes in a number

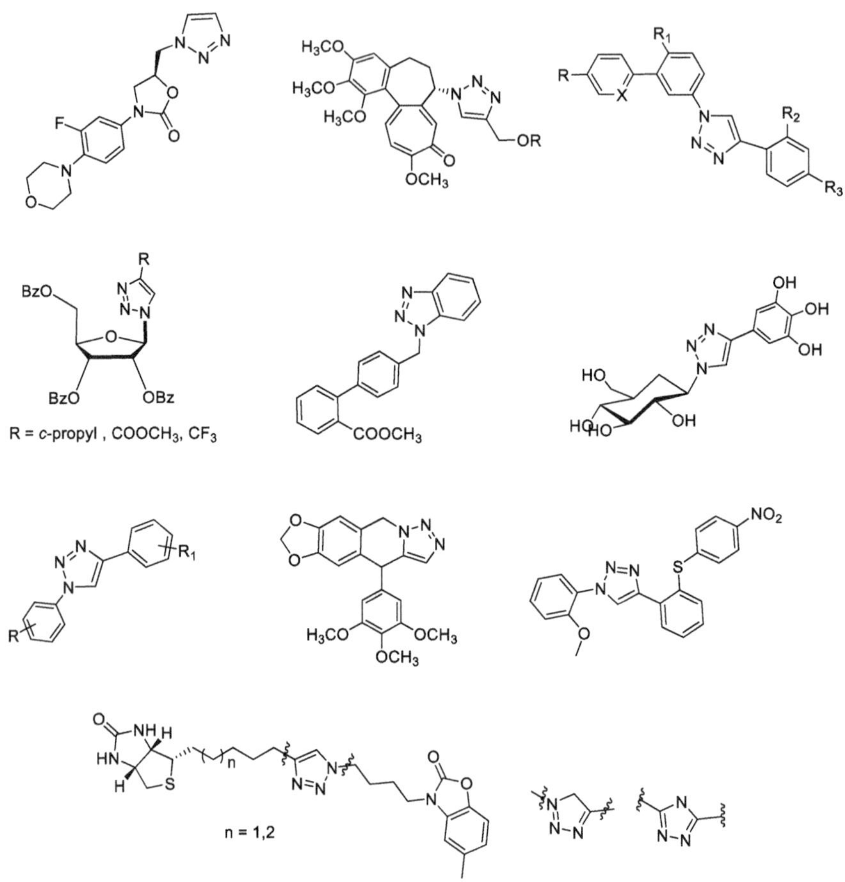

Figure 3 Known therapeutic agents with triazole as isosteres.

of cancers including colon, pancreatic, ovarian, and lung carcinoma.[7] In this context, a number of ST inhibitors have been developed to control the overexpression of ST6Gal 1. Furthermore, a number of strategies have also been developed to increase the efficacy and safety profiles of reported inhibitors.[8]

To this end, in a recent report, the 1,2,3-triazole and carbamate were used as neutral isosteres for the replacement of a charged phosphodiester linker in reported TS analogue of ST inhibitors (Figure 4),[9] and subsequently FEP calculations and molecular dynamics simulations were performed. The analogues with triazole and carbamate moieties showed better binding affinities toward human ST6Gal 1. The carbamate-carrying analogue exhibited a bit more effective binding. The interactions of ligands with active sites were found to be uninfluenced by the nature of the linker.

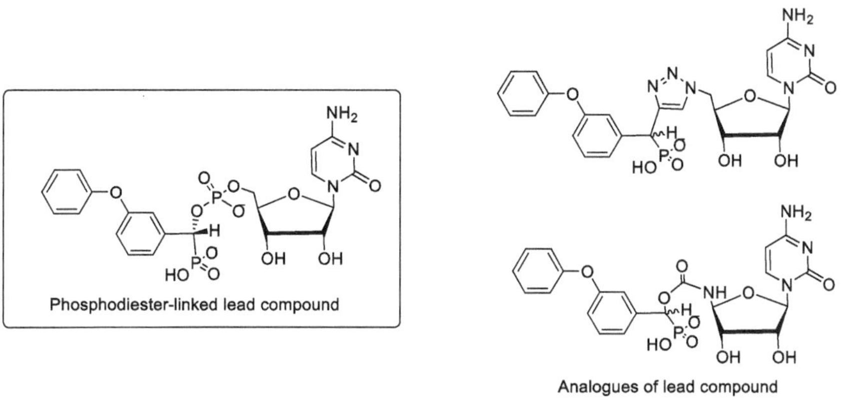

Figure 4 1,2,3-Triazole and carbamate as isosteres of charged phosphodiester.

These observations provided a new outlook about the perceived important role of phosphodiester linker. This study provided a rationale platform for the replacement of phosphodiester linker with carbamate and 1,2,3-triazole linkers. Furthermore, they have been suggested as potential isosteres for the development of new inhibitors of human ST6Gal 1.

A few heterocycles like isoxazoles, pyrazoles, oxadiazoles, and thiadiazoles have successfully been applied as isosteres of carboxylic acids.[10] The 4-hydroxy-1*H*-1,2,3-triazole has also been suggested as a promising isostere of the carboxylic acid group owing to its acidic properties at physiological pH. Very recently, Giarudo *et al.* prepared biomimetics of γ-aminobutyric acid (GABA), which is the main inhibitory neurotransmitter in the central nervous system (CNS). They prepared and pharmacologically characterize a series of N_1- and N_2-substituted 4-hydroxy-1,2,3-triazole analogues of known $GABA_AR$ ligands (Figure 5).[11] It was observed that the substituents like 3,3-diphenylpropyl or 2-naphthylmethyl at 5-position of the triazole influenced the binding affinities of the resulting analogues in low micromolar range. However, weak affinities in high to medium micromolar range were exhibited by the as-prepared analogues with unsubstituted Ns. These SAR studies provided more insight into the orthosteric GABAAR binding site using docking studies and electrostatic analyses. This pioneering study opened up new options for the development of diversely functionalized hydroxyl-triazole core for the synthesis of novel scaffolds.

Since the introduction of Cu-catalyzed azide alkyne cycloadditions for the facile preparation of 1,2,3-triazole, it has widely been used as an amide bond isostere due to its electronic characteristics and topological similarities with *trans*-amide. Over the years, a number of molecules with triazole as

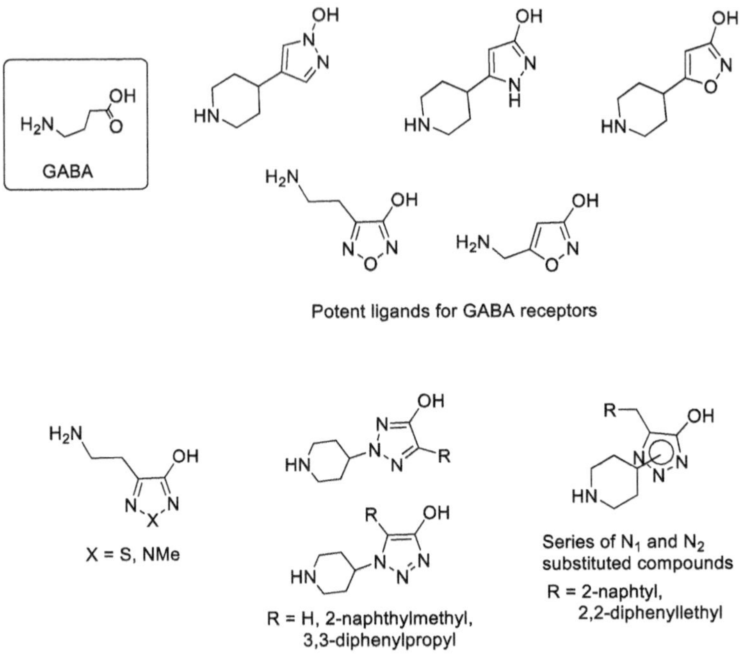

Figure 5 GABA mimics using 4-hydroxy-1*H*-1,2,3-triazole as isosteres of carboxylic acid.

an amide bond surrogate have shown therapeutic applications as antipsychotics, antibiofilms, antivirals, and chemotherapeutic agents. The amide replacement with triazole moiety enhances the biopotential and improves the pharmacokinetic profiles of potent biomolecules. However, it cannot be an ultimate option in all cases.[12,13] Very recently, Doiron *et al.* targeted amide containing small molecule modulators of the cystic fibrosis transmembrane conductance regulator (CFTR) proteins for bioisosteric replacement of amide with 1,2,3-triazole. The CFTR acts as HCO_3^- and Cl^- ion channel and helps to maintain the homeostasis in epithelial cells of reproductive tract, intestine, pancreases, sweat glands, and lungs. Any abnormalities or mutations in this channel cause cystic fibrosis disease, and currently more than 2000 such mutations have already been identified with phenylalanine 508 deletion (F508del) from the nucleotide binding domain as the most common one. Thus, the F508del-CFTR allele is found in more than 90% patients of cystic fibrosis (*CF*). This common mutation results in the weakening of protein stability at the plasma membrane and causes a gating defect of the ion channel.[14] This defect leads to the secretion of Cl^- ions in human bronchial epithelial cells, and its reduced transportation causes dehydration of the airway surfaces in *CF* patients. This leads to hyperinflammation and chronic lung infection causing lung failure, a major cause of death in *CF* patients.

As a combat strategy of *CF*, some small molecule modulation of mutation forms of CFTR has been identified as a major achievement in recent years. The CFTR modulators that counteract the folding deficiencies and enhance protein expression are categorized as correctors, while gating defect is counteracted by potentiators that increase Cl^- ion transportation. Both categories of modulators have received approval from the U.S. Food and Drug Administration (FDA). These groundbreaking therapeutic leads opened up new searches for CFTR modulators with more improved pharmacokinetic profiles.[15] Considering all these factors, Doiron *et al.* used 1,2,3-triazole as a bioisosteric replacement of amide in VX-770 and VX-809 (Figure 6).[16] Surprisingly, the as-prepared triazole-containing molecules exhibited a marked decrease in their biological activity and metabolic stability compared with their amide analogous. This study provided an important aspect of 1,2,3-triazole as an amide isostere suggesting extreme caution while considering such strategic synthesis, especially in case of medicinal compounds.

The privileged molecular architecture of gel-based materials have attracted applications across a number of domains including food industries,

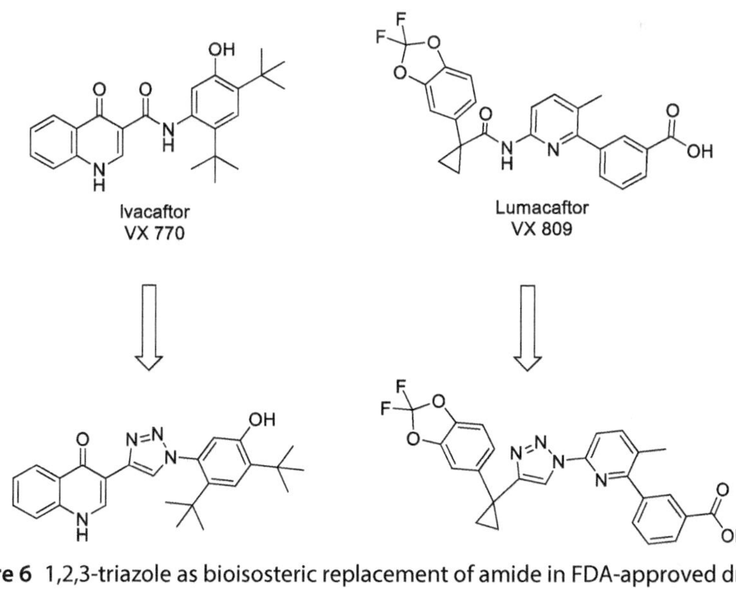

Figure 6 1,2,3-triazole as bioisosteric replacement of amide in FDA-approved drugs.

conductive scaffolds, cosmetics, catalysis and liquid crystals, etc.[17] In this context, the alkylamide derivatives of *trans*-1,2-diaminocyclohexane exhibited remarkable gelation ability in organic solvents. The compound (C_{12}-Cyc) and its enantiomers showed gelation in a range of solvents due to the antiparallel arrangement of amide-CO and amide-NH and perpendicular to Cyc ring supported by Van der Waals forces and hydrogen bonding.[18] Very recently, Taut *et al.* replaced both of the amide bonds with 1,2,3-triazole rings for the preparation of isosteric gelator to study the corresponding effect on gel properties and gelation ability of click-gelator (Figure 7).[19] The results displayed that the insertion of new isostere groups offered the same functional properties, thus it turned out be a reliable methodology for the fine-tuning of gelation properties. Theoretically, the triazole-substituted gelator was expected to show a variety of interacting patterns depending on the polar nature of the environment. The experimental results of gelation ability were found in agreement with theoretical calculations. Furthermore, the nontoxic click-gelator advantageously demonstrated phase-selective gelation of water-oil mixtures. The isosteric replacement strategy with robustness of triazolic moiety highlights its value for the practical preparation of functional materials beyond physical gels. In another very recent report, Häring *et al.* used 1,2,3-triazole as an isostere of 1,2,4-triazole and prepared a corresponding click-hydrogelator (Figure 8).[20] The stable supramolecular viscoelastic hydrogels with critical gelation concentration of 6 g/L were formed by click-TIA in water. Furthermore, morphological, mechanical,

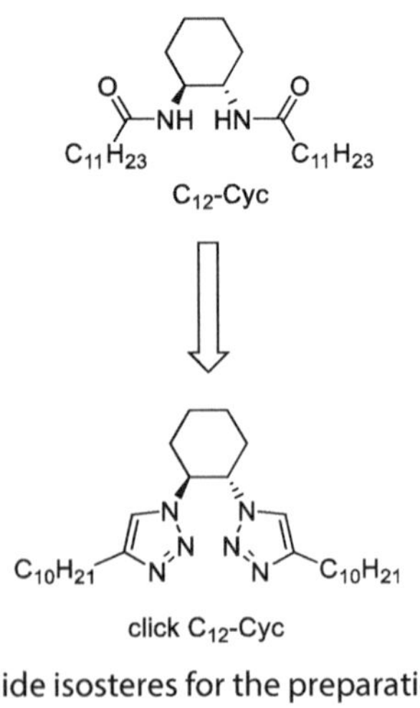

Figure 7 1,2,3-Triazole as amide isosteres for the preparation of isosteric gelator.

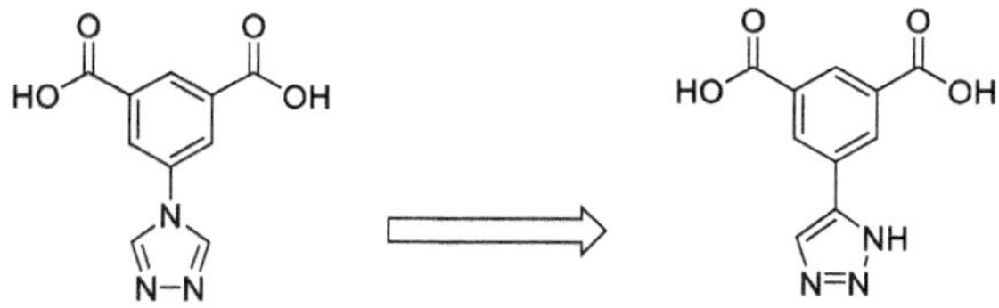

Figure 8 Use of 1,2,3-triazole as isostere of 1,2,4-triazole.

thermal, and rheological properties of both 5-TIA and click-TIA were studied using pure and hybrid hydrogels (click-TIA + 5-TIA in 1:0.2 M ratio). The first order kinetics was exhibited by oxytetracycline when the hydrogels were used for its encapsulation and *in vitro* controlled release. The click-TIA hydrogel exhibited a maximum drug release of about 60% after 8 h from pH 6.5 to 10, and the release rate reduced to half at lower pH range.

In humans, the replication of retroviruses is inhibited by an innate antiviral human DNA-editing enzyme APOBEC3G (A3G). This enzyme is targeted by HIV-1 accessary protein viral infectivity factor (Vif) for an *in vivo* replication. This enzyme destroys the retroviruses by catalyzing hypermutations in their DNA. Therefore, a functional Vif has always been an absolute requirement for viral replications in cells with active A3G. Thus, HIV-1 Vif has become an attractive target for antiviral therapy.[21] The known Vif antagonists like RN-18-based class of small molecules accelerate the A3G incorporation into virons, enhance cytidine deamination in via rus genome, and induce A3G-mediated degradation of Vif. The combination of all these mechanistic moves inhibit the HIV-1 infectivity.[22,23] The RN-18 demonstrated Vif-specific antiviral properties as it does not inhibit viral infection in MT4 cells even at 100 μM; however, it exhibited 6 and 4.5 μM IC_{50} values in H9 and CEM cells, respectively. Therefore, the Vif-A3G interaction has been focused as a vital target for the development of small molecule-based antiviral therapies. Very recently, Muhammad *et al.* focused metabolic stability and potency for the development of RN-18-based Vif antagonists. They adopted isosteric replacement strategy and replaced amide in RN-18 with metabolically stable, biocompatible, and conformationally restricted heterocycles including 1,2,3-triazole (Figure 9). The as-prepared Vif antagonist with 1,2,3-triazole exhibited IC_{50} of 1.2 μM suggesting triazole as a suitable bioisostere in RN-18 with improved activity and pharmacological profiles. The study provided an insight into the Vif-A3G axis as a potent antiviral target and also introduced a new potent class of anti-HIV candidates.

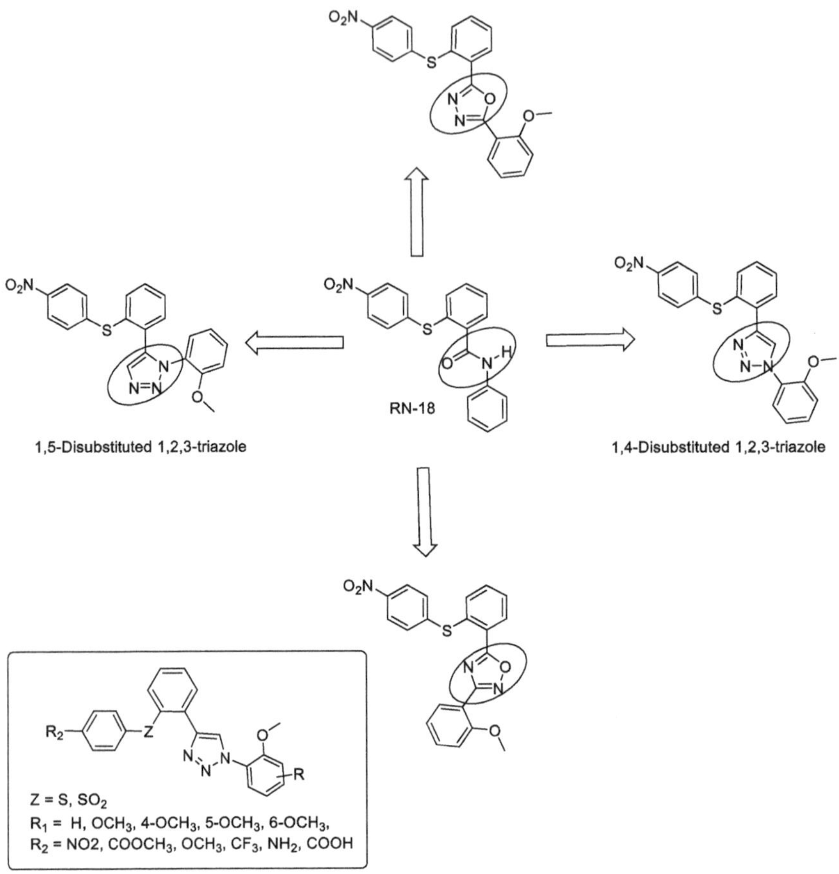

Figure 9 Heterocycles as isosteres of amide.

1,2,4-Triazoles as bioisosteres

The 1,2,4-triazoles have less common applications as bioisosteres of functional groups compared with 1,2,3-triazoles. However, herein a few recent examples have been presented to showcase their potential to act as bioisosteres of important functionalities and provide potent compounds.

Human African trypanosomiasis (HAT) and Chagas disease are the major infectious diseases in humans caused by protozoan parasites causing considerable life losses, especially in developing countries. The currently available therapeutic agents have been found unsatisfactory due to resistance development, serious side effects, and limited efficacy profiles (Figure 10). In search of potent alternatives, a number of lead compounds like nitroimidazole derivatives, quinolones, and triazoles have been under consideration

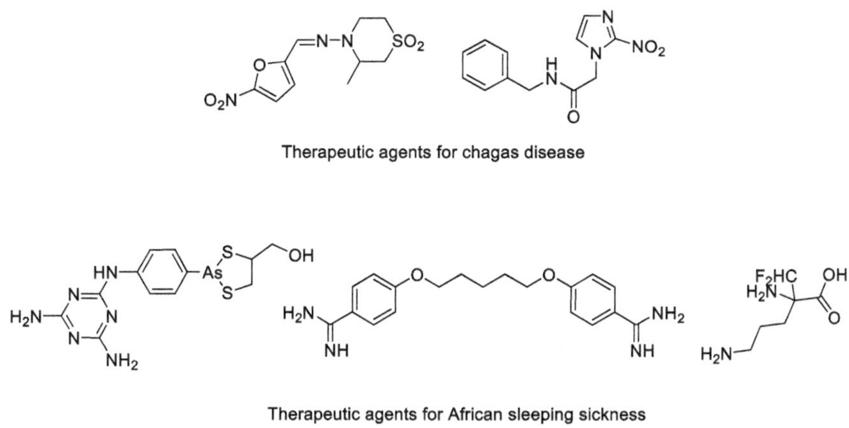

Figure 10 Known drugs for Chagas disease and African sleeping sickness.

over a decade leading to the identification of few promising prototypes.[24,25] In this context, a highly active lead compound megazol found limited success due to its genotoxicity and mutagenic properties, although it was found active against the causative agents of both said diseases. Adopting a bioisosteric replacement strategy, Carvalho *et al.* took megazol as a promising lead and substituted its heterocycle with 1,2,4-triazole to study the corresponding effects on antiparasitic activity (Figure 11).[26] Afterward, the Comet and fluorescein diacetate/ethidium bromide assays were employed to assess the cytotoxicity and genotoxicity in whole blood. The as-prepared compound was not genotoxic and readily metabolized, but it exhibited poorer antiparasitic activity. Furthermore, it exhibited greater binding affinity with active site of *Trypanosoma brucei* type 1 nitroreductase enzyme compared with megazol in docking studies.

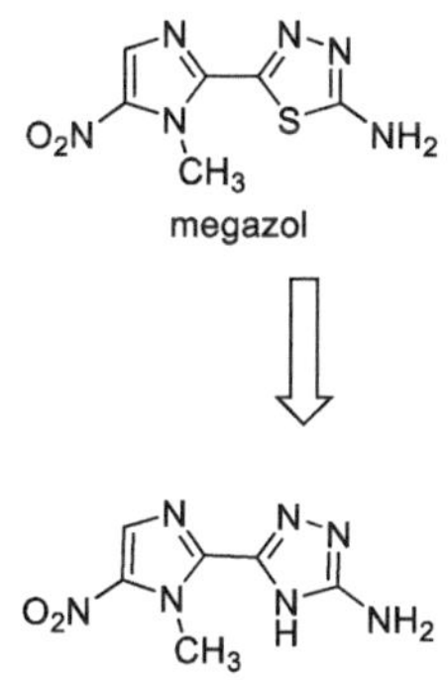

Figure 11 Preparation of 1,2,4-triazole analogue of megazol.

Somatostatin is a naturally occurring polypeptide that acts as a somatotropin release-inhibiting factor (SRIF) and exists in two bioactive forms, *viz.*, SRIF-14 and -28. They bind to structurally related receptors (five receptor subtypes sst_1-sst_5 belong to the G-protein-coupled superfamily) and activate them using core residues Trp, Lys, and β-turn for pharmacological actions.[27,28] The SIRF modulates many neuronal activities in the peripheral and CNS and also show inhibitory effects on exocrine and endocrine secretions. Thus, the respective receptors have been recognized as primary targets of the therapeutic agents. However, the sst_4 has been recognized as the main receptor involved in mediating pain mechanisms, inflammation, and CNS pathology. It has also become a main drug target due to its nonoffensive effects on insulin secretion, growth hormones, and glucagon production. The enzymatic instability and poor bioavailability of SRIF has reduced its therapeutic use, and alternatively preparation of its peptide analogues was perused actively over the past decades. Furthermore, attention has also been diverted for the development of na onpeptide, selective, and orally active sst_4 antagonist.[29] In this context, Daryaei *et al.* prepared a series of compounds using 1,2,4-triazole ring for high selectivity and affinity, and mimics of Lys^9 and Trp^8 were interchanged on 3, 4, and 5 position of the triazolic ring (Figure 12).[30] The previous SAR-studies helped in the selection of the side chains. Actually, they employed 1,2,4-triazole in place of *cis*-peptide bond mimicking the endogenous substrate within the binding pocket. The prepared compound 1 acted as an agonist and exhibited high selectivity and affinity toward sst_4. Furthermore, it was suggested as a promising lead for the development of sst_4 agonist with potential clinical applications against Alzheimer's disease and related disorders.

The uric acid transporters mediate the reabsorption of urates filtered by the human kidneys. The urate transporter 1 (URAT 1) maintains level of uric acid in serum. However, the reabsorption of uric acid could be blocked by the administration of known URAT 1 inhibitors.[31,32] The lesinuard was approved in 2015 as a selective URAT 1 inhibitor, but it exhibits a narrow therapeutic window and low efficacy profiles. Wu *et al.* initially developed a potent molecule 1 that showed better activity than lesinuard. Afterward, they developed further derivatives by using different isosteres of carboxylic acid functionality. They also used 1,2,4-triazole as a bioisostere replacement of carboxylic acid in one of the prepared derivatives (Figure 13).[33] The as-prepared derivative exhibited IC_{50} of $3.24 \pm 0.64\,\mu M$ compared with lesinuard (IC_{50} of $7.20\,\mu M$). This study helped in the development

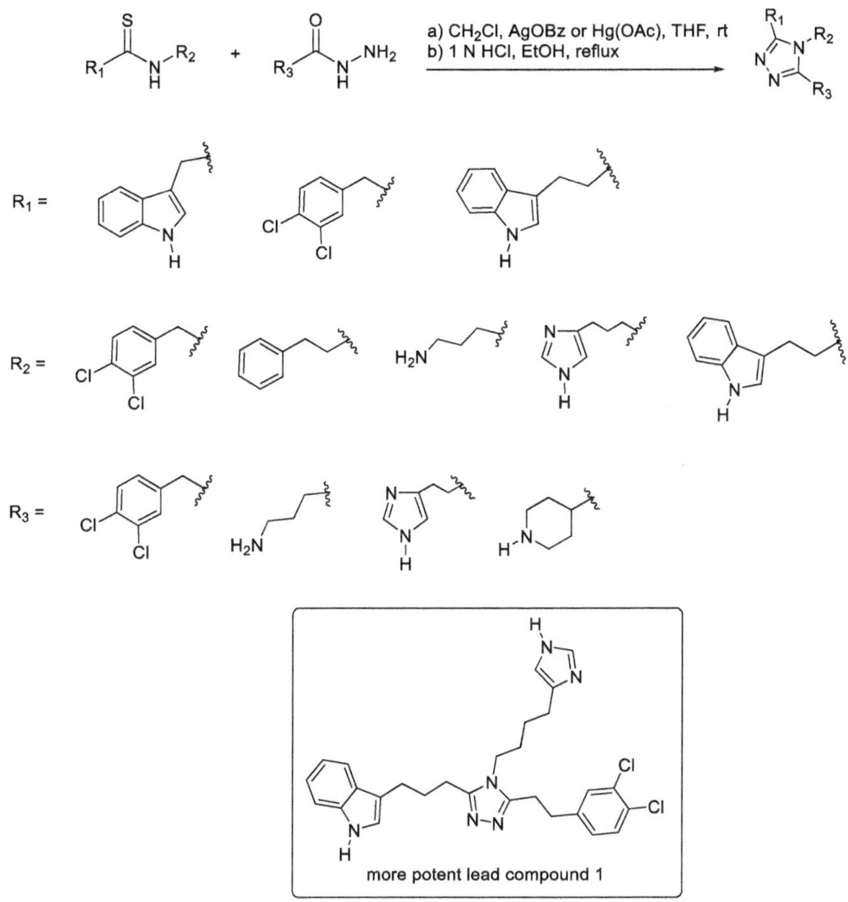

Figure 12 Series of compounds using 1,2,4-triazole and mimics of Lys[9] and Trp.[8]

of another highly potent candidate as URAT 1 inhibitor showing 225-fold higher activity than standard lesinuard.

A number of vital physiological functions are mediated by the nine isoforms of voltage-gated sodium channels ($Na_v1.1$ to $Na_v1.9$) in mammals. They are in fact transmembrane ion channels with different expressions in neuroendocrine cells, peripheral and CNS, and muscles. In humans, the $Na_v1.7$–1.9 shows their expression in nociceptors for the execution of pain signaling and amplifications. The $Na_v1.7$ was found to mediate a key role in pain sensation. Thus, it has become a promising target for the treatment of pain. Over the years, there has been great interest in the development of selective $Na_v1.7$ antagonism as novel analgesics.[34,35]

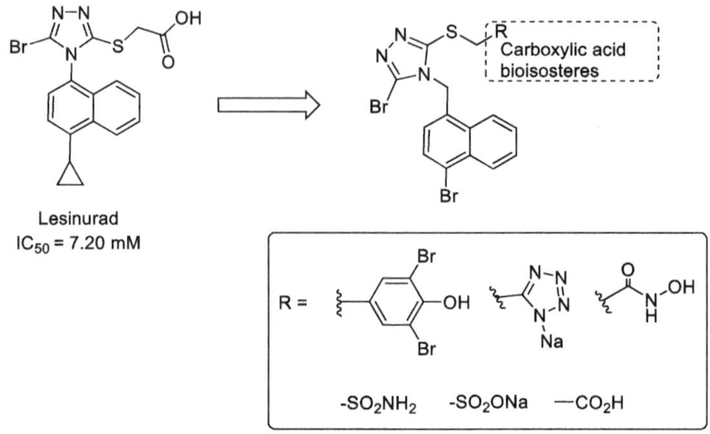

Figure 13 Known URAT 1 inhibitor and its bioisosteres.

After years of active research, Focken *et al.* identified compound 1 as a promising lead for the development of novel molecules for the treatment of pain. The lead molecule exhibited poor stability but good selectivity, thus it turned out to be a good starting point for optimization studies. They hypothesized to use five membered heterocycles to replace the aryl fluoride and carbonyl group of acyl sulfonamide in compound 1. Furthermore, they focused metabolic stability, improved physiochemical properties, and consequently introduced *N*-([1,2,4]triazolo[4,3-*a*]pyridine-3-yl) methanesulfonamides as a novel series of inhibitors with high selectivity (Figure 14).[36] The as-prepared molecules a, b, and c displayed low *in vivo* clearance, good *in vitro* metabolic stability, and high efficacy in model animals. In a transgenic mouse model, the lead b was also found highly effective against induced pain compared with compound c.

Conclusion

Bioisosteric replacement has emerged as a leading methodology for the rational modification of physiochemical properties of lead molecules. This strategy helps medicinal chemists to improve the efficacy, bioavailability, and selectivity of potent compounds. The bioisosteric replacement of different functionalities modulates drug resistance-related issues. Both 1,2,3- and 1,2,4-triazole have emerged as promising choices for bioisosteric replacement of a number of functionalities. Herein, recent noteworthy examples have been presented to highlight the role of triazoles as powerful bioisosteric options for practical applications.

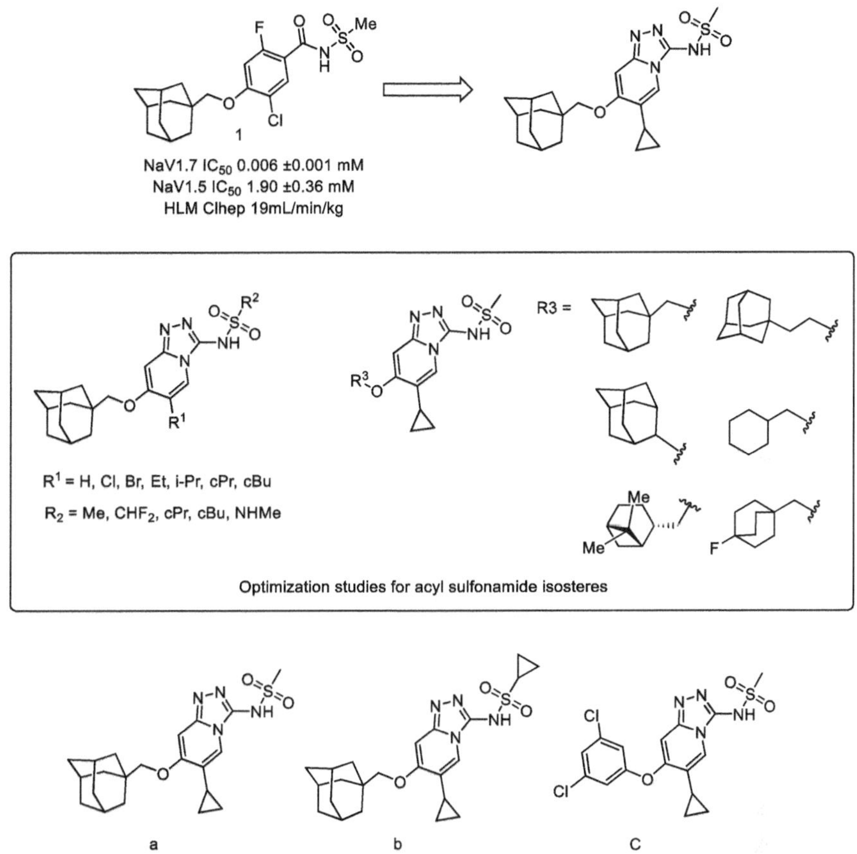

Figure 14 Optimization of acyl sulfonamide isosteres with 1,2,4-triazoles.

References

1. Pedersen DS, Abell A. 1,2,3-Triazoles in peptidomimetic chemistry. *Eur J Org Chem* 2011;**2011**(13):2399–411.
2. Diness F, Schoffelen S, Meldal M. Advances in merging triazoles with peptides and proteins. In: *Peptidomimetics I.* Springer; 2015. p. 267–304.
3. Malik MS, Ahmed SA, Althagafi II, Ansari MA, Kamal A. Application of triazoles as bioisosteres and linkers in the development of microtubule targeting agents. *RSC Med Chem* 2020;**11**(3):327–48.
4. Bonandi E, Christodoulou MS, Fumagalli G, Perdicchia D, Rastelli G, Passarella D. The 1,2,3-triazole ring as a bioisostere in medicinal chemistry. *Drug Discov Today* 2017;**22**(10):1572–81.
5. Rodrigues E, Macauley MS. Hypersialylation in cancer: modulation of inflammation and therapeutic opportunities. *Cancers* 2018;**10**(6):207.
6. Munkley J, Scott E. Targeting aberrant sialylation to treat cancer. *Fortschr Med* 2019;**6**(4):102.
7. Garnham R, Scott E, Livermore KE, Munkley J. ST6GAL1: a key player in cancer. *Oncol Lett* 2019;**18**(2):983–9.

8. Wang L, Liu Y, Wu L, Sun X-L. Sialyltransferase inhibition and recent advances. *Biochim Biophys Acta (BBA)-Proteins Proteom* 2016;**1864**(1):143–53.
9. Montgomery AP, Skropeta D, Yu H. Transition state-based ST6Gal I inhibitors: mimicking the phosphodiester linkage with a triazole or carbamate through an enthalpy-entropy compensation. *Sci Rep* 2017;**7**(1):1–11.
10. Ballatore C, Huryn DM, Smith III AB. Carboxylic acid (bio) isosteres in drug design. *ChemMedChem* 2013;**8**(3):385–95.
11. Giraudo A, Krall J, Nielsen B, Sørensen TE, Kongstad KT, Rolando B, et al. 4-Hydroxy-1,2,3-triazole moiety as bioisostere of the carboxylic acid function: a novel scaffold to probe the orthosteric γ-aminobutyric acid receptor binding site. *Eur J Med Chem* 2018;**158**:311–21.
12. Bi F, Ji S, Venter H, Liu J, Semple SJ, Ma S. Substitution of terminal amide with 1H-1,2,3-triazole: identification of unexpected class of potent antibacterial agents. *Bioorg Med Chem Lett* 2018;**28**(5):884–91.
13. Sun S, Jia Q, Zhang Z. Applications of amide isosteres in medicinal chemistry. *Bioorg Med Chem Lett* 2019;**29**(18):2535–50.
14. Gentzsch M, Mall MA. Ion channel modulators in cystic fibrosis. *Chest* 2018;**154**(2):383–93.
15. Habib A-RR, Kajbafzadeh M, Desai S, Yang CL, Skolnik K, Quon BS. A systematic review of the clinical efficacy and safety of CFTR modulators in cystic fibrosis. *Sci Rep* 2019;**9**(1):1–9.
16. Doiron JE, Le CA, Ody BK, Brace JB, Post SJ, Thacker NL, et al. Evaluation of 1,2,3-triazoles as amide bioisosteres in cystic fibrosis transmembrane conductance regulator modulators VX-770 and VX-809. *Chem Eur J* 2019;**25**(14):3662–74.
17. Owens GJ, Singh RK, Foroutan F, Alqaysi M, Han C-M, Mahapatra C, et al. Sol–gel based materials for biomedical applications. *Prog Mater Sci* 2016;**77**:1–79.
18. Zweep N, Hopkinson A, Meetsma A, Browne WR, Feringa BL, van Esch JH. Balancing hydrogen bonding and van der Waals interactions in cyclohexane-based bisamide and bisurea organogelators. *Langmuir* 2009;**25**(15):8802–9.
19. Tautz M, Torras J, Grijalvo S, Eritja R, Saldías C, Alemán C, et al. Expanding the limits of amide–triazole isosteric substitution in bisamide-based physical gels. *RSC Adv* 2019;**9**(36):20841–51.
20. Häring M, Rodríguez-López J, Grijalvo S, Tautz M, Eritja R, Martín VS, et al. Isosteric substitution of 4 H-1,2,4-triazole by 1 H-1,2,3-triazole in isophthalic derivative enabled hydrogel formation for controlled drug delivery. *Mol Pharm* 2018;**15**(8):2963–72.
21. Zhang R-H, Wang S, Luo R-H, Zhou M, Zhang H, Xu G-B, et al. Design, synthesis, and biological evaluation of 2-amino-N-(2-methoxyphenyl)-6-((4-nitrophenyl) sulfonyl) benzamide derivatives as potent HIV-1 Vif inhibitors. *Bioorg Med Chem Lett* 2019;**29**(24):126638.
22. Nathans R, Cao H, Sharova N, Ali A, Sharkey M, Stranska R, et al. Small-molecule inhibition of HIV-1 Vif. *Nat Biotechnol* 2008;**26**(10):1187–92.
23. Barce Ferro CT, Dos Santos BF, da Silva CD, Brand G, da Silva BAL, de Campos Domingues NL. Review of the syntheses and activities of some sulfur-containing drugs. *Curr Org Synth* 2020;**17**(3):192–210.
24. Singh Grewal A, Pandita D, Bhardwaj S, Lather V. Recent updates on development of drug molecules for human African trypanosomiasis. *Curr Top Med Chem* 2016;**16**(20):2245–65.
25. Scarim CB, Jornada DH, Machado MGM, Ferreira CMR, dos Santos JL, Chung MC. Thiazole, thio and semicarbazone derivatives against tropical infective diseases: Chagas disease, human African trypanosomiasis (HAT), leishmaniasis, and malaria. *Eur J Med Chem* 2019;**162**:378–95.
26. Carvalho ASd, Salomão K, Castro SLd, Conde TR, Zamith HPdS, Caffarena ER, et al. Megazol and its bioisostere 4H-1,2,4-triazole: comparing the trypanocidal, cytotoxic

and genotoxic activities and their in vitro and in silico interactions with the *Trypanosoma brucei* nitroreductase enzyme. *Mem Inst Oswaldo Cruz* 2014;**109**(3):315–23.
27. Hofland LJ, Lamberts SW. Somatostatin receptors and disease: role of receptor subtypes. *Bailliere Clin Endocrinol Metab* 1996;**10**(1):163–76.
28. Prévôt TD, Gastambide F, Viollet C, Henkous N, Martel G, Epelbaum J, et al. Roles of hippocampal somatostatin receptor subtypes in stress response and emotionality. *Neuropsychopharmacology* 2017;**42**(8):1647–56.
29. Kántás B, Börzsei R, Szőke É, Bánhegyi P, Horváth Á, Hunyady Á, et al. Novel drug-like somatostatin receptor 4 agonists are potential analgesics for neuropathic pain. *Int J Mol Sci* 2019;**20**(24):6245.
30. Daryaei I, Sandoval K, Witt K, Kontoyianni M, Crider AM. Discovery of a 3,4,5-trisubstituted-1, 2, 4-triazole agonist with high affinity and selectivity at the somatostatin subtype-4 (sst 4) receptor. *MedChemCommun* 2018;**9**(12):2083–90.
31. Tan PK, Ostertag TM, Miner JN. Mechanism of high affinity inhibition of the human urate transporter URAT1. *Sci Rep* 2016;**6**(1):1–13.
32. Pan Y, Kong L-D. Urate transporter URAT1 inhibitors: a patent review (2012–2015). *Expert Opin Ther Pat* 2016;**26**(10):1129–38.
33. Wu J-w, Yin L, Liu Y-q, Zhang H, Xie Y-f, Wang R-l, et al. Synthesis, biological evaluation and 3D-QSAR studies of 1, 2, 4-triazole-5-substituted carboxylic acid bioisosteres as uric acid transporter 1 (URAT1) inhibitors for the treatment of hyperuricemia associated with gout. *Bioorg Med Chem Lett* 2019;**29**(3):383–8.
34. Dib-Hajj SD, Black JA, Waxman SG. Voltage-gated sodium channels: therapeutic targets for pain. *Pain Med* 2009;**10**(7):1260–9.
35. Bennett DL, Clark AJ, Huang J, Waxman SG, Dib-Hajj SD. The role of voltage-gated sodium channels in pain signaling. *Physiol Rev* 2019;**99**(2):1079–151.
36. Focken T, Chowdhury S, Zenova A, Grimwood ME, Chabot C, Sheng T, et al. Design of conformationally constrained acyl sulfonamide isosteres: identification of N-([1,2,4] triazolo [4,3-a] pyridin-3-yl) methane-sulfonamides as potent and selective h NaV1.7 inhibitors for the treatment of pain. *J Med Chem* 2018;**61**(11):4810–31.

CHAPTER 3

Site-specific Incorporation of Triazole for Functionalization of Nucleotides, Oligonucleotides, and Nucleic Acids

Arruje Hameed[a], Amjad Hameed[b], Ali Raza[c], and Tahir Farooq[c,*]
[a]Department of Biochemistry, Government College University, Faisalabad, Pakistan
[b]Nuclear Institute for Agriculture and Biology (NIAB), Faisalabad, Pakistan
[c]Department of Applied Chemistry, Government College University, Faisalabad, Pakistan
*Corresponding author. E-mail: tahirfarooqfsd@gmail.com

Introduction

Over the last few decades, chemically modified nucleic acids and their oligomers have established their worth as promising research tools in the emerging forefronts of biology, molecular diagnostics, and pharmaceutics.[1,2] Chemical modifications of nucleic acids are suggested to overcome some issues related to their polyanionic backbone and to induce resistance against nucleobase-mediated cleavage.[3] The site-specific chemical modifications of nucleotides, oligonucleotides, and nucleic acids furnish them with novel functionalities and biopotentials.[4] Such site-specific modifications are performed at nitrogenous bases, phosphodiester linkage, or ribose sugar through various standard chemical reactions. Out of various types of chemical modifications, click chemistry has emerged as a leading approach for site-specific incorporation of triazole on the three previously mentioned main areas in nucleic acid analogs.[5] The physiological inertness of both alkyne and azide functionalities has made them attractive entities for expedient, high-yielding, and regiocontrolled Cu-catalyzed alkyne azide cycloadditions (CuAAC) under various environment conditions. For the site-specific incorporation of 1,2,3-triazole, either azido-functionality is introduced on the desired position of aforementioned three main areas and made to react with alkyne containing molecules or the reverse approach is adopted (Figures 1–3).[6]

The site-specific insertion of triazole moiety does help manipulate structural and functional properties of nucleic acids, thus it has become an emerging tool for gene function analyses and development of micro-RNA- and small interfering RNA-based therapeutic agents.[7] Both siRNA and

Advances in Triazole Chemistry
https://doi.org/10.1016/B978-0-12-817113-4.00006-8

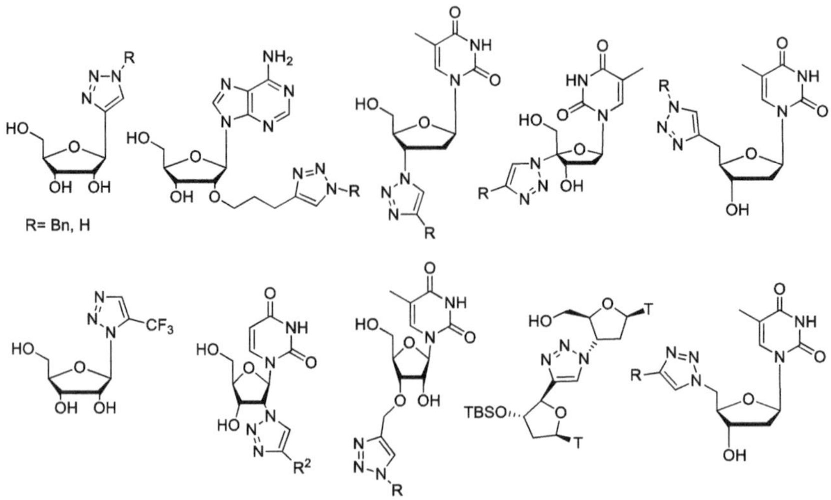

Figure 1 Point-specific incorporation of triazole on sugar starting from azide- or alkyne-functionalized entity.

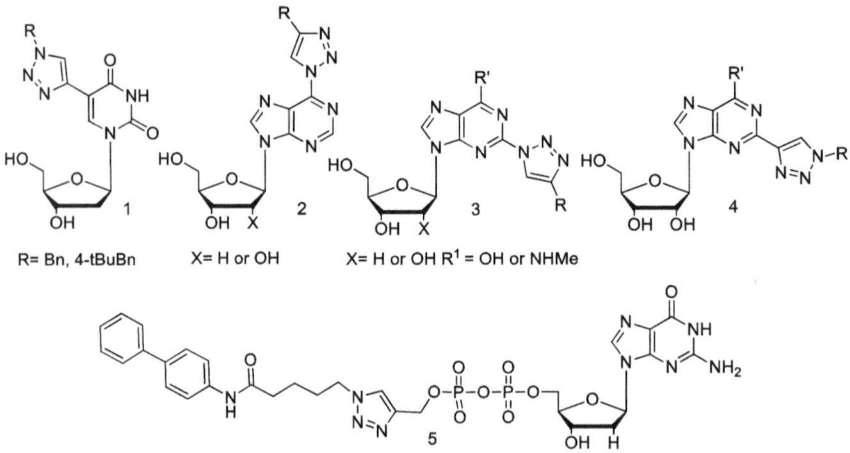

Figure 2 Point-specific incorporation of triazole on nucleobases and phosphodiester.

miRNA are employed as mechanistic tools for post-transcriptional gene silencing in an RNA interface. However, the potential applications of siRNA in gene silencing and as therapeutic agents had always been challenged by delivery and cell membrane penetration issues related to their negative charge and size.[8] In this context, triazole-based chemical modifications of siRNA are recommended for the improvement of pharmacokinetic profiles, avoidance of off-target interactions and immune activation, and enhancement of nuclease resistance.[9,10] The subsequent sections of this chapter

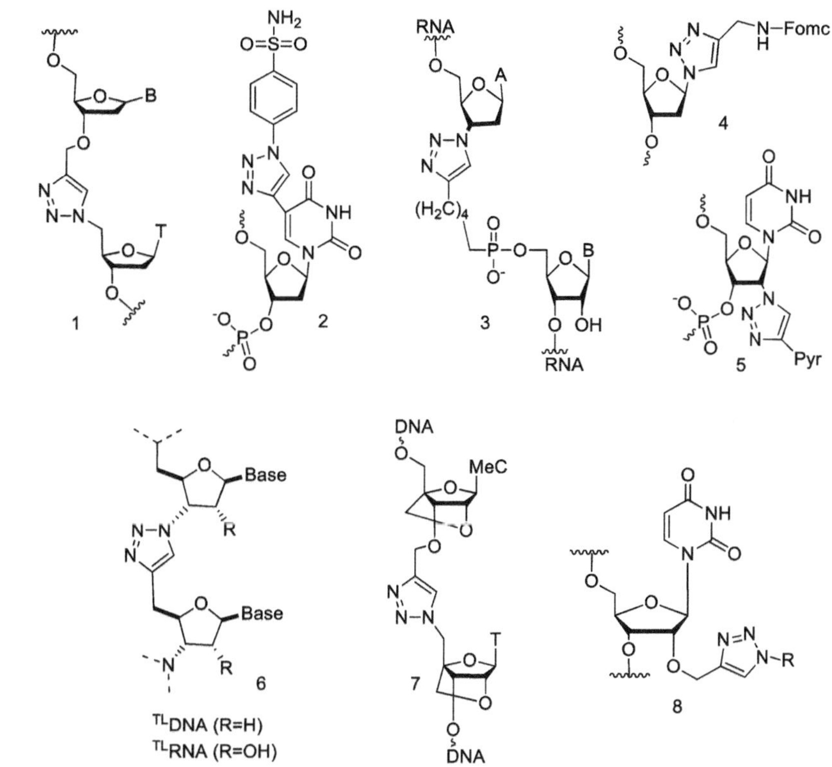

Figure 3 Site-specific triazoles for functionalization of nucleic acids.

highlight the most notable examples of point-specific incorporation of 1,2,3-triazoles on nucleotides, oligonucleotides, or nucleic acids starting with both azide-functionalized and alkyne-functionalized entity.

Point-specific incorporation of 1,2,3-triazole on sugar moiety

Most commonly, the chemical modifications are performed at ribose moiety; however, nucleobases and the backbone are also targeted for such modifications.[11] The siRNAs with 2′-F and 2′-OCH_3 substituents performed relatively better for targeting site-specifically positioned genes. However, such chemical alterations have suffered limitations of tolerance. Aigner and his co-researchers postulated 2′-N_3 as an efficient substituent for potential enhancement of siRNA considering the polarity and size of the group. They described efficient synthesis of 2′-azido containing RNA using a solid-phase approach (Scheme 1 and Figure 4).[12] The 2′-azido substituted ribose potentiated the applications of such nucleoside analogs in bioconjugation and siRNA with additional biorthogonal reactivity.

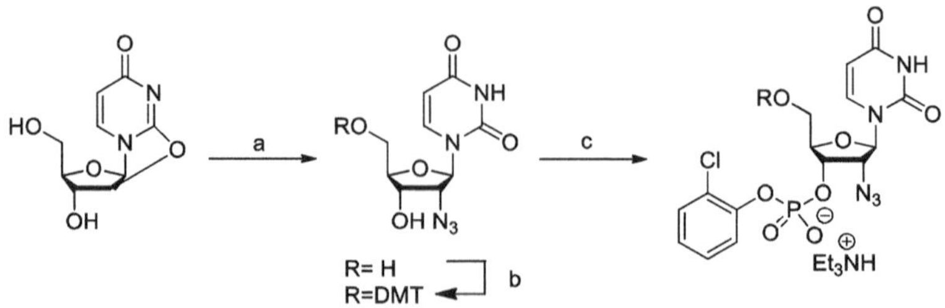

Reagents and conditions:
a) 1.8 eq LiF, 1.8 eq $(CH_3)_3SiN$, in (TMEDA/DMF 1:1), 110°C
b) 1.5 eq 4,4'-dimethoxytriphenylmethyl-chloride in pyridine, rt 24 h
c) 4 eq N-methylimidazole, in THF, rt, 5min ii: 2.5 eq 2-chlorophenyl phosphorodichloridate, 5.5 eq 1,2,4-triazole, 5 eq Et_3N, in THF, rt, 10 min, 86%

Scheme 1 Synthesis of 2′-azido functionalized building block for solid-phase synthesis of RNA.

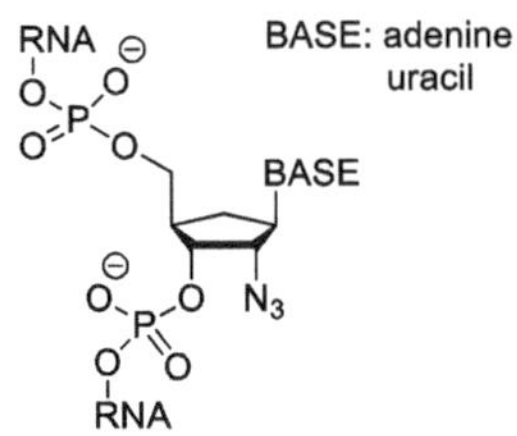

Figure 4 Structure of RNA with site-specifically 2′-azido-modified nucleosides.

Fauster and his colleagues revealed the facile synthesis of 2′-azido guanosine and 2′-azido cytidine as building blocks and site-specifically incorporated them into RNA (Scheme 2).[13] Furthermore, they introduced

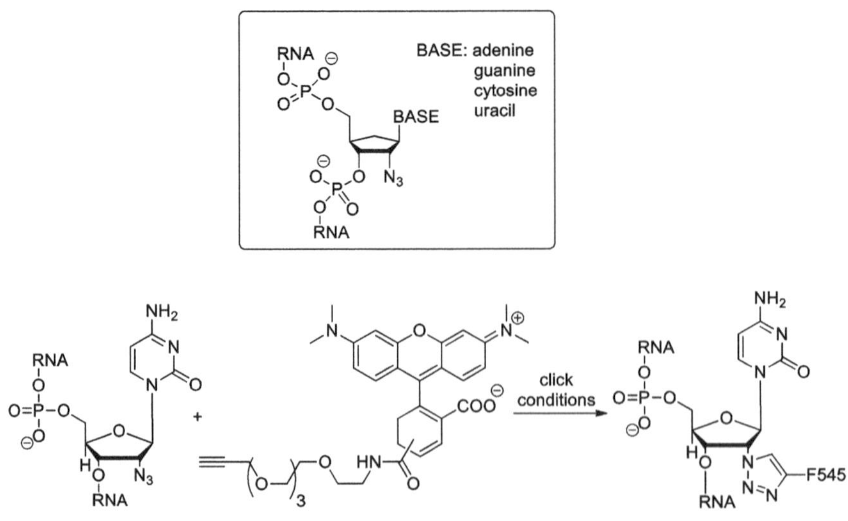

Scheme 2 Facile introduction of florescent dye on 2-azido building blocks.

florescent dye through click chemistry protocol thus establishing the feasibility of reverse attachment of desired substituents compared with previously used alkyne substituted nucleic acids with limited flexibility.[14,15] This example signified the selective labeling of nucleic acid with multiple dyes.

This explained 2′-azido substituents were also employed for the development of double-headed nucleosides with two nucleobases using CuAAC (Scheme 3).[16] The introduction of such double-headed nucleosides improve interaction in DNA duplex. Furthermore, they also provide extra stabilization to three-way junctions both in DNA:RNA and DNA especially in (+1)-zipper motif. Jørgensen and his colleagues envisioned that 1,2,3-triazole containing double-headed nucleosides could be used to synthesize artificial nucleic acids.

The site-specific bis-labeling of long RNA was considered a challenging task until recently when Hall *et al.* came up with a novel strategy (adapted Schemes 4 and 5).[17] They employed 2-*O*-propargyl nucleoside precursor to prepare pre-miRNA. The labeling was used as a sense-probe to access *in vitro* miRNA maturation. The hetero-bis-labeling was performed using click/reverse click strategy.

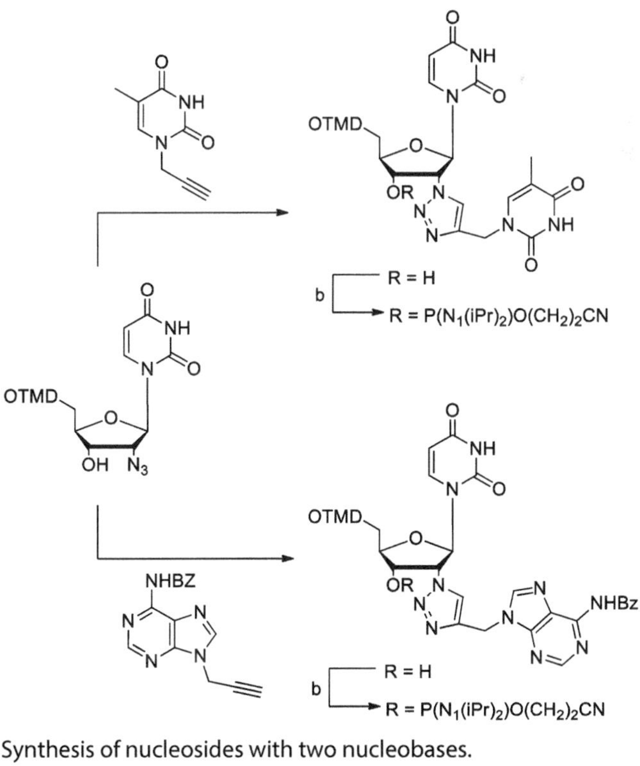

Scheme 3 Synthesis of nucleosides with two nucleobases.

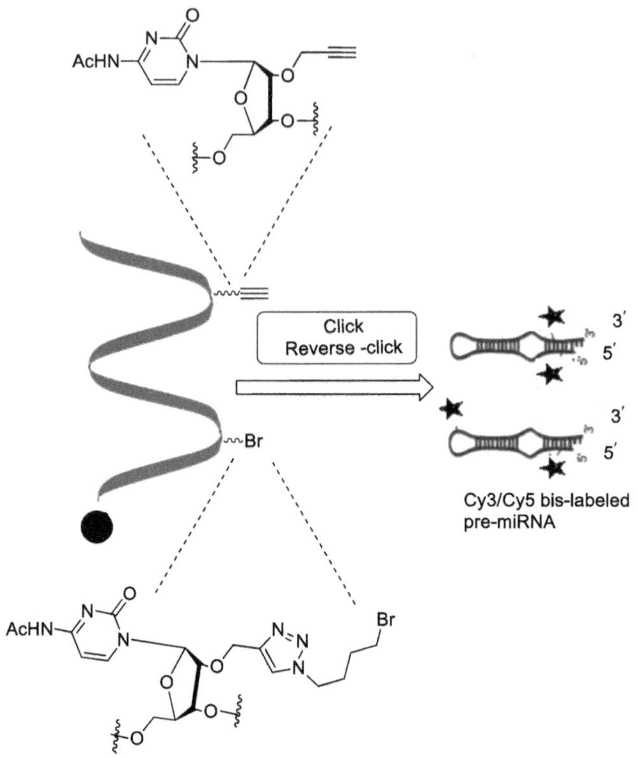

Scheme 4 Synthesis of pre-miRNA.

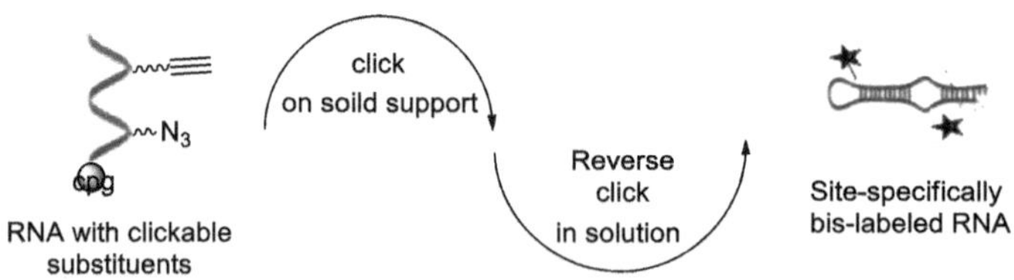

Scheme 5 Click strategy for bis-labeling of nucleic acids.

Astakhova *et al.* reported a facile methodology for the synthesis of oligonucleotides double-labeled with internally positioned cyanine and xanthene fluorescent dyes. The automated solid-phase approach involved the click coupling of different cyanine and xanthene-based azides with 2′-*O*-propargyl uridine (Scheme 6).[18] Promising photophysical properties were demonstrated by these novel probes applicable in *in vivo* imaging and molecular diagnostics. The 2′-*O*-propargyl substituent has played a key role for the internal attachment of a dye pair, which is usually a challenging task.

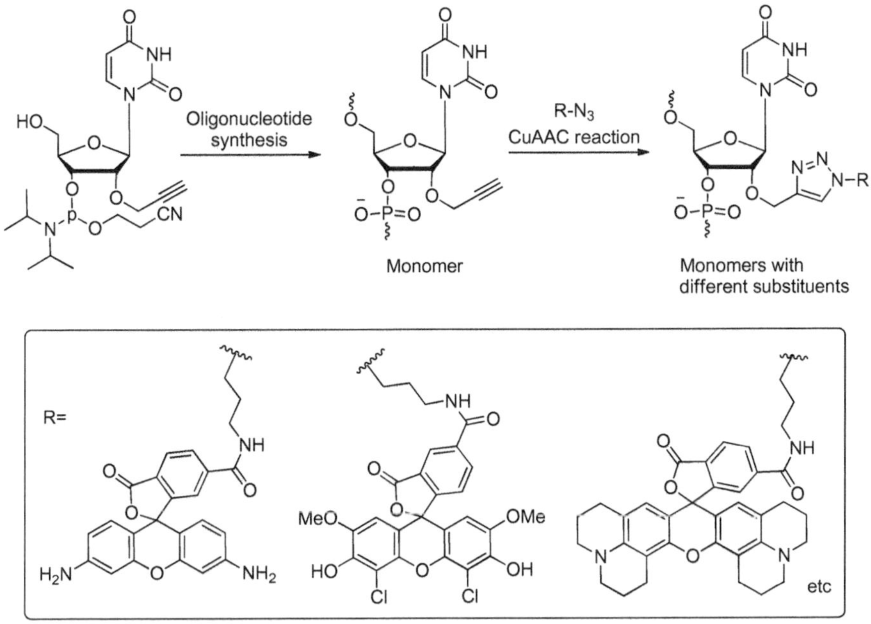

Scheme 6 Oligonucleotides double-labeled with fluorescent dye.

The C3-azido substituted nucleosides were used for the synthesis of anti-mycoplasma nucleosides. These C3-triazole containing nucleosides were developed as substrate for important activation enzyme TK1 in bacteria and humans. The testing of these analogs helped to identify active site's sequence diversity and differentiated the structural and functional features of bacterial and human TK1 (Scheme 7).[19]

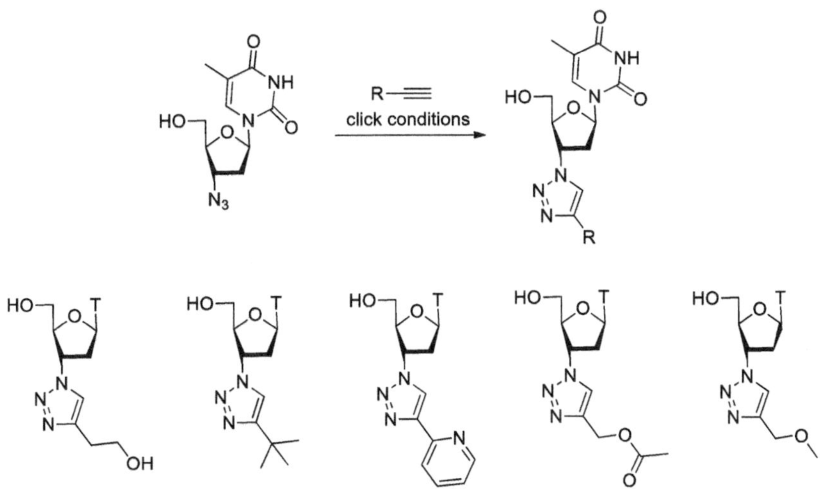

Scheme 7 Synthesis of C3-triazole containing anti-mycoplasma nucleosides.

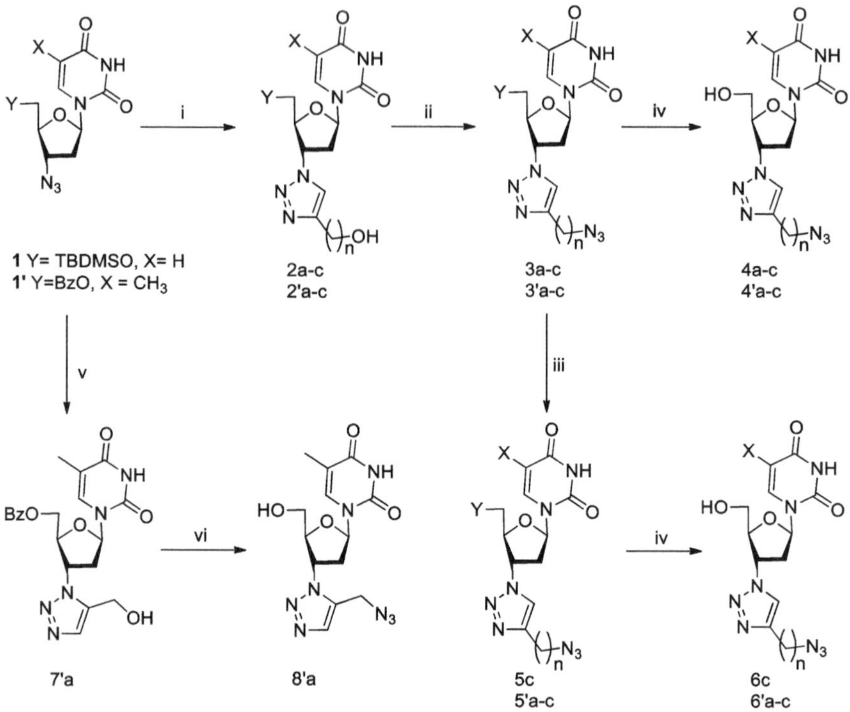

Scheme 8 Synthesis of C3′-1,2,3-triazolo-nucleosides as anti-HIV agents.

Roy and his colleagues reported the synthesis of C3′-1,2,3-triazolo nucleosides as anti-HIV agents. In fact, the nucleosides carried azido-alkyl-chain functionalized-[1,2,3]-triazole substituent at 3′-position (Scheme 8).[20]

The biocompatibility of triazolic-linkage in chemically modified oligonucleotides has become a point of attraction in the emerging field of synthetic nucleic acids (Figure 5). The triazole-backbone mimics the functions of phosphodiester containing DNA in human cells and bacteria; however, the duplex loses the stability with complementary DNA/RNA when compared with original DNA setup. Such binding affinity issues make triazole-linked oligonucleotides misfit to act as antisense, which requires high binding potential to target NA.

Oligonucleotides containing locked nucleic acid (LNA) displayed better selectivity and enhanced affinity to target complementary RNA. The melting point of DNA:RNA duplex increased around 10°C even when a single LNA was incorporated into it.[21] The LNA provides more stability and greater binding potential to oligonucleotides, but they are

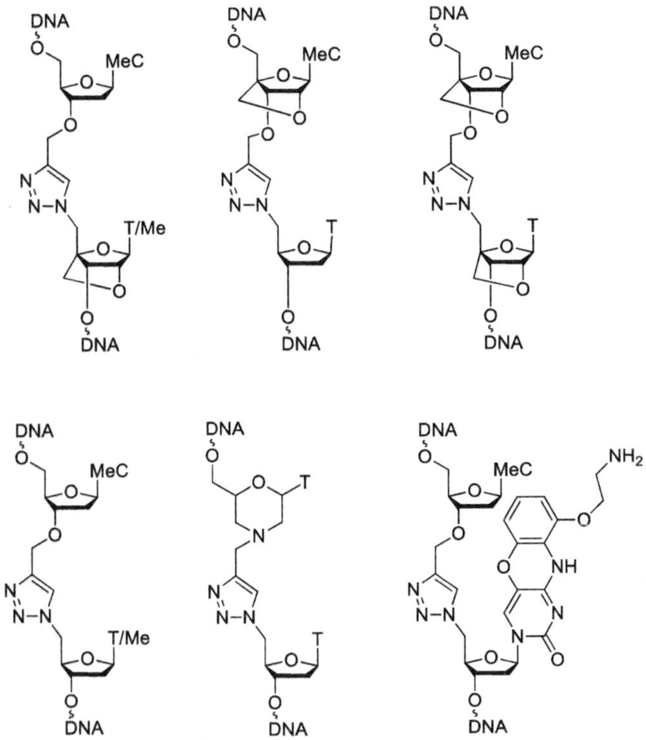

Figure 5 DNA with modified backbone.

still non-resistant to nucleases due to the presence of a phosphodiester backbone.[22,23]

Kumar *et al.* noticed that the incorporation of a triazole backbone into the LNA containing oligonucleotide could serve well to resist *in vivo* degradation without compromising binding affinity. They synthesized triazole-linked LNA using LNA sugars substituted with 3′-*O*-propargyl group under click conditions (Schemes 9 and 10).[24] They concluded that oligonucleotides containing both features were more resistant to degradation with high affinity for complementary RNA compared with those that contain either triazole or LNA only.

Last year, Wu *et al.* disclosed a new strategy for mapping oxidative damages in DNA at single-nucleotide resolution. The strategy involved the utilization of repair enzymes, polymerase-mediated insertion of 3′-*O*-propargyled nucleoside, and click-labeling of a damaged site in a genome (Scheme 11).[25] The click-code-seq high-resolution mapping of the damaged site helps to

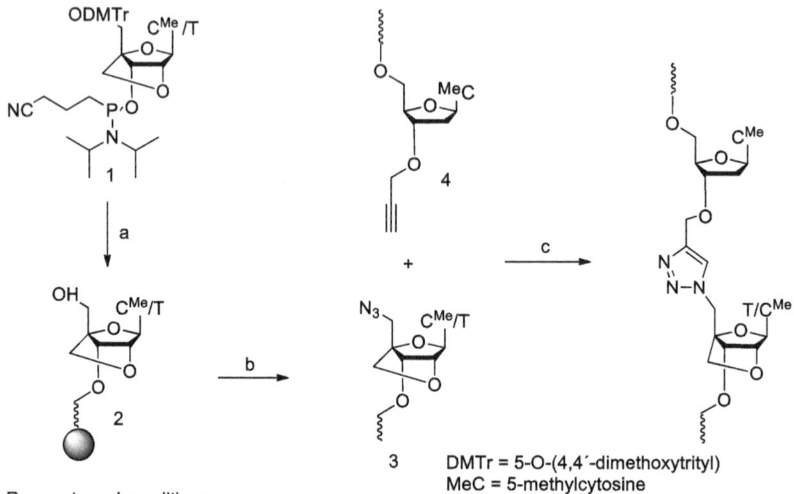

Reagents and conditions:
a) Automated solid phase DNA synthesis. b) i) Methyltriphenoxyphosphonium iodide, DMF, ii) NaN_3, DMF, 55 °C; 3. aq NH_3, 55 °C. c) $CuSO_4$, sodium ascorbate, tris(3-hydroxypropyltriazolylmethyl)amine

Scheme 9 Synthesis of triazole 3′-LNA.

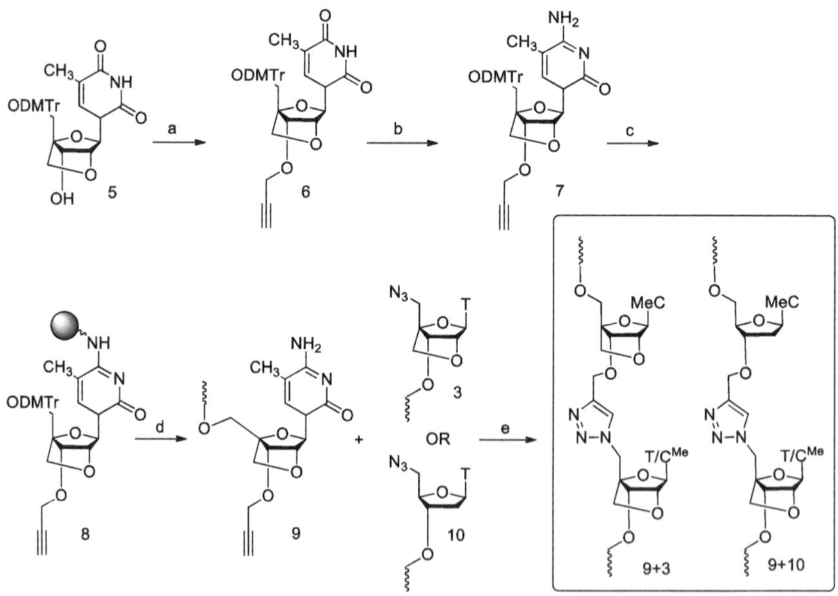

Reagents and conditions:
a) i) NaH, DMF, 0 °C, 30 min, rt, 1 h; ii) propargyl bromide, 0 °C, 30 min, rt, 16 h. b) 1. N-Methylimidazole, 0 °C, 15 min; 2. $POCl_3$, 0 °C, 30 min. 3. Conc. NH_3/H_2O, rt, 16 h. (c) i) ammino resin, succinic anhydride, DMAP, pyridine, rt, 20 h; ii) DIC, HOBT, pyridine, rt, 20 h iii) pentachlorophenol, pyridine, rt, 3 h; 4. piperidine (10% in DMF), rt, 5 min; 5. rt, 1 h. d) Automated solid phase DNA synthesis. (e) $CuSO_4$, sodium ascorbate, tris(3-hydroxypropyltriazolylmethyl)amine.

Scheme 10 Synthesis of 5′-LNA and 5′,3′-LNA.

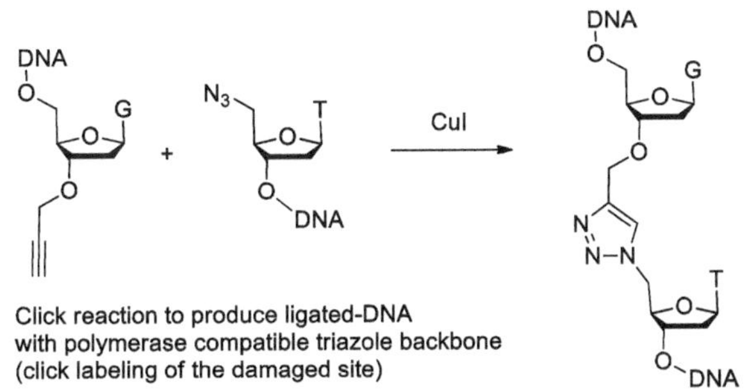

Scheme 11 Polymerase-mediated insertion of 3′-*O*-propargyled nucleoside and click labeling.

understand sequence context, DNA-protein interaction, and patterns of chemical modification.

In a quite recent report, the click DNA ligation approach has been employed for one-pot chemical synthesis of oligonucleotides to further their assembly to produce a gene. Kukwikila *et al.* used oligonucleotides containing 3-alkyne and 5-azide units for the synthesis of a base-pair gene (adapted Scheme 12).[26] This gene is capable of encoding green fluorescent protein iLOV. This newly synthesized gene carries eight 1,2,3-triazole units at a chemical ligation site, and even then showed biocompatibility and underwent replication by DNA polymerases and encoded a functioning iLOV protein *in vitro* in *Escherichia coli*. This one-pot strategy was found to be a powerful approach for assembly of genes because previously the enzyme-mediated assembly of oligonucleotides into a gene and genome was considered a challenging task.

Over the last few years, the study of enzymatic activities has greatly been influenced by the applications of activity-based probes (ABPs) in enzyme inhibitor screenings, proteomic profiling, and *in vivo* imaging. Many classes of enzymes have been targeted using ABPs. An ABP generally contains a tag reporter of fluorophores for detection, a linker, and a reactive group for covalent labeling of an active site on an enzyme. Kim and his colleagues studied enzymatic activities by parallel DNA sequencing or quantitative PCR, a new immunoassay approach involving DNA-linked ABP.

As a model study, oligonucleotide-probe (conjugate of 3′-alkynyl-5′-fluorescein 20-mer oligonucleotide and an azido-fluorophosphonate) was used for labeling of an active site of serine hydrolase (adapted Scheme 13).[27] The lower sample requirements, throughput, and improved sensitivity are the advantages of this DNA-based approach.

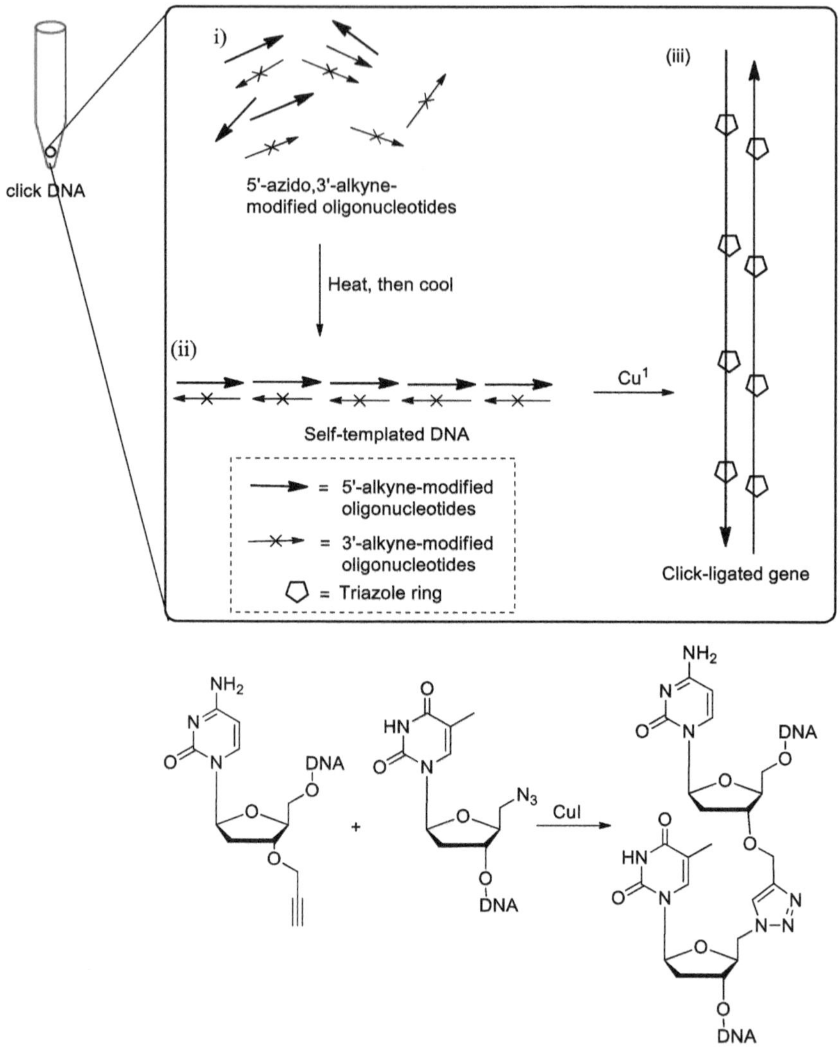

Scheme 12 Gene synthesis by one-pot methodology.

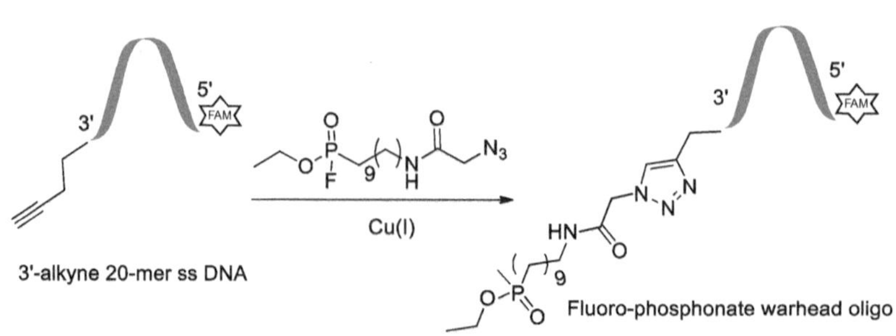

Scheme 13 Synthesis of oligonucleotide probe.

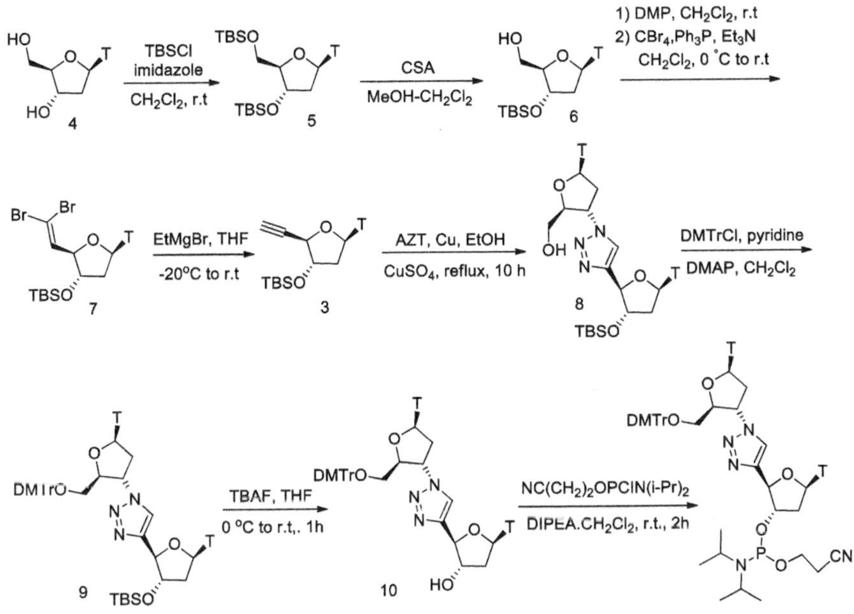

modified dinner (TT') 1

Scheme 14 Synthesis of dinucleoside phosphoramidite with triazole as linker.

Idris and his colleagues described the synthesis of dinucleoside phosphoramidite with triazole as linker (Scheme 14). In a solid-phase synthetic approach, they incorporated it into 10- and 12-mer oligonucleotides. From a therapeutic perspective, such modifications are required for improved pharmacokinetic and extra stability of oligonucleotides (Scheme 15).[28] The synthesis involved the CuAAC of C4-alkyne substituted thymidine.

The 4′-azidothymidine reacted with different alkynes under Cu(I)- and Ru-(II)-catalysis to generate novel nucleosides with corresponding 1,2,3-triazoles. However, the steric hindrance failed the Ru-catalyzed cycloadditions and the 4′-triazolo nucleosides received from Cu-catalysis showed no activity against HCV and moderate activity against influenza A virus and HIV-1 (Schemes 16 and 17).[29] Wang and his colleagues suggested

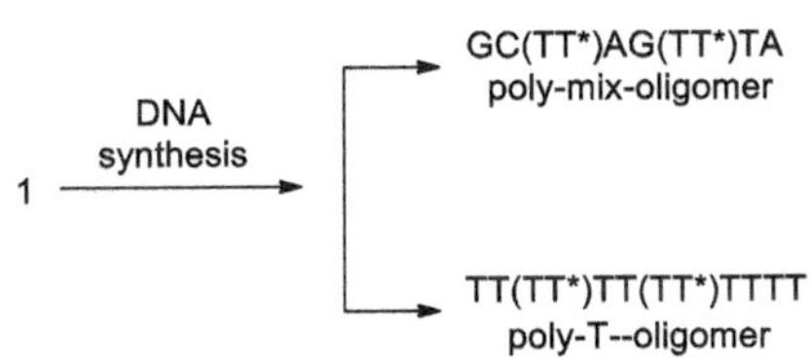

Scheme 15 Oligomer synthesis from modified dimer.

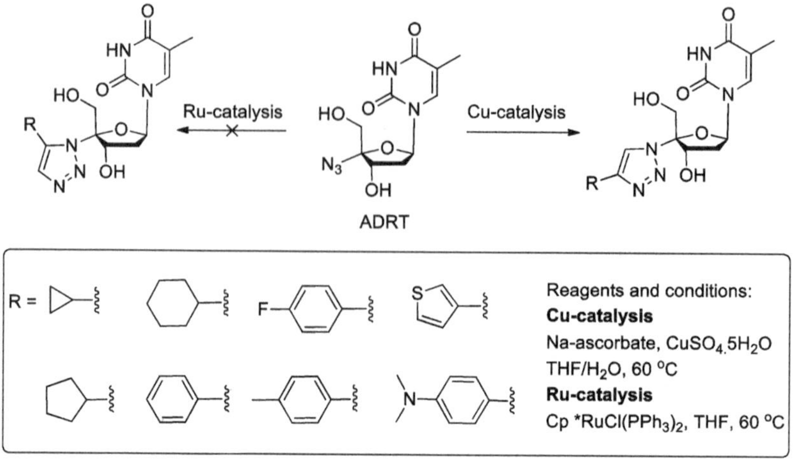

Scheme 16 Cu- and Ru-catalysis for synthesis of 4-triazolo nucleosides.

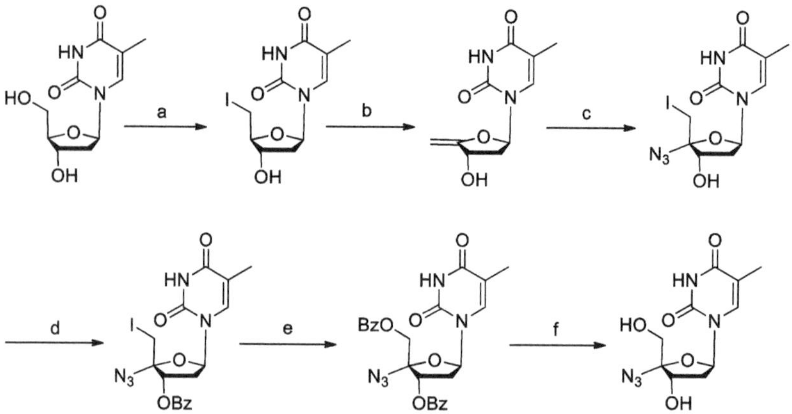

Reagents and conditions:
a) I_2, PPh_3, imidazole, THF, 25 °C, 18 h, 77%; b) NaOMe, MeOH, 60 °C,15 h, 81%; c) $[Bn(Et)_3N]Cl$, NaN_3, NMO, I_2, THF, 5 °C, 2 h, 52%; d) BzCl, DMAP, pyridine, rt, 3 h; e) $(Bu_4N)HSO_4$, K_2HPO_4, m-chlorobenzoic acid, m-CPBA, DCM/H_2O (4:1), rt, 15 h, 84%; f) 7N NH_3/MeOH, rt, 12 h, 79%.

Scheme 17 Synthesis of ADRT.

that the chemical modification of 4′-azido compromised the antiviral potential of such nucleosides.

Potent anti-HIV1 activity was shown by novel 4′-*C*-triazole containing nucleoside analogs without any cytotoxic effect. Wu *et al.* recommended them as suitable candidates to develop potent NTRIs (Scheme 18).[30]

Since 1990, interest has been shown for the synthesis of oligonucleotides containing triazolic-moiety instead of phosphate with improved

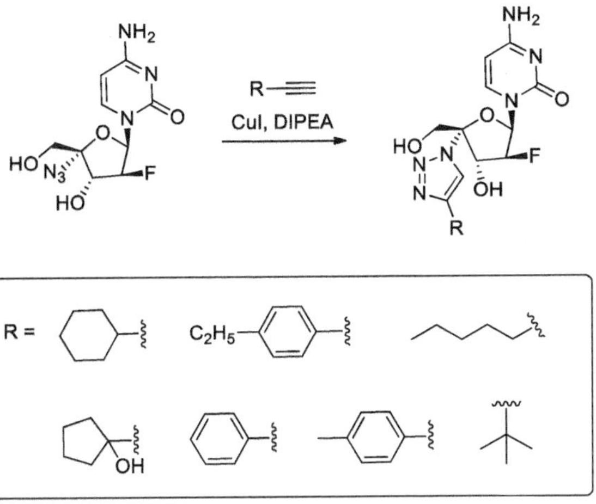

Scheme 18 Synthesis of 4′-*C*-triazole containing nucleoside analogs.

therapeutic profiles (Figure 6). The advent of CuAAC has boosted the applications of such analogs in nanotechnology, biotechnology, and synthetic biology over the last few years. However, triazole containing such analogs exhibited issues of water solubility and low binding affinity in complex biological environment.

Sharma *et al.* described the novel synthesis of locked nucleotides linked through 1,2,3-triazole replacing phosphate as a linker. The synthesis utilized C4-alkynyl substituted nucleosides for CuAAC reactions (Scheme 19).[31]

The incorporation of triazole-linker at internal position of duplex reduced binding affinity of locked nucleic acid (LNA). The dimers with locked sugars and triazole substituents at 3′ or 5′-termini displayed excellent nuclease stability and better binding affinity representing the flexibility of duplex at specific positions.

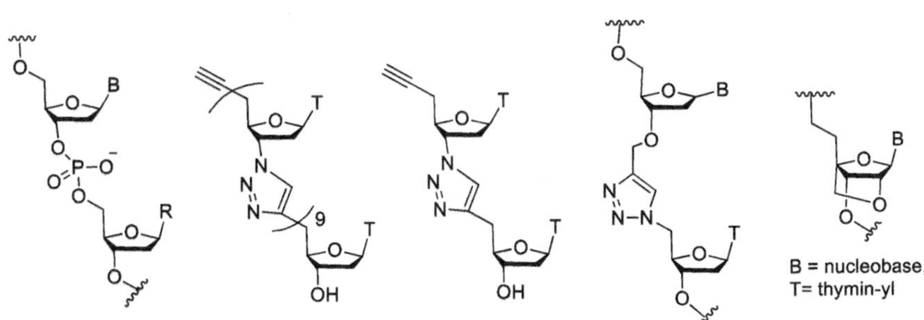

1 2 3 4 5

DNA 1, triazole-linked decamer 2, triazole-linked dimer 3, biocompatible triaxole linkage 4, and LNA 5

Figure 6 Structurally modified DNA.

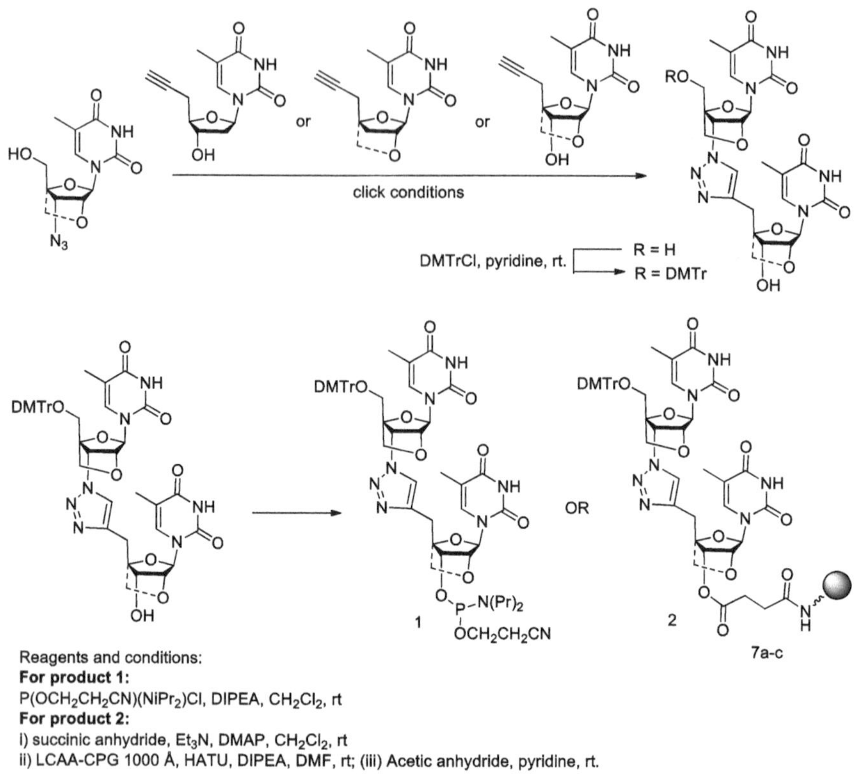

Scheme 19 Preparation and solid support derivatization of triazole-linked LNA.

More than a decade ago, a novel solid-phase synthesis of ribonucleic acid analog containing triazole-linker (^{TL}RNA) was disclosed starting from D-xylose (Schemes 20 and 21).[32] The triazole linker, being a non-degradable entity, was suggested to provide extra integrity to ^{TL}RNA, which mimics the structural and biological features of RNA (Figure 7).

Very recently, Malkowski and his research group introduced azido group at 5′-*C*-position of the nucleosides employing tosyl-azide (Scheme 22).[33] Furthermore, propargylglycine and alkylated serine were made to react with 5′-azido-nucleosides under CuAAC conditions furnishing novel 5′-triazole-nucleoside-amino acid conjugates (Figure 8). The amino acids were used due to their tunable properties.

The 5′-triazole-nucleoside with adenosines and cytidines were prepared and evaluated as α-2,3-sialyltransferase inhibitors. These structural analogs of cytidine diphosphate were designed as sialyltransferase inhibitors as part of an emerging cancer treatment strategy (Figure 9). The elevated activity

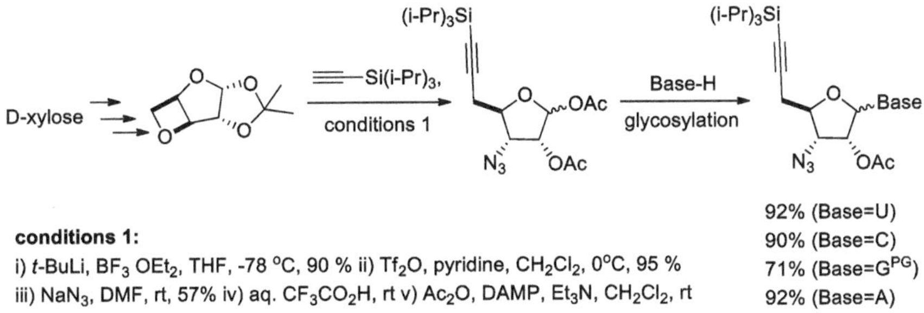

Scheme 20 Synthesis of ribonuleosides.

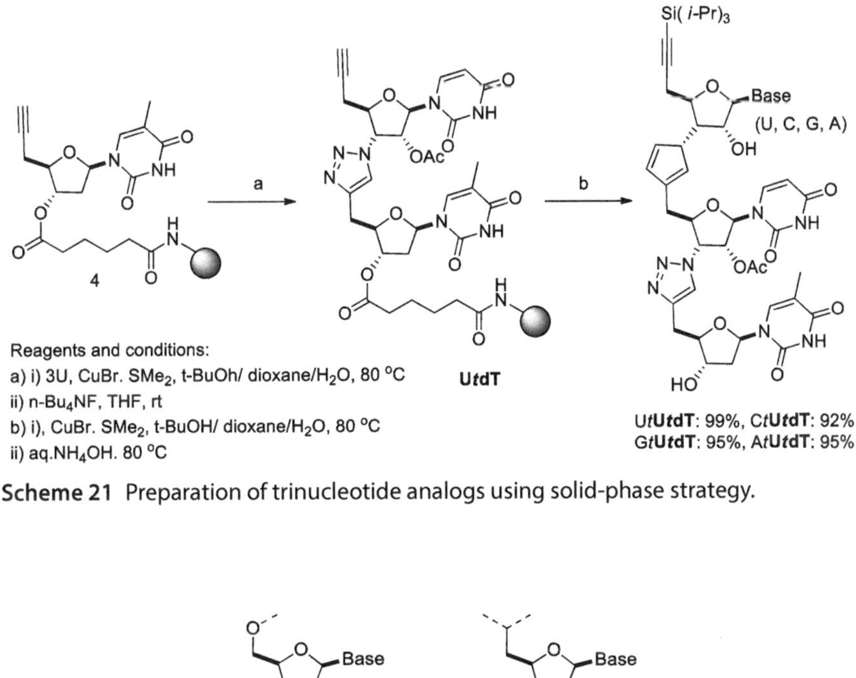

Scheme 21 Preparation of trinucleotide analogs using solid-phase strategy.

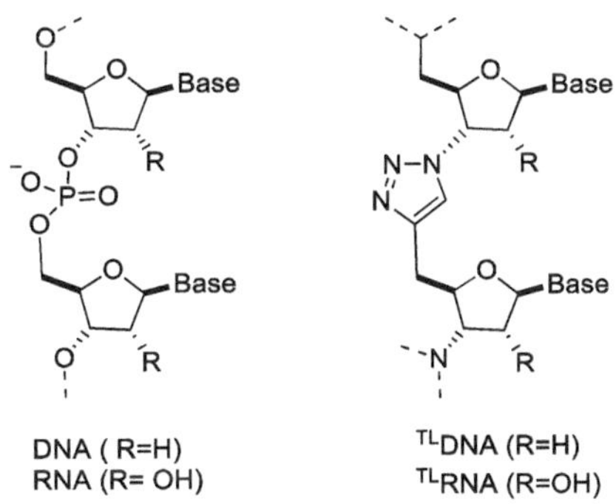

Figure 7 Structures of DNA, RNA, ^{TL}DNA, ^{TL}RNA.

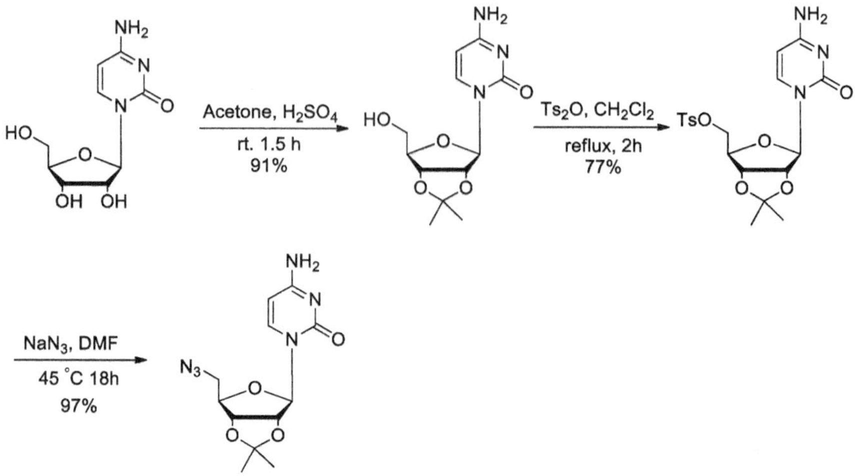

Scheme 22 Synthesis of 5′-azido-nucleosides.

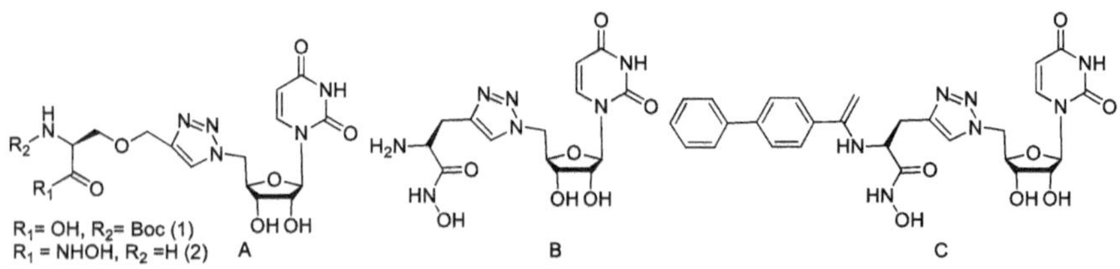

Figure 8 5′-Triazole-nucleoside-amino acid conjugates.

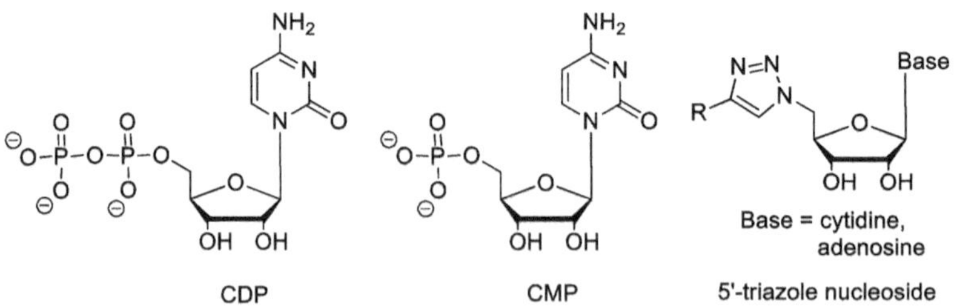

Figure 9 Structures of CDP, CMPn and their triazole mimics.

of enzyme is a potential marker of tumorogenesis especially for breast carcinomas. The synthesized 5′-triazole-cytidine analogs were found to be more effective inhibitors compared with 5′-triazole-adenosines (Scheme 23).[34]

Point-specific Incorporation of triazole on nucleobases

Due to the availability of modifiable sites, nucleobases are relatively more attractive areas for chemical modifications to produce variously functionalized derivatives of nucleotides, oligonucleotides, and nucleic acids.

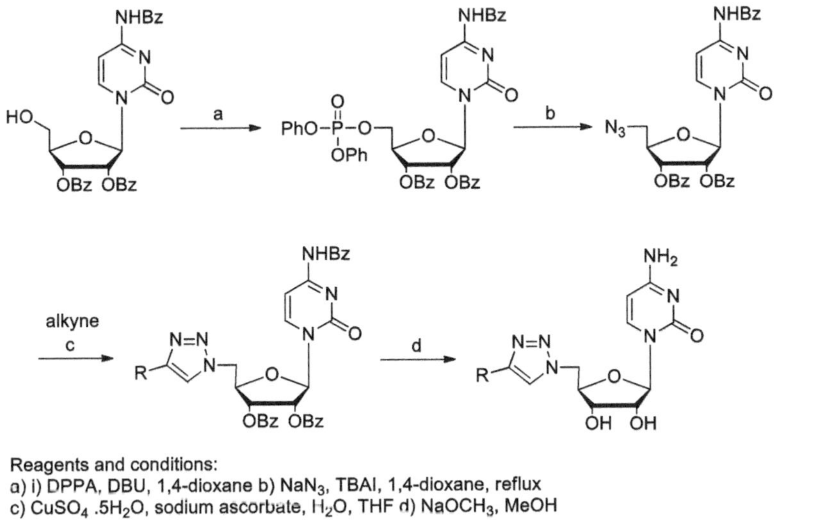

Scheme 23 The 5′-triazole-adenosines as α-2,3-sialyltransferase inhibitors.

The site-specific incorporation of such functionalized nucleobases does not disturb the natural Watson-crick interactions in RNA and DNA duplex.[35,36]

Chemical modifications of nucleobases induce fluorescent properties in nucleic acids that are otherwise weakly fluorescent in nature. The fluorescent properties due to triazole-based modifications on nucleobases depends on the attachment site of a triazole and the substituent(s) attached to it. So, site-specific incorporation of differently substituted triazoles on nucleobases is being used as an emerging approach to study structures, activities, and interactions of nucleic acids.[37]

In the last few years, fluorogenic DNA probes have been well developed by mainly exploiting the chemopotential of 7-alkynyl deazapurine nucleosides. However, unavailability of facile synthetic procedures for 8-alkynyl- and 8-azido-7-deazapurine nucleosides made them less attractive entities until recently. Kavoosi *et al.* disclosed the facile conversion of 8-halo-7-deazapurine nucleosides to 8-azido-7-deazapurine nucleosides, which subsequently underwent strain-promoted click reactions (Schemes 24–26).[38] Furthermore, C8-H of 7-deazapurine was functionalized with benzotriazoles through I_2-mediated strategy furnishing benzotriazole containing nucleosides.

Contrary to this, Wen and his research group utilized 8-alkynyl-guanine and adenosine nucleosides for the development of monosubstituted-1,2,3-triazole containing fluorescent probes (Scheme 27).[39] These nucleosides displayed excellent fluorescent properties, and the DNA polymerase β was used for incorporation of synthesized 8-(1*H*-1,2,3-triazol-4-yl)-2′-deoxyadenosine 5′-phosphate (8-TrzdATP) into one nucleotide-gap containing DNA duplex (Scheme 28).

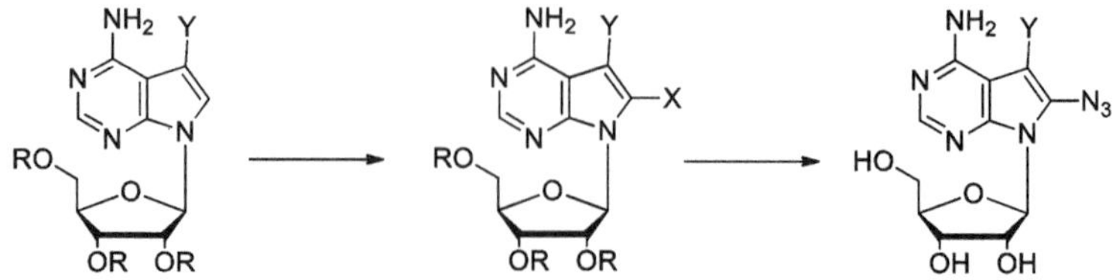

1a: R= Ac, Y= H
1b: R= H, Y= H
1c: R= H, Y= CN
1d: R= H, Y= $CONH_2$

2a: R= Ac, Y= Br, X= H
2b: R= Ac, Y= Br, X= Br
2c: R= H, Y= Br, X= H
2d: R= H, Y= Br, X= Br
2e: R= H, Y= H, X= Br
2f: R= H, Y= CN, X= Br
2g: R= H, Y= $CONH_2$, X= Br

3a: R= H
3b: R= CN
3c: R= $CONH_2$

conditions and reagents:
a) i) NBS or DBH ii) NH_3/ MeOH b) NaN_3/ DMF, rt, 16 h

Scheme 24 Preparation of 8-azido-7-deazapurine nucleosides.

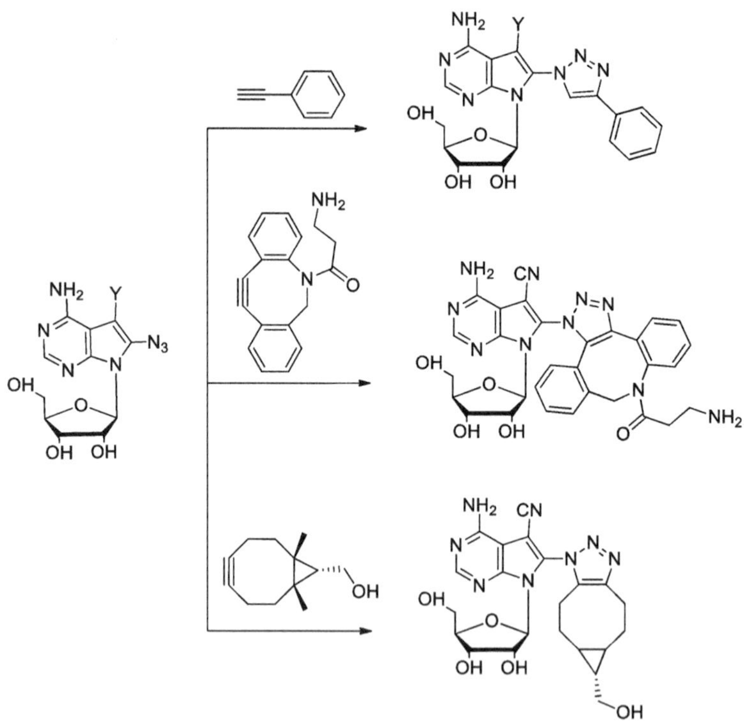

Scheme 25 Cycloaddition reactions of 8-azido-7-deazapurine nucleosides.

Cosyn *et al.* synthesized 2-(1,2,3-triazolyl)adenosine derivatives as A_3 receptor antagonist, partial agonist, and agonists employing click chemistry for site-specific incorporation of triazoles (Scheme 29).[40] Two series of compounds were synthesized starting with 2-azido or 2-alkynyl substituted nucleobases. Few novel compounds were found as promising leads for further pharmacological studies.

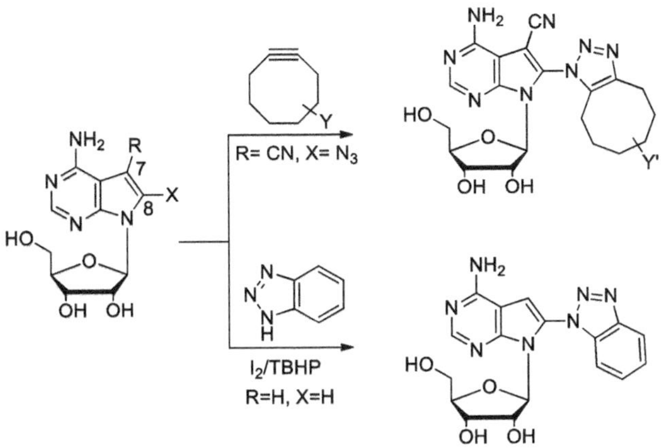

Scheme 26 Strain-promoted click reaction and attachment of benzotriazole.

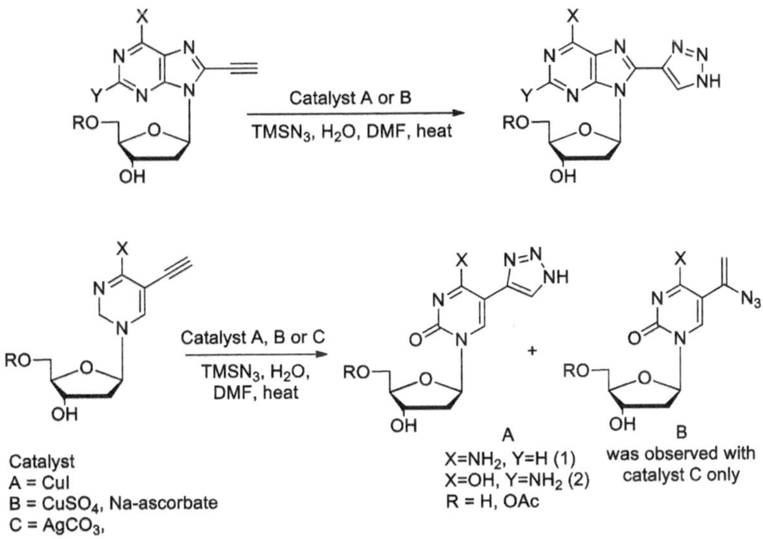

Scheme 27 Synthesis of nucleosides with monosubstituted-1,2,3-triazole.

In 2016, Mayer and his research group used nucleobase-modified aptamer to develop a cost-effective, rapid, and novel production strategy for alkyne-modified nucleic acids on a large scale.This high fidelity solid-phase approach avoids the production of an oxidative by-product, which generally appears due to alkaline work-up. This quantitative, modular, and post-functionalization route has enabled the inexpensive preparation of DNA-based nano-architectures with wide scope of applications (Schemes 30 and 31).[35]

In the last year, Liu and his co-workers used 5-alkyne functionalized nucleobases and 5-azido sugar moieties to synthesize nucleoside macrocycles (Scheme 32).[41] The click chemistry protocol provided high yields

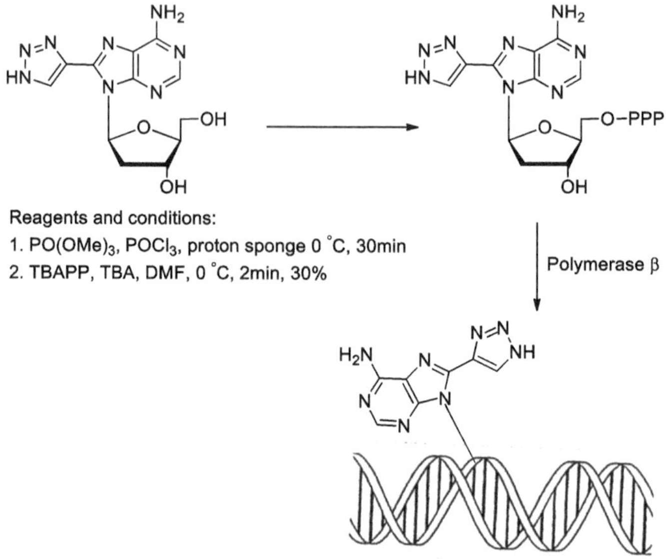

Scheme 28 Preparation and incorporation of 8-TrzdATP into DNA.

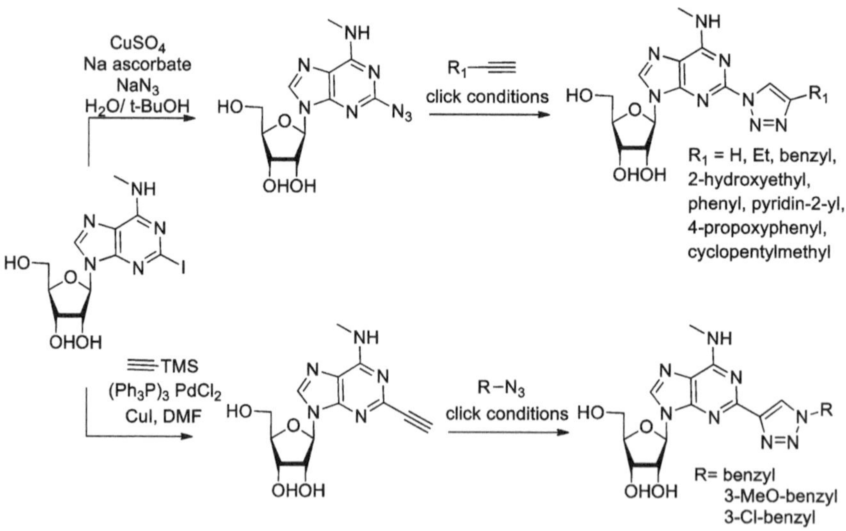

Scheme 29 Synthesis of triazole containing adenosine derivatives.

of macrocycles without involving any protecting group. The Watson-Crick recognition sites were easily available on these macrocycles for base pairing with proteins and nucleic acids. Such macrocyclic nucleosides with high lipophilic nature were proposed as effective transmembrane delivery agents for nucleotides and oligonucleotides.

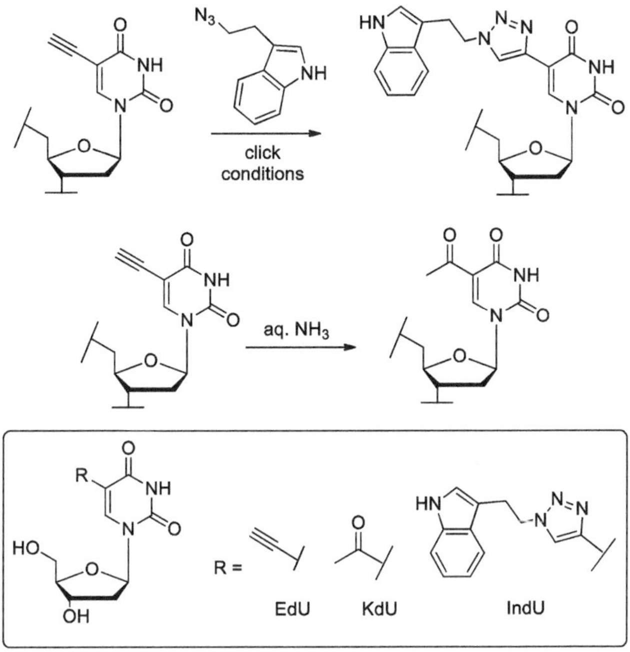

Scheme 30 Functionalization of DNA and nucleosides and by-product formation.

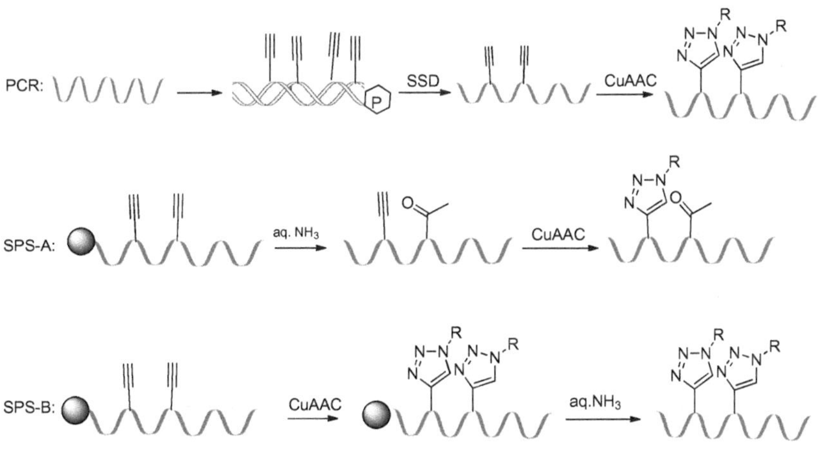

Scheme 31 Different routes for the preparation of triazole functionalized DNA.

Conclusion

The physiological inertness and ease of synthesis of 1,2,3-triazoles has made them instrumental for the chemical modification of nucleic acids. The CuAAC reactions have widely been used for attaching ligands, lipophobic,

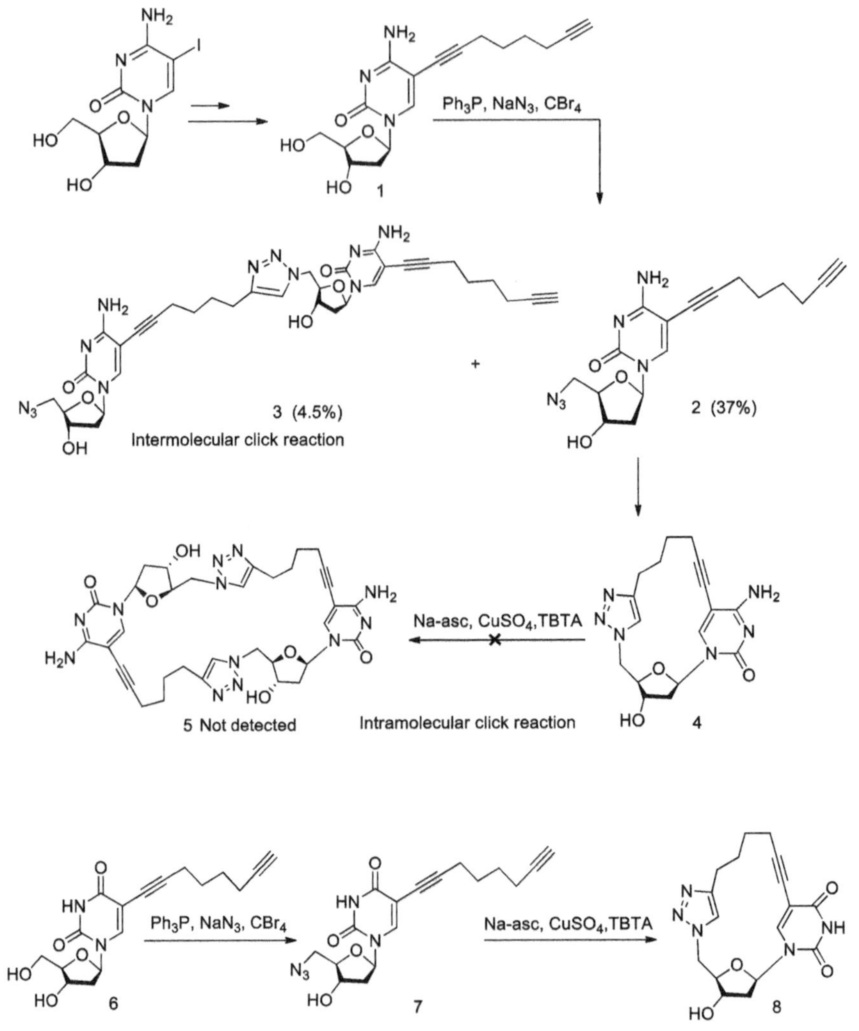

Scheme 32 1,2,3-Triazole containing synthetic nucleoside macrocycles.

lipophilic, and fluorophore functionalities to nucleosides, nucleotides, and oligonucleotides. The availability of clickable sites on natural nucleosides create opportunities to develop novel triazole-derivatives with applications in therapeutics, gene-silencing, bioconjugation, and nucleic acid labeling through a modular click chemistry approach. The triazole-based chemical modifications of nucleic acids hold a promising future for the development of novel derivatives due to the ease of site-specific introduction of triazole on a variety of available clickable sites. This chapter highlighted some remarkable 1,2,3-triazole-based site-specific functionalization of nucleotides, oligonucleotides, and nucleic acids starting from azide- or alkyne-functionalized entities.

References

1. Schrum JP, Siddiqi S, Ejebe K. *Modified nucleosides, nucleotides, and nucleic acids, and uses thereof.* Google Patents; 2016.
2. Ni S, Yao H, Wang L, Lu J, Jiang F, Lu A, et al. Chemical modifications of nucleic acid aptamers for therapeutic purposes. *Int J Mol Sci* 2017;**18**(8):1683.
3. Hagen G, Peel BJ, Samis J, Desaulniers J-P. Synthesis and in vitro assessment of chemically modified siRNAs targeting BCL2 that contain 2′-ribose and triazole-linked backbone modifications. *MedChemComm* 2015;**6**(6):1210–5.
4. Kuwahara M, Sugimoto N. Molecular evolution of functional nucleic acids with chemical modifications. *Molecules* 2010;**15**(8):5423–44.
5. El-Sagheer AH, Brown T. Click chemistry with DNA. *Chem Soc Rev* 2010;**39**(4):1388–405.
6. Krishna H, Caruthers MH. Alkynyl phosphonate DNA: a versatile "click" able backbone for DNA-based biological applications. *J Am Chem Soc* 2012;**134**(28):11618–31.
7. Kurreck J. RNA interference: from basic research to therapeutic applications. *Angew Chem Int Ed* 2009;**48**(8):1378–98.
8. Deng Y, Wang CC, Choy KW, Du Q, Chen J, Wang Q, et al. Therapeutic potentials of gene silencing by RNA interference: principles, challenges, and new strategies. *Gene* 2014;**538**(2):217–27.
9. Efthymiou TC, Huynh V, Oentoro J, Peel B, Desaulniers J-P. Efficient synthesis and cell-based silencing activity of siRNAS that contain triazole backbone linkages. *Bioorg Med Chem Lett* 2012;**22**(4):1722–6.
10. Selvam C, Mutisya D, Prakash S, Ranganna K, Thilagavathi R. Therapeutic potential of chemically modified si RNA: recent trends. *Chem Biol Drug Des* 2017;**90**(5):665–78.
11. Shukla S, Sumaria CS, Pradeepkumar P. Exploring chemical modifications for siRNA therapeutics: a structural and functional outlook. *ChemMedChem* 2010;**5**(3):328–49.
12. Aigner M, Hartl M, Fauster K, Steger J, Bister K, Micura R. Chemical synthesis of site-specifically 2′-azido-modified RNA and potential applications for bioconjugation and RNA interference. *ChemBioChem* 2011;**12**(1):47–51.
13. Fauster K, Hartl M, Santner T, Aigner M, Kreutz C, Bister K, et al. 2′-Azido RNA, a versatile tool for chemical biology: synthesis, X-ray structure, siRNA applications, click labeling. *ACS Chem Biol* 2012;**7**(3):581–9.
14. Paredes E, Evans M, Das SR. RNA labeling, conjugation and ligation. *Methods* 2011;**54**(2):251–9.
15. Motorin Y, Burhenne J, Teimer R, Koynov K, Willnow S, Weinhold E, et al. Expanding the chemical scope of RNA: methyltransferases to site-specific alkynylation of RNA for click labeling. *Nucleic Acids Res* 2010;**39**(5):1943–52.
16. Jørgensen AS, Shaikh KI, Enderlin G, Ivarsen E, Kumar S, Nielsen P. The synthesis of double-headed nucleosides by the CuAAC reaction and their effect in secondary nucleic acid structures. *Org Biomol Chem* 2011;**9**(5):1381–8.
17. Pradère U, Hall J. Site-specific difunctionalization of structured RNAs yields probes for microRNA maturation. *Bioconjug Chem* 2016;**27**(3):681–7.
18. Astakhova IK, Wengel J. Interfacing click chemistry with automated oligonucleotide synthesis for the preparation of fluorescent DNA probes containing internal xanthene and cyanine dyes. *Chem Eur J* 2013;**19**(3):1112–22.
19. Lin J, Roy V, Wang L, You L, Agrofoglio LA, Deville-Bonne D, et al. 3′-(1, 2, 3-Triazol-1-yl)-3′-deoxythymidine analogs as substrates for human and Ureaplasma

parvum thymidine kinase for structure–activity investigations. *Bioorg Med Chem* 2010;***18***(9):3261–9.

20. Roy V, Obikhod A, Zhang H-W, Coats SJ, Herman BD, Sluis-Cremer N, et al. Synthesis and anti-HIV evaluation of 3′-triazolo nucleosides. *Nucleosides Nucleotides Nucleic Acids* 2011;**30**(4):264–70.
21. Kaur H, Babu BR, Maiti S. Perspectives on chemistry and therapeutic applications of locked nucleic acid (LNA). *Chem Rev* 2007;**107**(11):4672–97.
22. Hagedorn PH, Persson R, Funder ED, Albæk N, Diemer SL, Hansen DJ, et al. Locked nucleic acid: modality, diversity, and drug discovery. *Drug Discov Today* 2018;**23**(1):101–14.
23. Soler-Bistué A, Zorreguieta A, Tolmasky ME. Bridged nucleic acids reloaded. *Molecules* 2019;**24**(12):2297.
24. Kumar P, El-Sagheer AH, Truong L, Brown T. Locked nucleic acid (LNA) enhances binding affinity of triazole-linked DNA towards RNA. *Chem Commun* 2017;**53**(63):8910–3.
25. Wu J, McKeague M, Sturla SJ. Nucleotide-resolution genome-wide mapping of oxidative DNA damage by click-code-seq. *J Am Chem Soc* 2018;**140**(31):9783–7.
26. Kukwikila M, Gale N, El-Sagheer AH, Brown T, Tavassoli A. Assembly of a biocompatible triazole-linked gene by one-pot click-DNA ligation. *Nat Chem* 2017;**9**(11):1089.
27. Kim D, Jetson RR, Krusemark CJ. A DNA-assisted immunoassay for enzyme activity via a DNA-linked, activity-based probe. *Chem Commun* 2017;**53**(68):9474–7.
28. Chandrasekhar S, Srihari P, Nagesh C, Kiranmai N, Nagesh N, Idris MM. Synthesis of readily accessible triazole-linked dimer deoxynucleoside phosphoramidite for solid-phase oligonucleotide synthesis. *Synthesis* 2010;**2010**(21):3710–4.
29. Vernekar SKV, Qiu L, Zacharias J, Geraghty RJ, Wang Z. Synthesis and antiviral evaluation of 4′-(1, 2, 3-triazol-1-yl) thymidines. *MedChemComm* 2014;**5**(5):603–8.
30. Wu J, Yu W, Fu L, He W, Wang Y, Chai B, et al. Design, synthesis, and biological evaluation of new 2′-deoxy-2′-fluoro-4′-triazole cytidine nucleosides as potent antiviral agents. *Eur J Med Chem* 2013;**63**:739–45.
31. Sharma VK, Singh SK, Krishnamurthy PM, Alterman JF, Haraszti RA, Khvorova A, et al. Synthesis and biological properties of triazole-linked locked nucleic acid. *Chem Commun* 2017;**53**(63):8906–9.
32. Fujino T, Endo K, Yamazaki N, Isobe H. Synthesis of triazole-linked analogues of RNA (TLRNA). *Chem Lett* 2012;**41**(4):403–5.
33. Malkowski S, Dishuck C, Lamanilao G, Embry C, Grubb C, Cafiero M, et al. Design, modeling and synthesis of 1, 2, 3-triazole-linked nucleoside-amino acid conjugates as potential antibacterial agents. *Molecules* 2017;**22**(10):1682.
34. Lee L, Chang KH, Valiyev F, Liu HJ, Li WS. Synthesis and biological evaluation of 5′-triazole nucleosides. *J Chin Chem Soc* 2006;**53**(6):1547–55.
35. Tolle F, Rosenthal M, Pfeiffer F, Mayer Gn. Click reaction on solid phase enables high fidelity synthesis of nucleobase-modified DNA. *Bioconjug Chem* 2016;**27**(3):500–3.
36. Xu W, Chan KM, Kool ET. Fluorescent nucleobases as tools for studying DNA and RNA. *Nat Chem* 2017;**9**(11):1043.
37. Das S, Samanta PK, Pati SK. Watson–Crick base pairing, electronic and photophysical properties of triazole modified adenine analogues: a computational study. *New J Chem* 2015;**39**(12):9249–56.
38. Kavoosi S, Rayala R, Walsh B, Barrios M, Gonzalez WG, Miksovska J, et al. Synthesis of 8-(1, 2, 3-triazol-1-yl)-7-deazapurine nucleosides by azide–alkyne click reactions and direct CH bond functionalization. *Tetrahedron Lett* 2016;**57**(39):4364–7.
39. Wen Z, Tuttle PR, Howlader AH, Vasilyeva A, Gonzalez L, Tangar A, et al. Fluorescent 5-pyrimidine and 8-purine nucleosides modified with an N-unsubstituted 1, 2, 3-triazol-4-yl moiety. *J Org Chem* 2019;**84**(6):3624–31.

40. Cosyn L, Palaniappan KK, Kim S-K, Duong HT, Gao Z-G, Jacobson KA, et al. 2-Triazole-substituted adenosines: a new class of selective A3 adenosine receptor agonists, partial agonists, and antagonists. *J Med Chem* 2006;**49**(25):7373–83.
41. Liu J, Leonard P, Müller SL, Daniliuc C, Seela F. Nucleoside macrocycles formed by intramolecular click reaction: efficient cyclization of pyrimidine nucleosides decorated with 5′-azido residues and 5-octadiynyl side chains. *Beilstein J Org Chem* 2018;**14**(1):2404–10.

CHAPTER 4

Triazoles in Synthesis and Functionalization of Polymers

Tahir Farooq* **and Ali Raza**
Department of Applied Chemistry, Government College University, Faisalabad, Pakistan
*Corresponding author. E-mail: tahirfarooqfsd@gmail.com

Introduction

The advances in triazole chemistry have revolutionized almost all areas of synthetic chemistry and the preparation of macromolecules including polymer synthesis. Especially, the emergence of Cu-catalyzed triazole synthesis has greatly influenced the polymerization, post-functionalization, and surface modifications of natural and synthetic polymers over the last few years. Polymer chemists can easily synthesize, functionalize, and control polymer structures at desirable length scales by employing robust approaches of triazole chemistry that do not involve harsh conditions and separation techniques.[1,2] Both main classes of triazoles, viz., 1,2,4-triazole and 1,2,3-triazole have been well exploited for synthesis, functionalization, and surface modification of polymers (Figure 1).

1,2,4-Triazoles and polymer synthesis

The advancements in the synthetic strategies of 1,2,4-triazoles have made them interesting heterocycles with applications in various scientific disciplines including materials and polymer synthesis.[3]

In recent years, the synthesis of conjugated polymers with alternative sulfur–nitrogen and heterocyclic moieties have become an attractive research area. Such efforts have been made to develop new [SN]x analogues of well-known poly(sulfur nitride), non-metallic polymeric superconductors having safety issues with their synthesis. In this connection, Mingfeng and his *co*-researcher reported an efficient synthesis of a conjugated polymer with alternative sulfur-nitrogen and 1-alkyl-1,2,4-triazolic moieties (Scheme 1).[4] The polymer exhibited a broad absorption band range with high solubility in a number of non-polar solvents including tetrahydrofuran, chloroform, dichloromethane, and chlorobenzene.

Advances in Triazole Chemistry
https://doi.org/10.1016/B978-0-12-817113-4.00003-2

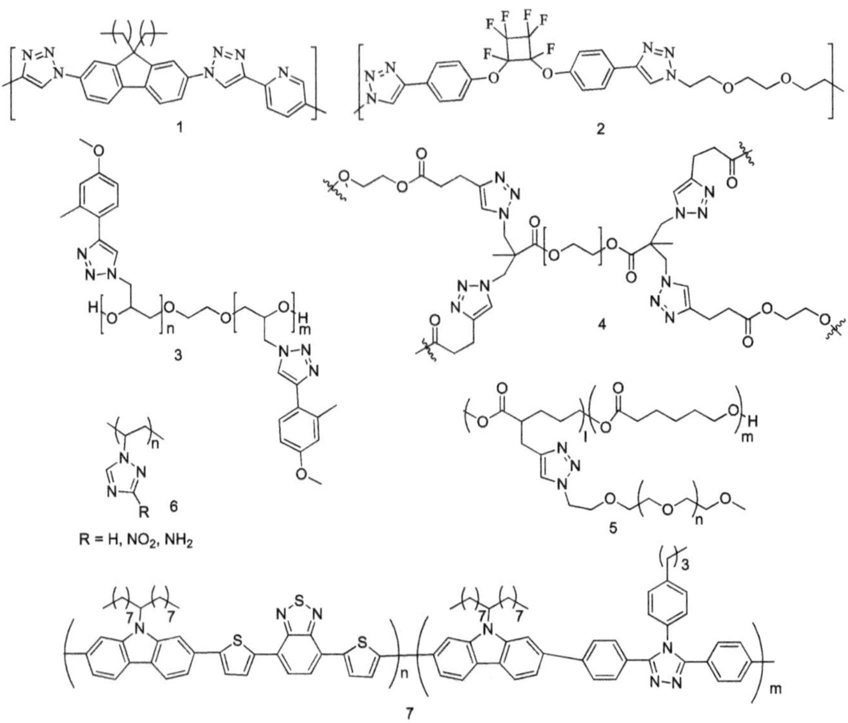

Figure 1 1,2,4- or 1,2,3-Triazole containing known polymers.

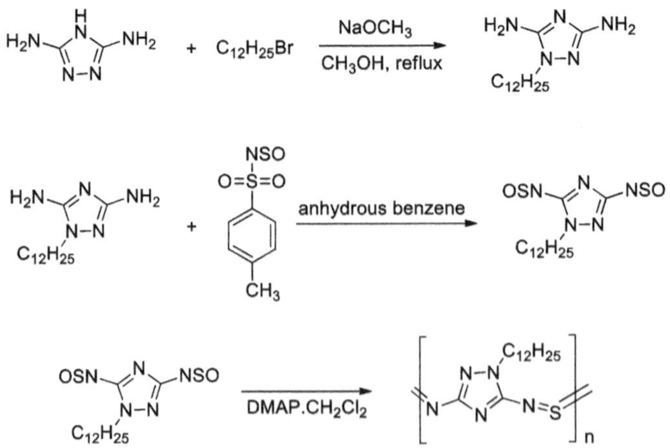

Scheme 1 Synthesis of conjugate polymers with alternative S–N and 1-alkyl-1,2,4-triazoles.

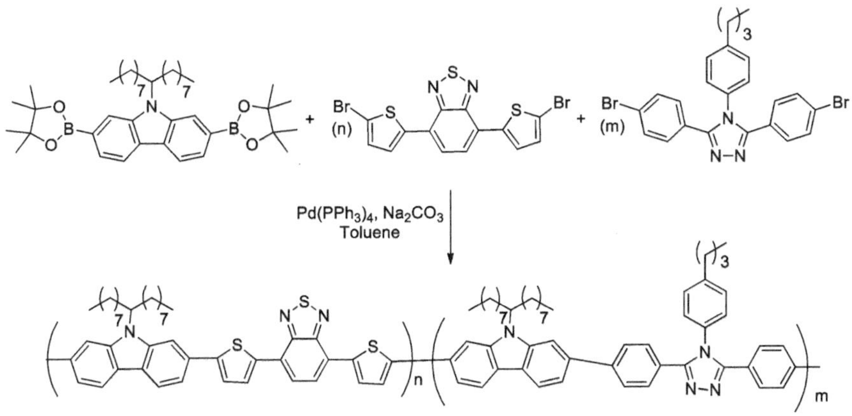

Scheme 2 Synthesis of carbazole-benzothiadiazole-triazole-based copolymer.

Eunhee Lim employed the Suzuki-coupling polymerization process to synthesize carbazole-benzothiadiazole-triazole-based copolymers with tunable electrochemical and optical properties (Scheme 2).[5] The introduction of triazole as an electron-transporting unit into the carbazole-benzothiadiazole-based copolymer improved the photovoltaic and optical properties, film morphology, and solubility of the resulting polymer.

The positions of the donor atoms and the bridging capacity of N1,N2 or N2,N4 have made the 1,2,4-triazole or its derivatives attractive candidates for the construction of metal complexes or the synthesis of coordination polymers (CPs) in different geometrical forms.

The systematic applications of CPs have revolutionized the fields of drug delivery, catalysis, light-emitting devices, chemosensors, and gas/liquid detection. However, the controlled synthesis of CPs with predictable properties has always been a challenging task due to the lack of understanding about structure-property relationships. The rational selection of organic linkers and inorganic metal connectors serves well to construct CPs with desired properties. The *N*-heterocycles with N or O atoms as donor sites are common organic linkers that bind to soft transition metals forming highly stable structures. In this connection, the 1,2,4-triazole ligands have established their versatile worth to generate simple and high symmetry structures due to the bridging fashion of μ_3-1κN:2κN:4κN, μ_2-1κN:2κN, and μ_2-2κN:4κN. Novel open porous CPs with new topological nets were synthesized using 3-amino-1*H*-1,2,4-triazole as three connecting anionic or two connecting neutral ligands. Liu and his colleagues used hydrothermal conditions to synthesize 3D structure of [Cd5(atr)7(Ac)3(H_2O)2] atr-containing Cd conducting polymer.[6]

The thieno[3,2-b]thiophene-2,5-dicarboxaldehyde and 3,5-diamino-1,2,4-triazole underwent condensation polymerization and furnished (poly(*N*-thieno[3,2-b]thiophen-2-yl)methylene)-1*H*-1,2,4-triazol-5-amine) poly(TTMA)) with a low optical band gap (Scheme 3).[7] An increase in molarity reduced the absorption band edge values and band gap of ply(TTMA) making it a suitable candidate for the development of optoelectronic devices. Furthermore, the thieno[3,2-b]thiophene structures make polymers suitable for the construction of organic semiconductors while triazole rings are favored for high performance electrical memory polymers.

Bozkurt and his colleagues reported the triazole and tetrazole functionalized methacrylamide and methacrylates as anhydrous polymer electrolytes via post-polymerization functionalization (Scheme 4).[8] Initially, the PMAC was synthesized by free radical polymerization then subsequent functionalization by triazole. Furthermore, the anhydrous polymer electrolytes were obtained when stoichiometric rations (1, 2, and 4 with respect to

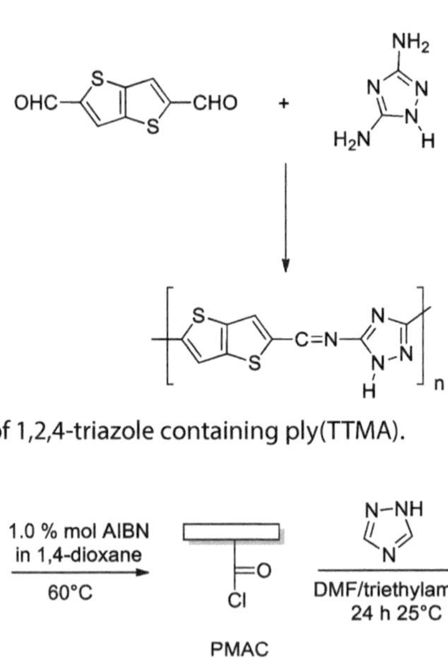

Scheme 3 Synthesis of 1,2,4-triazole containing ply(TTMA).

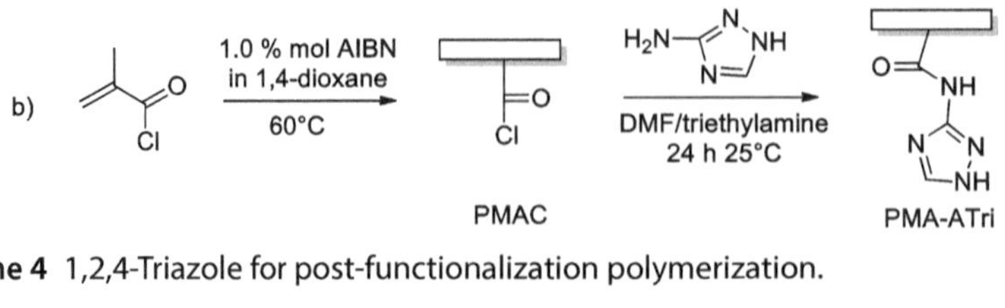

Scheme 4 1,2,4-Triazole for post-functionalization polymerization.

triazole) of triflic acid (TA) were added as a doping agent. Under anhydrous conditions, the poly(methacryloyl-1,2,4-triazole) (PMA-tri-$(TA)_4$) and poly(methacryloyl-3-amino-1,2,4-triazole) (PMA-ATri-$(TA)_4$) exhibited maximum proton conductivities of 8.7×10^{-4} S cm^{-1} and 0.02 S cm^{-1}, respectively, at 150°C, and the samples showed thermal stability up to 200°C.

Liu and his fellows copolymerized 1-vinyl-1,2,4-triazole (VTZ) and *N*-acryloyl glycinamide (NAGA) without involving any cross-linking agent and produced a supramolecular polymer, the poly(*N*-arolyoyl glycinamide-*co*-1-vinyl-1,2,4-triazole) (PNAGA-PVTZ) hydrogels to treat degenerated soft supporting tissues (Scheme 5).[9] The hydrogel demonstrated anti-inflammatory, anti-microbial, and *in vivo* biocompatibility potential. Furthermore, remarkable mechanical properties including excellent compressive strength (H 11 MPa), large stretchability (H 1300%), and high tensile strength (H 1.2 MPa) were observed at swelling equilibrium state. The properties like reprocessability over a temperature range, thermoplasticity, and self-repairability of the hydrogel were linked with the presence of triazole substituents. Thus, the multifunctional hydrogels were found as suitable candidates as implantable biomaterials.

Very recently, Mahata and his colleagues reported the synthesis of lignin-graft-polyoxazoline copolymer hydrogel functionalized with 1,2,4-triazole to achieve anti-biofilm and anti-microbial properties (Scheme 6).[10] The hydrogel reduced the production of iNOS significantly, and also the expression level of IL-1β in LPS induced macrophage cells. Furthermore, the hydrogel exhibited anti-inflammatory, healing potential, and free radical scavenging activity. This lignin-based hydrogel with multifunctional effects was found to be a suitable candidate to develop ointment formulations for wound therapy.

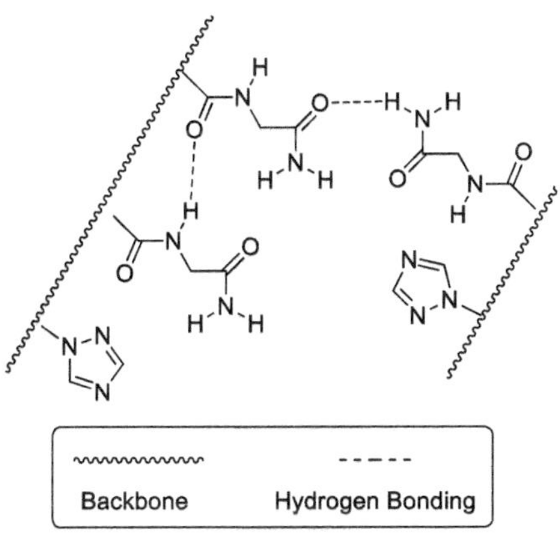

Scheme 5 1,2,4-Triazole containing functional hydrogels.

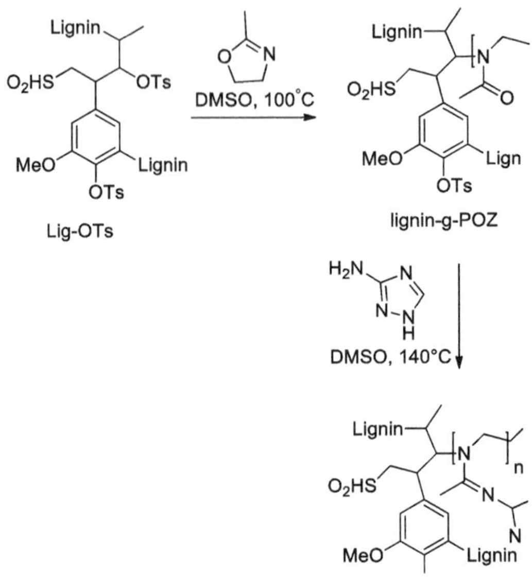

Scheme 6 Synthesis of potent hydrogels functionalized with 1,2,4-triazole.

1,2,3-Triazole and polymer synthesis

The Cu(I) catalyzed alkyne azide cycloaddition (CuAAC), the relatively newer version of Huisgen 1,3-dipolar additions with characteristics like quantitative yields under facile experimental conditions, tolerance of diverse functionalities, and selectivity, has become a versatile tool greatly influencing the paradigm of polymer synthesis. The construction of 1,2,3-triazole through CuAAC has widely been used for the synthesis of linear polymers using monomers equipped with alkyne and azide functionalities. To this end, synthesis of a variety of clickable monomers have been synthesized over the past few years. As an example of addition polymerization, the bis-alkyne and bis-azide were reacted under CuAAC conditions to achieve linear polymer (Scheme 7).[11] The same approach was used for the synthesis of fluorene-based conjugated polymers (Scheme 8) and 1,2,3-triazole-based polymer organogelator (Scheme 9).[12,13] According to another report, the dipropargyl ether was reacted with bis-azido polystyrene under CuAAC conditions to achieve polymer (Scheme 10).[14] These Cu(I)-catalyzed cycloadditions have been used in different ways for the synthesis of a variety of polymers.

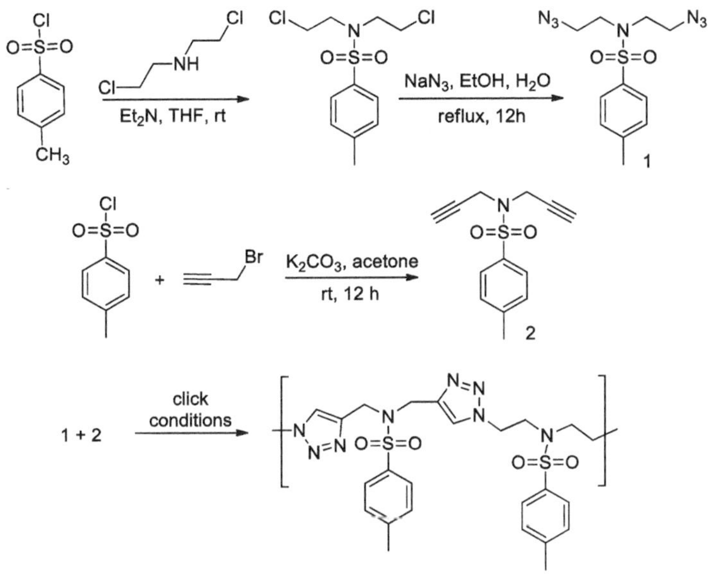

Scheme 7 Synthesis of polymer from clickable monomers.

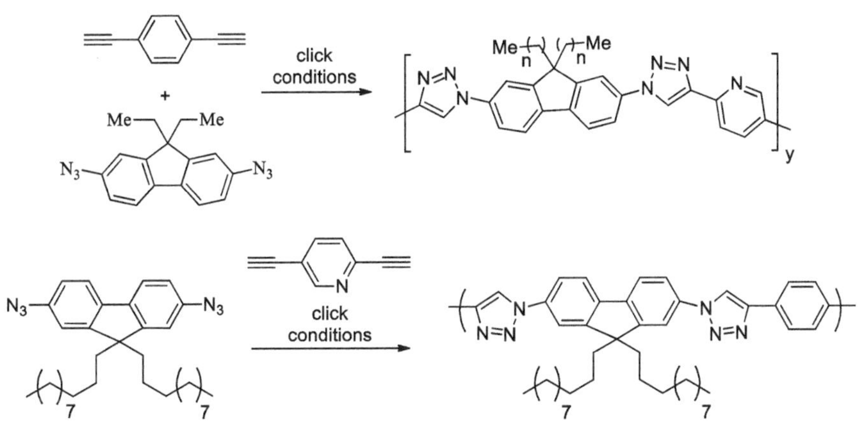

Scheme 8 Synthesis of conjugated polymer with triazolic backbone.

Combining polymerization reactions with CuAAC

Polymers with controlled composition and architecture, narrow molecular weight distribution, pre-determined molecular weight, and chain-end functionality could be prepared by controlled radical polymerization techniques (CRP). A variety of macromolecular architectures have been generated by combining controlled polymerization reactions with click chemistry due to its complementary attitude. Living polymerization is a

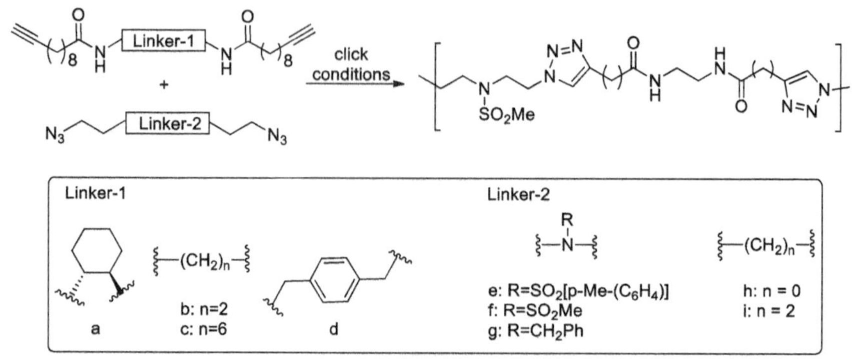

Scheme 9 Synthesis of polymeric organogelator from clickable monomers.

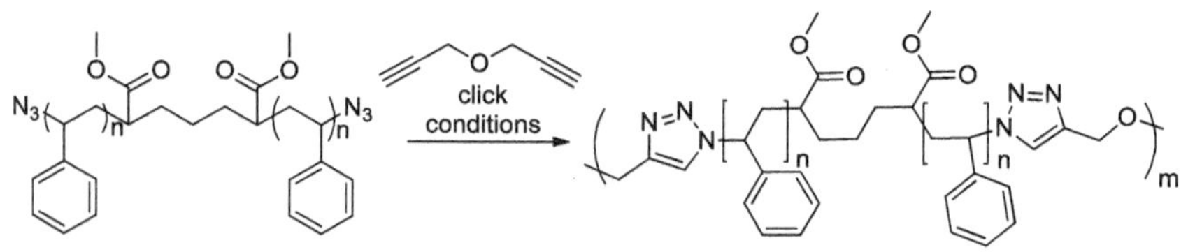

Scheme 10 CuAAC polymerization of macromonomer.

chain polymerization that proceeds in the absence of termination or chain transfer reactions. The propagating polymer chain end remains active as the monomer is consumed. The use of appropriate initiators and terminating agents could help to prepare chain-end functionalized polymer of choice.[15]

Functional polymers could be prepared by combining CuAAC reactions with the following polymerization reactions:

- Ring opening polymerization (ROP)
- Ring opening metathesis polymerization (ROMP)
- Radical polymerization

The ROP has been used to synthesize functionalized aliphatic polymers of high molecular weight with controlled polydispersity index and physical characteristics. This approach has also been used to introduce alkyne and azide functionalities at the chain ends or along the chains. The ABA-type block copolymers of poly(ε-caprolactone-b-ethylene glycol-b-ε-caprolactone) were synthesized by combining ROP and click chemistry. The ROP of ε-caprolactone and 3-azido-1-propanol provided PCL-N_3 while PEG reacted with propargyl chloride and furnished propargyl-PEG. Both of these counterparts reacted under click conditions and produced block copolymers (Scheme 11).[16]

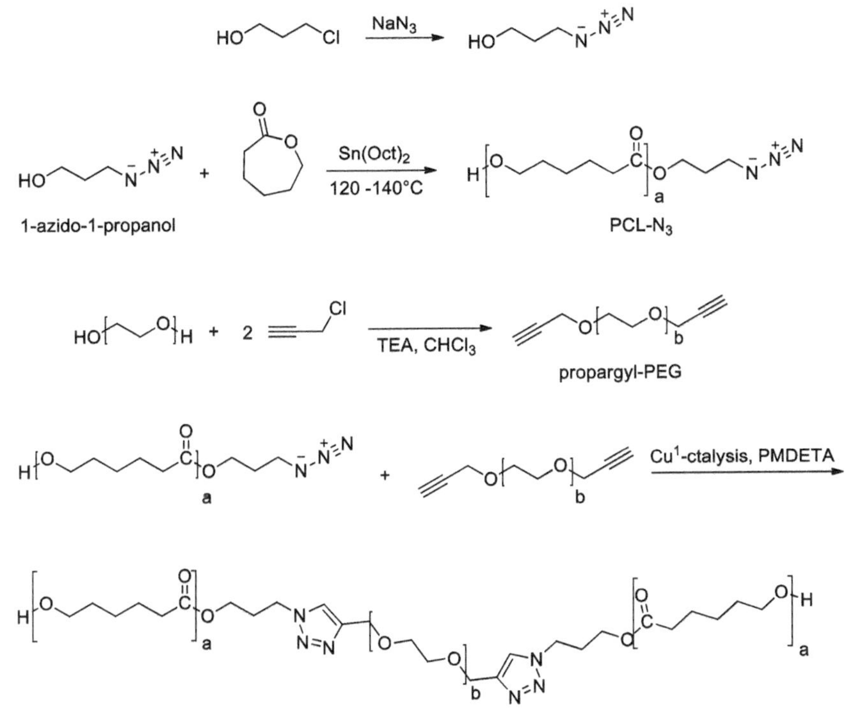

Scheme 11 Combination of ROP and click chemistry for synthesis of block copolymer.

Uyar and his research fellows employed the same combination for the synthesis of bisbenzoin group end-functionalized poly(ε-caprolactone) macrophotoinitiator (PCL-$(PI)_2$). Initially, the ROP of ε-CL provided PCL with a cyclohexene end-chain group (PCL-CH) that subsequently underwent bromination and azidation to furnish PCL-$(N_3)_2$. Under separate reaction conditions, the propargyl bromide reacted with benzoin and furnished alkyne-functionalized benzoin, which then reacted with PCL-$(N_3)_2$ under click conditions providing (PCL-$(PI)_2$) (Scheme 12).[17]

Similarly, the combined application of ROMP and click chemistry has received considerable attention over the last few years for the synthesis of well-defined functional polymers. In one such attempt, Choi and his research team used alkyne containing *endo*-tricyclo[4.2.2.$0^{2,5}$]deca-3,9-diene (TD) monomers for the synthesis of polymers via controlled ROMP. Furthermore, these polymers were subjected to post-polymerization functionalization employing CuAAC reactions at terminal alkyne moieties (Scheme 13). A broad range of functional polymers were prepared through this synthetic strategy.[18]

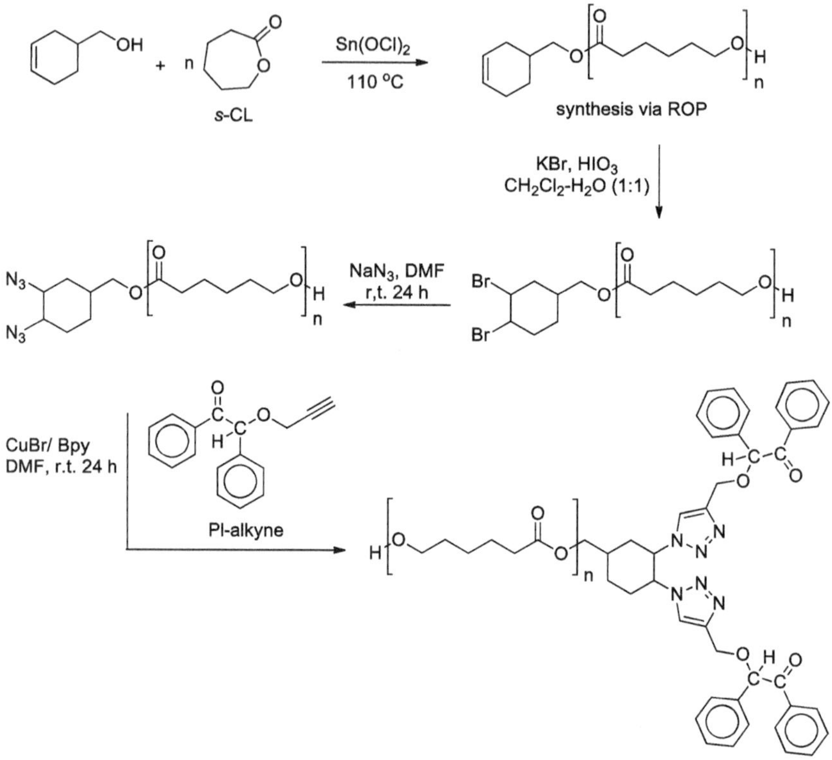

Scheme 12 Combination of ROP and click chemistry for end functionalization of polymers.

Kiessling and his *co*-workers employed ROMP and click chemistry to generate degradable polymers with a range of functionalities. They used azide containing bicyclic oxazinones for CuAAC reaction with 1-propargyl-α-D-mannose-2,3,4,6-tetraacetate to achieve functionalized monomers that underwent ROMP to afford polymers with carbohydrate substituents (Scheme 14).[19]

Functional polymers have efficiently been prepared by employing CRP techniques such as reversible addition fragmentation transfer polymerization (RAFT), atom transfer radical polymerization (ATRP), and nitroxide mediated polymerization (NMP) during the last few decades. Such techniques due to the mechanistic approach results in polymer chains having a dormant unit as the end cap. The dormant unit could be a dithioester moiety in RAFT, an alkoxyamine moiety in NMP, and a halogen atom in ATRP. Hence, diverse functional groups including alkyne and azide could be introduced at a dormant unit in a post-polymerization reaction.

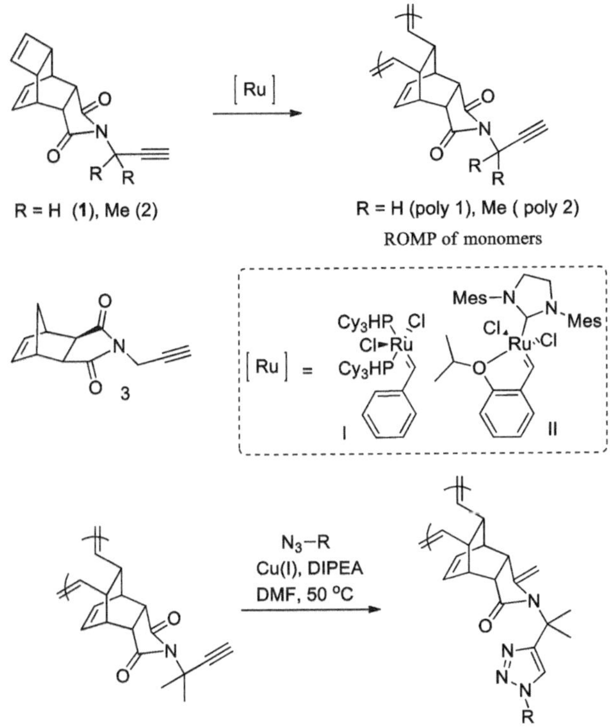

Scheme 13 ROMP of monomer and post-polymerization functionalization using CuAAC.

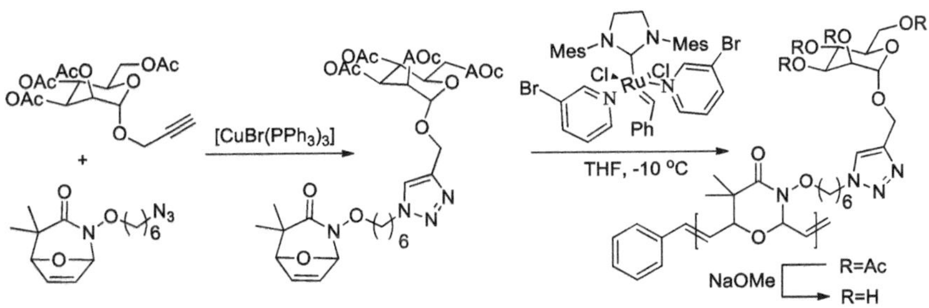

Scheme 14 Click chemistry for functionalization of monomers before ROMP.

The combination of CuAAC reaction and ATRP technique has widely been adopted for the synthesis of functional polymers as both are Cu-mediated approaches. The ATRP technique results in polymer chains with halogen ends that could easily be transformed to azides. Functionalized biocompatible polymers and sequence-defined bioconjugates were prepared by the facile and versatile combination of ATPR and click chemistry (Scheme 15). This synthetic strategy reported by Lutz and co-workers came out to be a universal approach to achieve novel biomaterials like functional biosurfaces and tailor-made bioconjugates.[20]

As a second strategy, the initiators used for ATRP were incorporated with clickable azide functionality. A variety of azido-initiators with different size and nature could be prepared by this general methodology, which could undergo ATRP to furnish bioconjugates, functional surfaces, and grafted polymers (Scheme 16).[21] However, the chances of copper contaminations have always restricted the applications of such polymers as biomaterials because ATRP techniques utilize Cu-catalyst in relatively high amounts.

Many reports described the synthesis of functional polymers and biomaterials employing the combination of RAFT and CuAAC strategy.[22,23] Ming and his colleagues combined RAFT, grafting-to approach, and CuAAC reaction to prepare responsive polymer-protein conjugates using bovine serum albumin (BSA) as model protein. The propargyl maleimide was reacted with cysteine residue of the BSA to access alkyne functionalized protein, and the RAFT was employed for azido-terminated

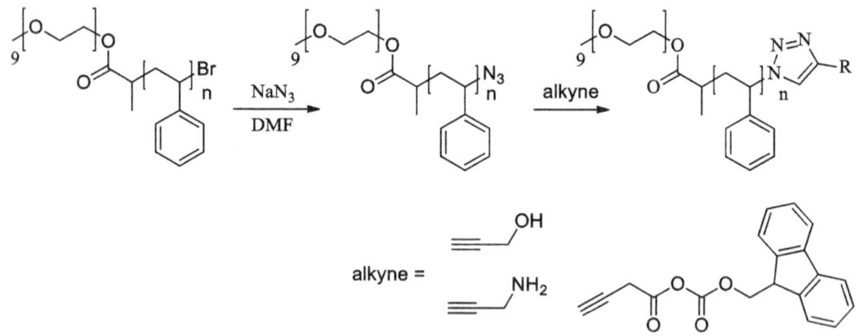

Scheme 15 Combination of ATPR and click chemistry for functional polymers.

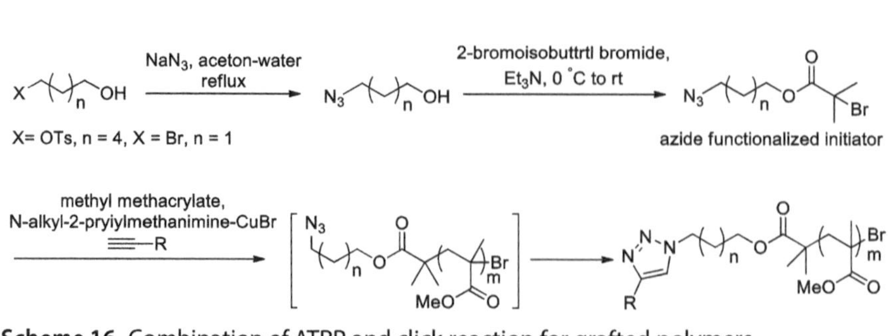

Scheme 16 Combination of ATRP and click reaction for grafted polymers.

poly(*N*-isopropylacrylamide) (PNIPAM-N3) preparation. Furthermore, they were coupled under click conditions to achieve responsive PNIPAM-BSA bioconjugates (Scheme 17).[24]

The NMP approach utilizes nitroxides for reversible capturing of propagating species resulting in the formation of alkoxyamines as a dormant unit. Various copolymers have been synthesized by combining NMP strategy with CuAAC reactions. Hawker and his colleagues employed this combination for the synthesis of water-soluble terpolymers by managing simultaneous and cascade functionalization (Scheme 18).[25]

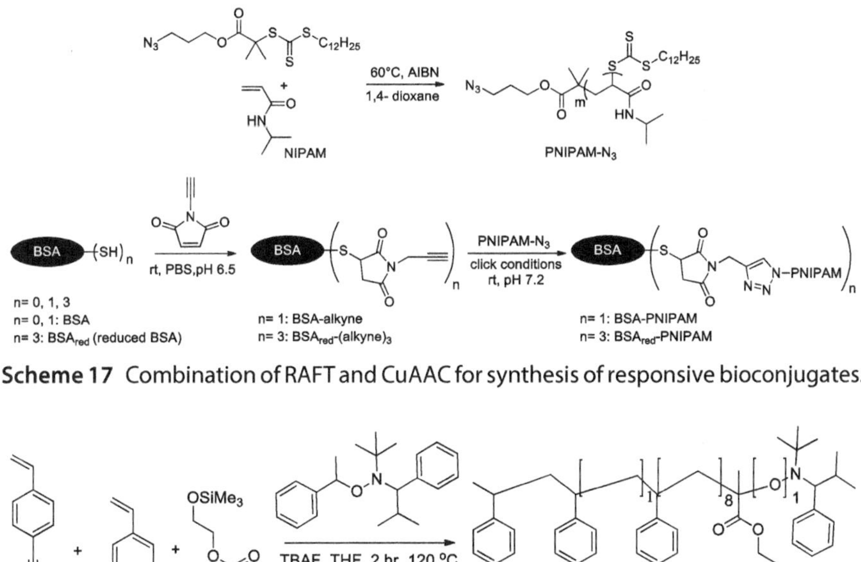

Scheme 17 Combination of RAFT and CuAAC for synthesis of responsive bioconjugates.

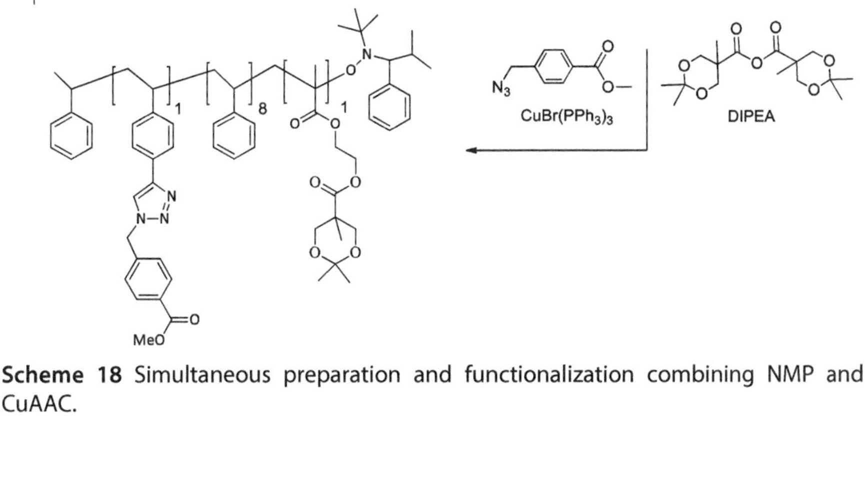

Scheme 18 Simultaneous preparation and functionalization combining NMP and CuAAC.

Synthesis of cross-linked polymers constructing 1,2,3-triazole

The synthesis of cross-linked polymers has also been revolutionized by the introduction of click chemistry concepts. A number of cross-linked polymers have been synthesized using CuAAC strategy.

The poly(vinyl alcohol) functionalized with alkyne were reacted with PVA-azides under click conditions to produce cross-linked polymeric material (Scheme 19).[26]

Baker *et al.* used acetylene-containing polyglycolides for CuAAC functionalization to produce cross-linked polymers with interesting water-soluble and biodegradable properties for biomedical applications (Scheme 20).[27]

Another report described the use of CuAAC reaction for the synthesis of star polymers. These polymers contained four arms of different chemical nature. Initially, different polymerization techniques were used to prepare block copolymers like PtBA-b-PEG and PMMA-b-PEG alkyne functionality, and PS-b-PCL with azide substituent, which subsequently underwent click reaction to afford quarter polymer (Scheme 21).[28]

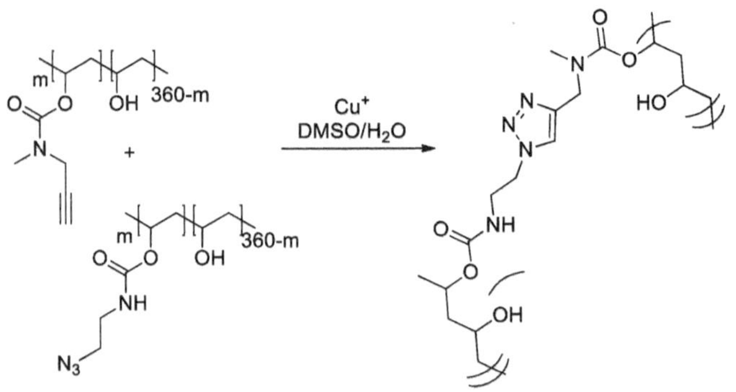

Scheme 19 Cross-linking of polymers using triazole.

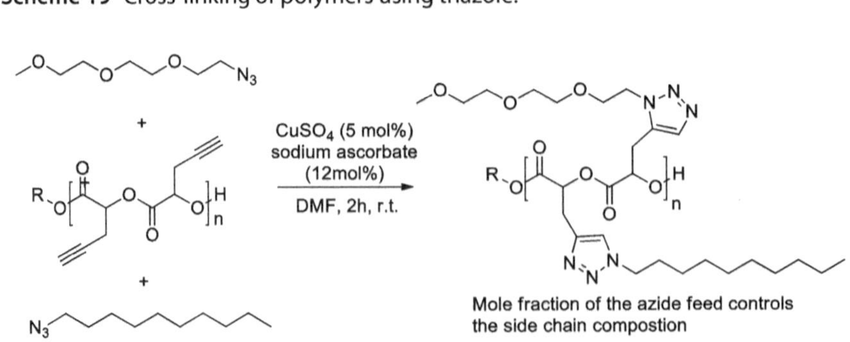

Scheme 20 Triazole-linked thermoresponsive degradable polymers.

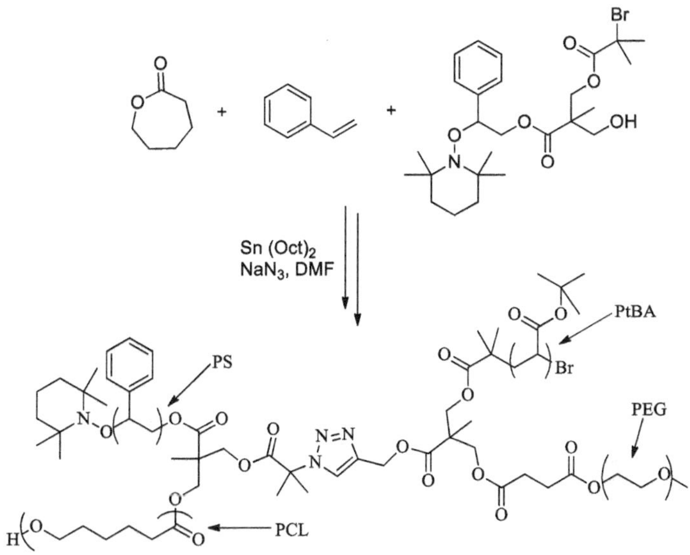

Scheme 21 Triazole-linked star quarterpolymer.

Turro and his colleagues reported the synthesis of photocleavable star polymers by employing ATRP and CuAAC strategy. The ATRP technique was used to synthesize photocleavable azide-macromonomer that underwent end-group modifications subsequently. Furthermore, insoluble polymeric materials were obtained by cross-linking the azide-macromonomers with tetra-functional acetylene under click conditions (Scheme 22).[29]

Accelerated synthesis of polymers using one-pot and click chemistry strategies

For the last few decades, one-pot strategy has become an exciting tool for the construction of small as well as complex synthetic and natural molecules. Likewise, the efficiency and robustness of highly specific CuAAC reaction has revolutionized the synthetic chemistry including polymer synthesis.[2] The chemists have developed combination of click chemistry and various polymerization techniques as one-pot multi-step strategies for the facile synthesis of a variety of polymers.[30]

A one-pot strategy involving ROP, NMP, and CuAAC reactions was employed for the synthesis of 3-miktoarm star terpolymer (Scheme 23). This high fidelity approach controlled each chemical transformation selectively thus avoiding the purification requirements.[31]

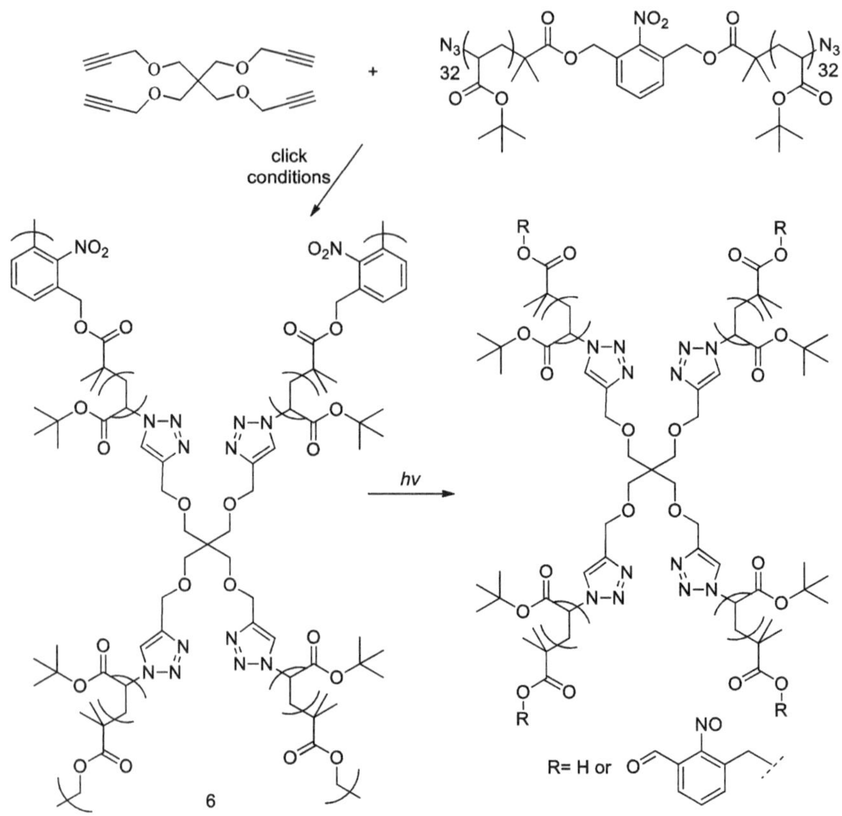

6

Scheme 22 Synthesis of photodegradable cross-linked polymeric materials.

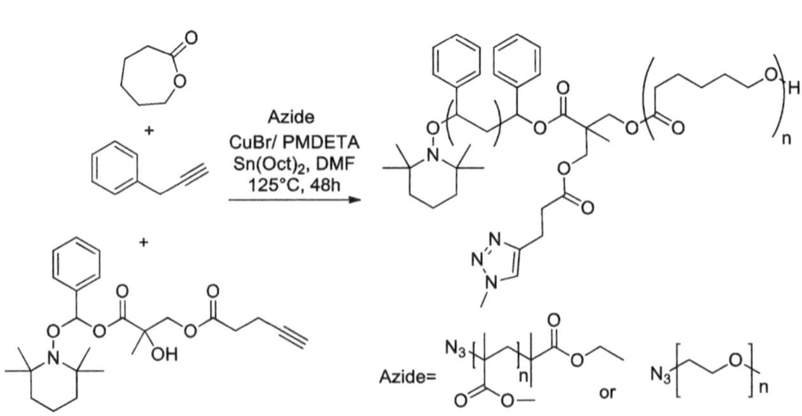

Scheme 23 Preparation of 3-miktoarm star terpolymer by one-pot strategy.

Another report described the synthesis of pH- and thermoresponsive block copolymer using ROP, CuAAC, and ATRP polymerization in a one-pot strategy (Scheme 24).[32]

Telechelic polystyrene derivatives were also synthesized using one-pot three-step strategy involving ATRP, in situ azidation, and click reactions (Scheme 25).[33]

The triblock copolymers with amphiphilic characteristics were synthesized involving two click reactions, i.e., the Diels-Alder and CuAAC in a one-pot approach. For this purpose, a polystyrene block with an anthracene substituent at one end and an azide group at the other was prepared as a difunctional entity, which subsequently underwent tandem reaction with alkyne functionalized PCL or PEG and PMMA with maleimide functionality. Resultantly, the triblock copolymers PMMA-PS-PCL and PMMA-PS-PEG were received in high yields (85–90%) (Scheme 26).[34] In previous

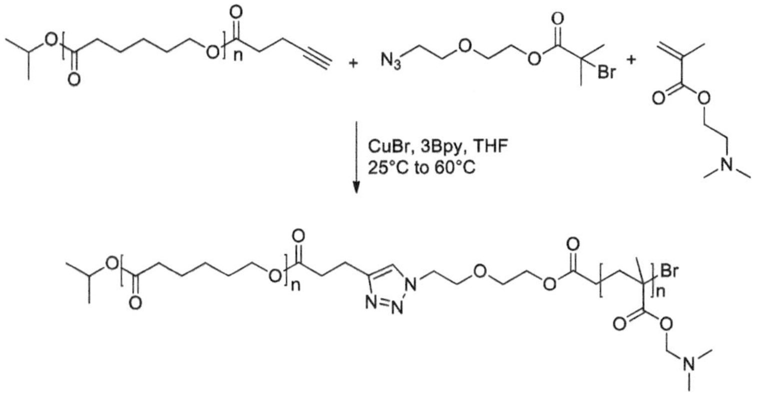

Scheme 24 One-pot strategy for amphiphilic and adaptative block copolymers.

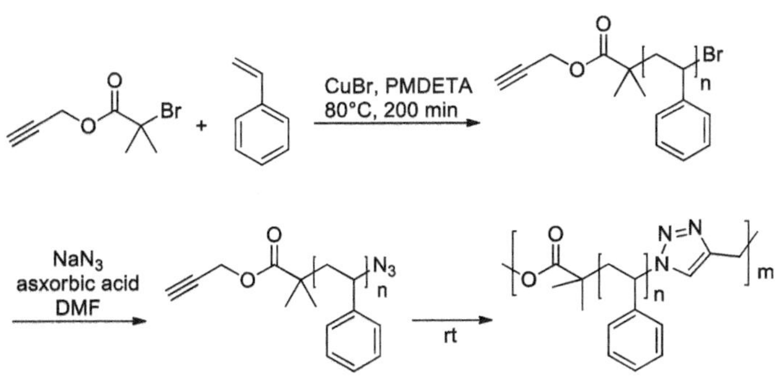

Scheme 25 Synthesis of telechelic polymer using one-pot strategy.

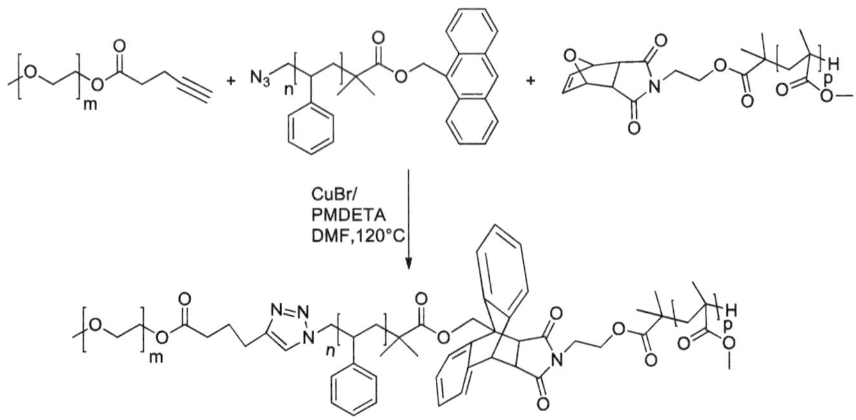

Scheme 26 Diels-Alder and CuAAC in one-pot synthesis of triblock copolymers.

cases, the one-pot approach involved click reaction along with traditional polymerization reactions.

Functionalization of biopolymers using click triazoles

The biopolymers like RNA, DNA, polypeptides, and polysaccharides, etc. could be functionalized by structural modifications to achieve new materials. Especially, the recent advances in triazole chemistry and developments in synthetic chemistry have revolutionized the synthesis of glycopolymers with a broad spectrum of applications in various scientific disciplines.

Reineke and his co-workers described the synthesis of trehalose containing stable, water-soluble, and biocompatible glycopolymers using click polymerization. They were designed to achieve facile delivery of nucleic acids in serum containing biological media.

Initially, the dialkyne-oligoamine and diazido-trehalose monomers were synthesized before polymerization reaction under click conditions (Scheme 27).[35]

Incorporation of an azido functionality at 6-position in the cellulose helps it to undergo CuAAC reactions with various alkynes resulting in biopolymers with modified charge and solubility, and properties like charge transfer, electrochemical, and fluorescence (Scheme 28).[36]

A cyclic peptide scaffold was functionalized with different peptides using click chemistry. The sequential method involved the deprotection of amines, propiolic acid coupling, and CuAAC with azidoacetylated peptides (Scheme 29).[37]

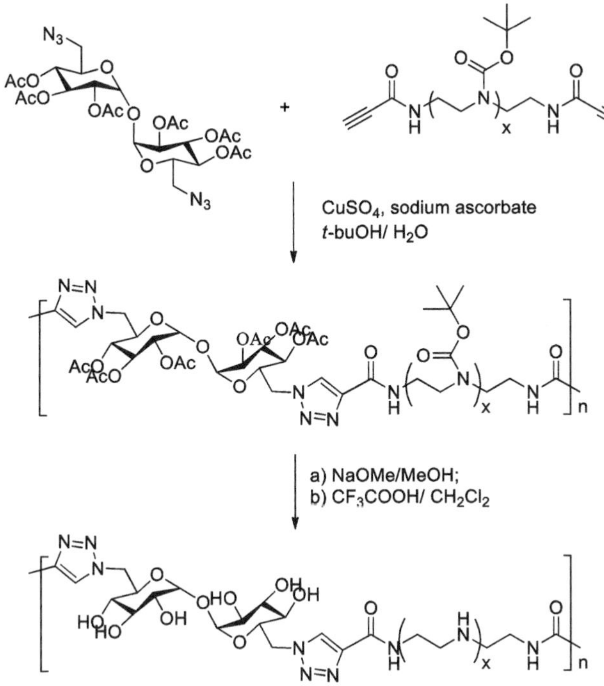

Scheme 27 Trehalose containing glycopolymers.

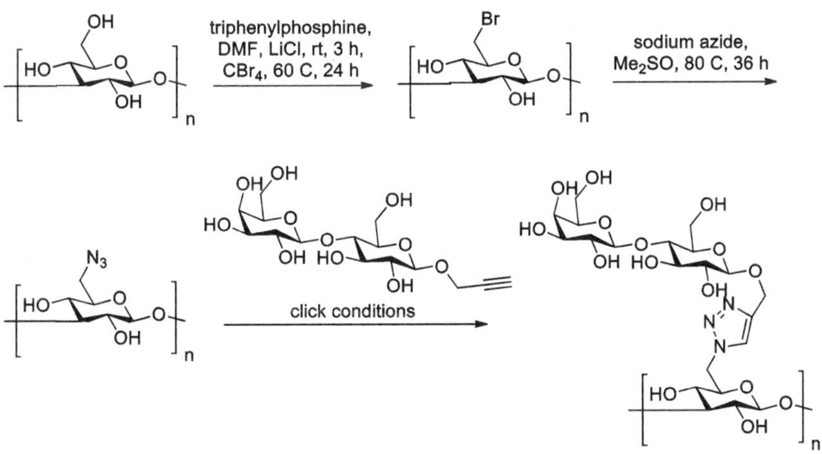

Scheme 28 Click chemistry on polysaccharides.

Schatz and his colleagues reported the synthesis of polysaccharide-block-polypeptide copolymers employing CuAAC reactions (Scheme 30). The poly(γ-benzyl L-glutamate) (PBLG) and dextran were selected as model blocks due to their biocompatible nature. The azide functionality was introduced at PBLG, and the alkyne substituent was incorporated at

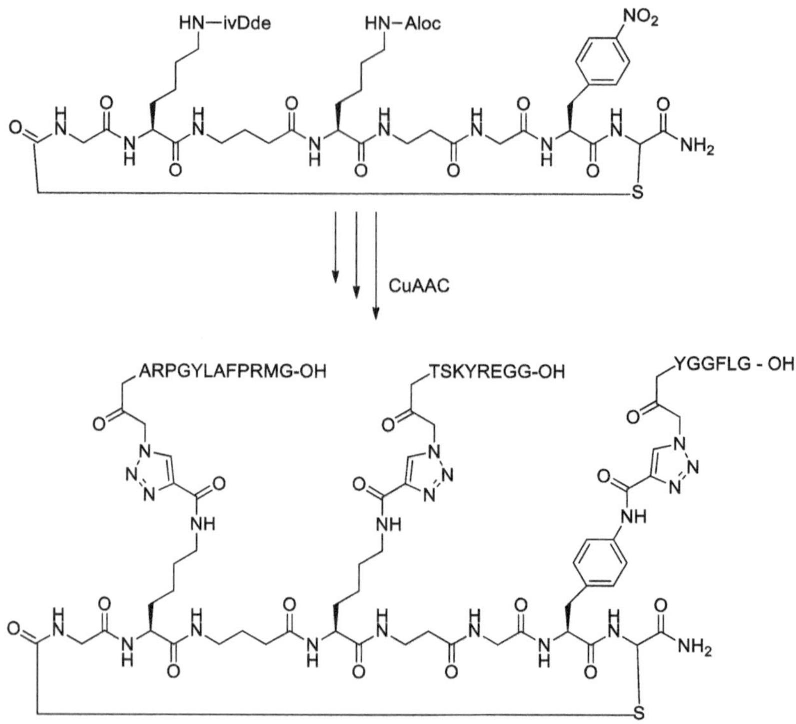

Scheme 29 Functionalization of peptides with peptides using CuAAC.

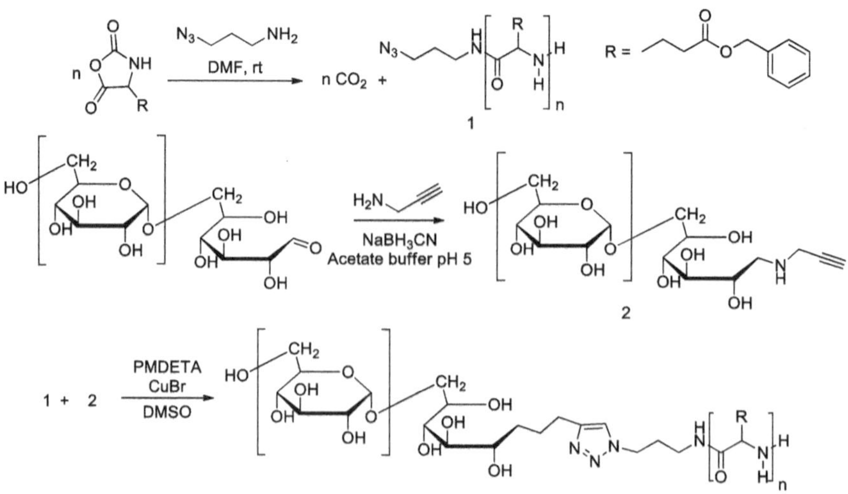

Scheme 30 CuAAC for synthesis of polysaccharide-block-polypeptide copolymers.

dextran. Furthermore, they were made to react under CuAAC to produce a block copolymer with the ability to self-assemble into vesicle in aqueous media. These polysaccharide-block-polypeptide copolymers find wide applications in drugs and gene-delivery systems.[38]

The high efficiency, orthogonality, versatility, and simplicity of CuAAC has made it a successful approach for ligation, cross-linking, site-specific labeling, and modification of DNA. The oligodeoxynucleotieds (ODNs) modified by click reactions found wide applications in material science, nanotechnology, bioconjugation, and drug discovery.

Seela and his colleagues reported a highly efficient and quantitative "bis-click" reaction for cross-linking of DNA (Scheme 31). Furthermore, ODNs were made to undergo cross-linking to establish the scope of this CuAAC-mediated reaction (Scheme 32).[39] Cancer thermotherapy exploits such cross-linking property of DNA for the arrest of cell replications resulting in cell death.

The ODNs could be functionalized with alkyne substituents using methyltransferases (MTases). Literature suggested that DNA could undergo regioselective and predictable site-specific alkylation with adenosine-derived *N*-mustard acting as a cofactor for MTases. Subsequently, the ODNs with alkyne functionality underwent CuAAC with azide in a sequence-selective manner (Scheme 33).[40]

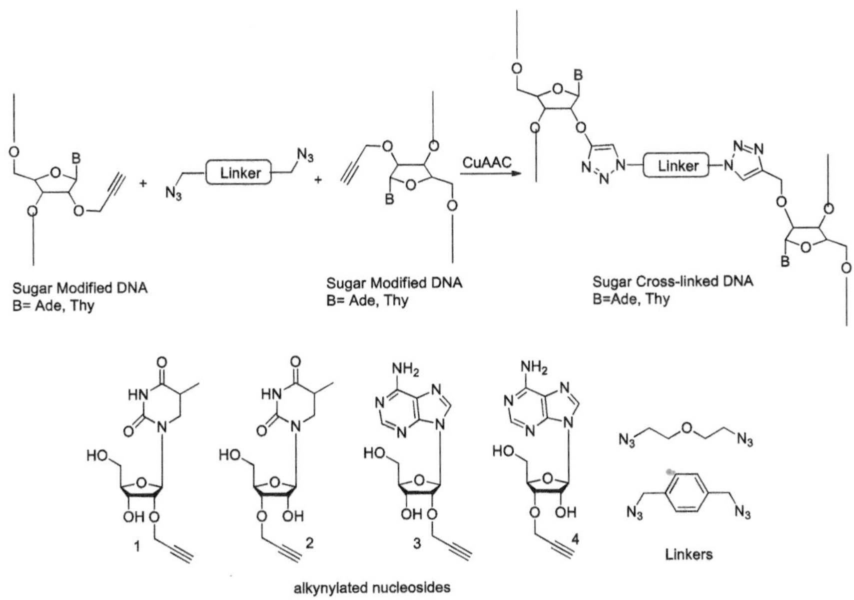

Scheme 31 CuAAC for cross-linking of sugar modified DNA.

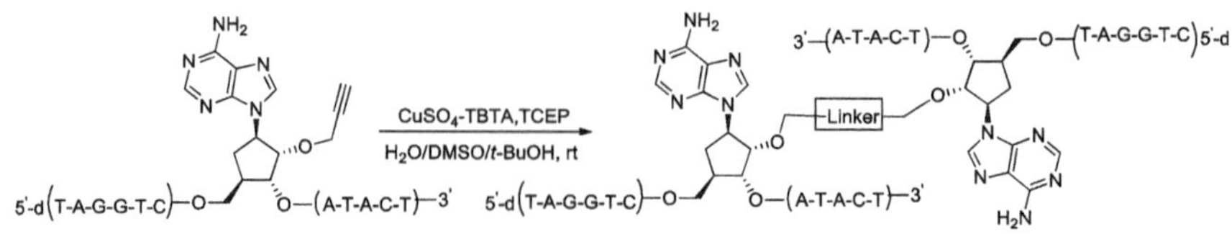

Scheme 32 CuAAC for cross-linking of ODN.

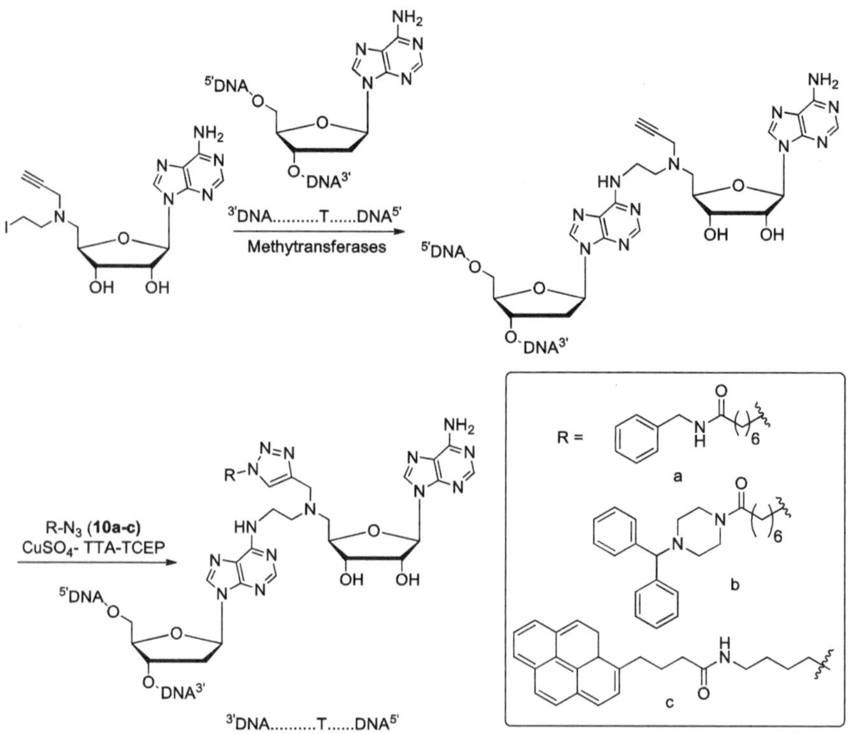

Scheme 33 Functionalization of ODNs using CuAAC.

The basic phosphodiester linkages were replaced with triazoles during the click synthesis of a number of RNA analogues of approximately 100 nucleotides. Alkyne and azide functionalized deoxy- and ribonucleotide polymers were used in CuAAC reactions to generate DNA:RNA hybrids and catalytically active RNA ribozymes (Scheme 34).[41] The triazole-based backbone linkages were found potentially compatible as the hammerhead and hairpin ribozymes efficiently cleaved their substrates.

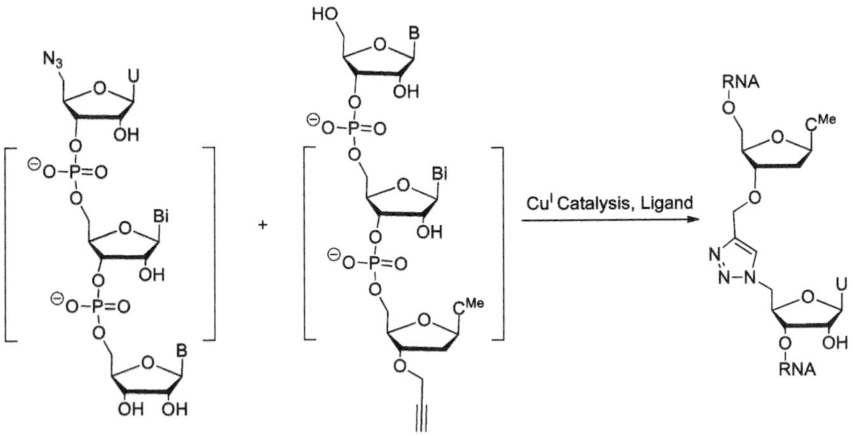

Scheme 34 Replacement of phosphodiester linkage with triazole.

1,2,3-Triazloes for the synthesis of dendritic polymers

A stepwise approach is employed to synthesize monodisperse polymeric materials with regular branching and symmetrically controlled layered structures throughout the assembly, which are called dendrimers. Either convergent or divergent procedure is adapted for the synthesis of dendrimers. In a convergent approach, the dendron arms are first constructed then connected to a multifunctional core representing in a chain-end growth while a divergent procedure starts with preparation of multifunctional cores and their subsequent sequence-wise attachments, which are called core-out growth.

Highly efficient reactions with the capacity of quantitative conversions are preferred to maintain a geometrical increment of functionalities ensuring the symmetrical growth of dendrimers. In this connection, the CuAAC reaction has extensively been used for the construction of dendritic structures due to the atom's economical, chemoselective, and environmentally benign nature. This click chemistry concept was first introduced in the field of dendrimer synthesis in 2004 by Hawker and his research group.[42] After that, a number of groups employed CuAAC as convergent, divergent or the combination of both approaches for the synthesis of dendrimeric architecture.

Peng and his research group prepared poly(aminoester) dendrimers through an integrated strategy combining CuAAC and convergent approach on divergently prepared dendrons. This integrated concept reduced the synthetic steps to achieve poly(aminoester) dendrimer in better yield (Scheme 35).[43]

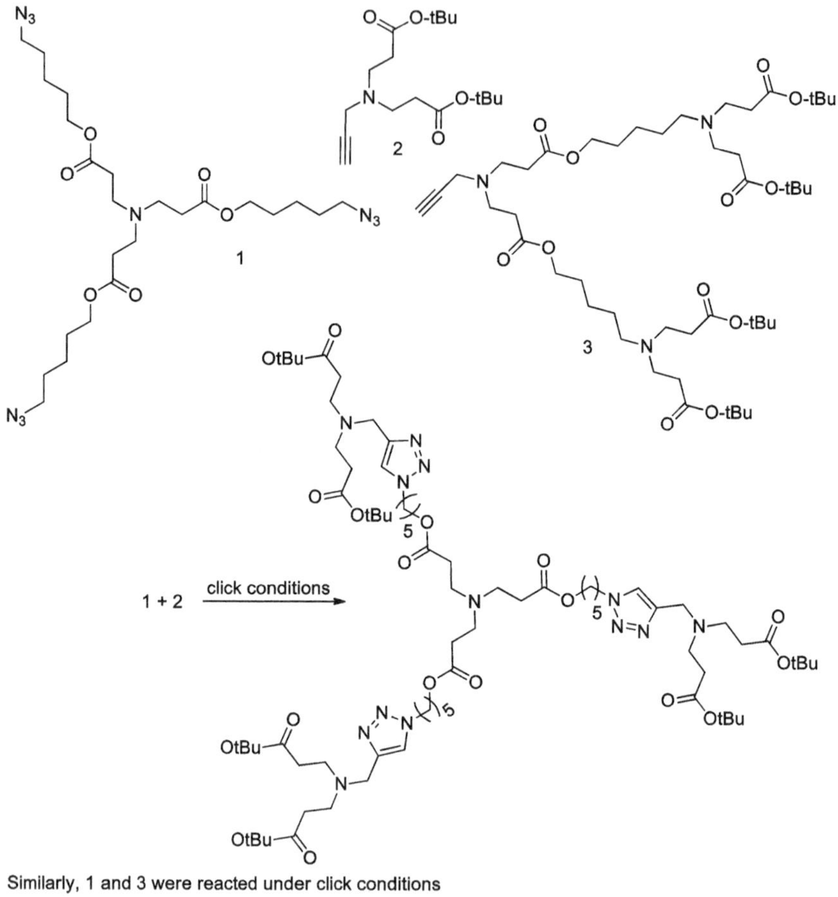

Scheme 35 Applications of click chemistry for synthesis of poly(aminoester) dendrimers.

The second generation poly(aryl ester) and Percec-type poly(benzyl ether) dendrons carrying cyanobiphenyl mesogens were used to prepare liquid-crystalline dendrimers under CuAAC conditions (Scheme 36).[44]

Quite recently, the click chemistry conditions were applied on β-cyclodextrin to generate heptavalent MUC1 glycopeptide dendrimers with possible biomedical applications (Scheme 37).[45] Such multivalent molecules are key synthetic targets as they are widely involved in lectin-glycan, receptor-ligand interactions, and cell-cell interactions in biological systems. They are constructed for use in tissue-engineering, anti-cancer drug carriers, and anti-inflammatory drugs. Glycopeptide dendrimers were found as promising candidates for developing synthetic vaccines and inhibitors.[46] So this facile synthesis of MUC1 glycopeptide dendrimers is a welcome advance in this regard.

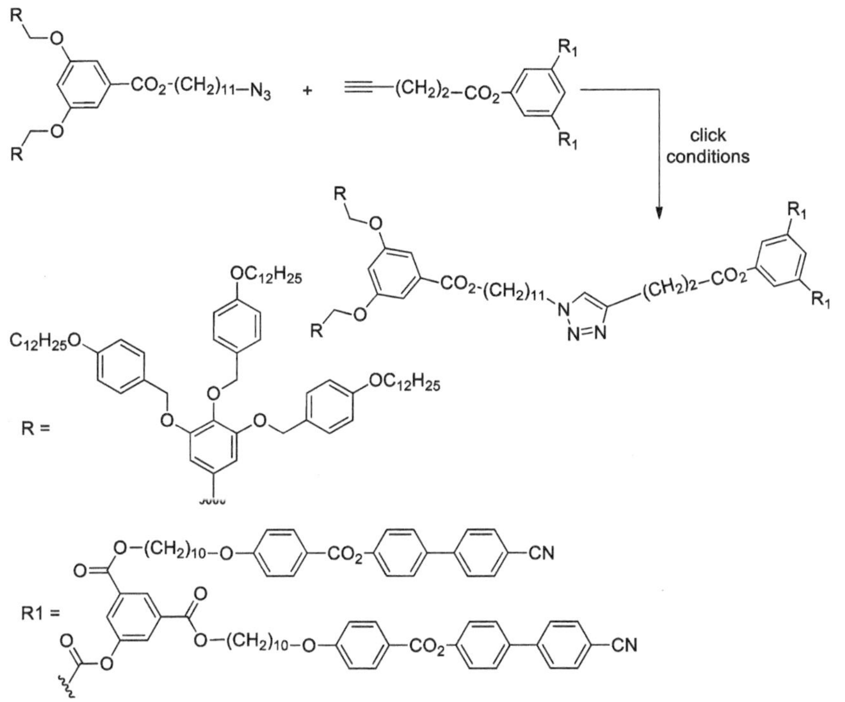

Scheme 36 Application of click triazole for synthesis of liquid-crystalline dendrimers.

Development of polymer-based gels using CuAAC reactions

It has been observed that the physical and chemical properties of the resulting polymeric materials are greatly influenced when triazole ring acts as a linker between two molecular fragments. Many interesting conformational and supramolecular properties including gelation properties were displayed by oligo- and polytriazolic polymers in this regard.[47] The gelation refers to the entrapping and immobilization of solvent molecules by the addition of a gelator in little quantity. When gelators develop a reversible three-dimensional self-assembly through non-covalent interactions like Van der Waals forces, p-p stacking interactions, or hydrogen bonding in sol-gel process, the resulting network structures are regarded as physical gels. But most polymer gelators create 3D networks through covalent interactions during polymerization thus produce irreversible chemical gels.[48] The stimuli responsive physical gels are considered to be more versatile because their properties can be tuned easily. However, the examples of polymer physical gels are rare compared with other low molecular weight gels (LMWG) due

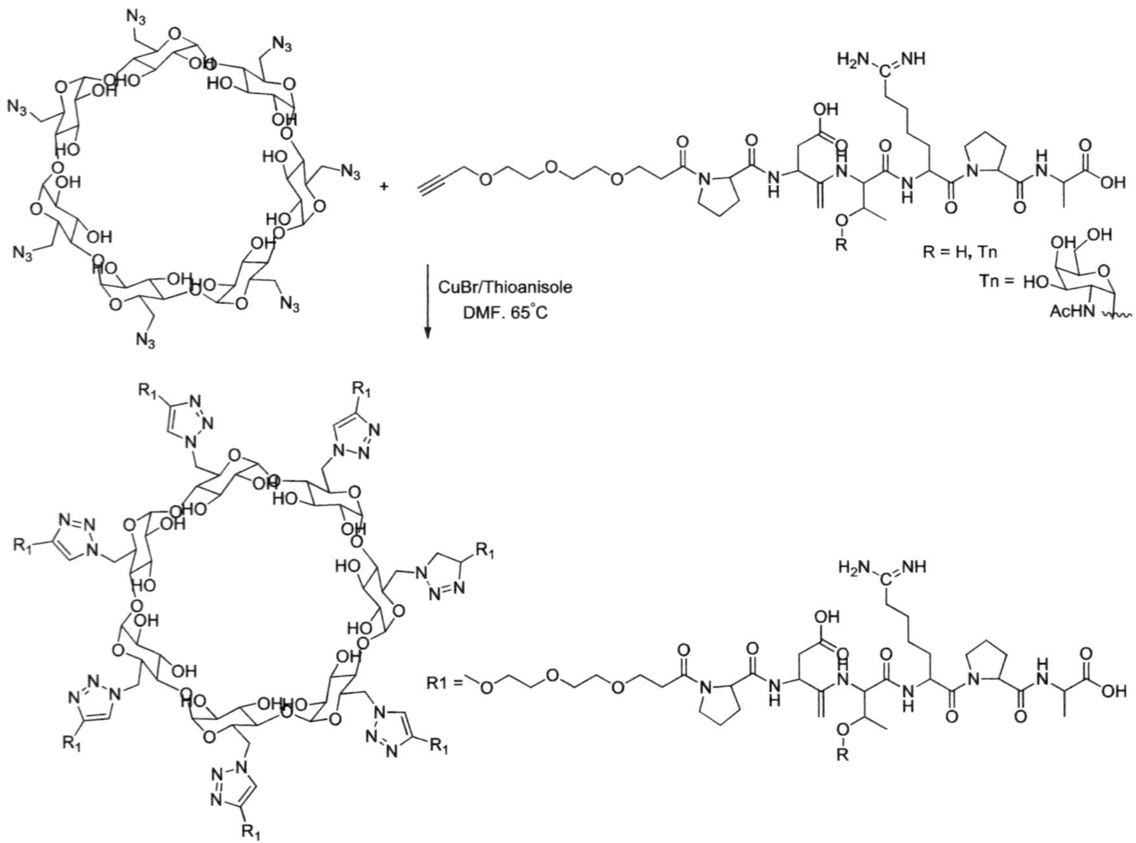

Scheme 37 Synthesis of MUCI glycopeptide dendrimers using CuAAC.

to some intrinsic structural issues. In fact, polymer chain mobility, backbone conformations, side chain, and multiple functional group interactions greatly influence the gelation property of the complex polymer system.[49,50]

In 2006, the poly(amide-triazole) polymers were presented for the first time as better organogelators compared with their constituent monomers. The compound 3 showed much better gelating properties compared with diazide and diacetylene as they (3%) did not form a gel in acetonitrile. However, no gel formation was observed even with compound 3 when 2,6-lutidine-acetonitril (1:20) was employed as a mixed solvent. When diazide and diacetylene were made to polymerize with their counterparts under click conditions, the resulting poly(amide–triazole) demonstrated much improved gelation properties in the mixed system, while the in situ click product 6 of diazide 2 and diacetylene 1 was found as relatively weak organogelator (Scheme 38).[51]

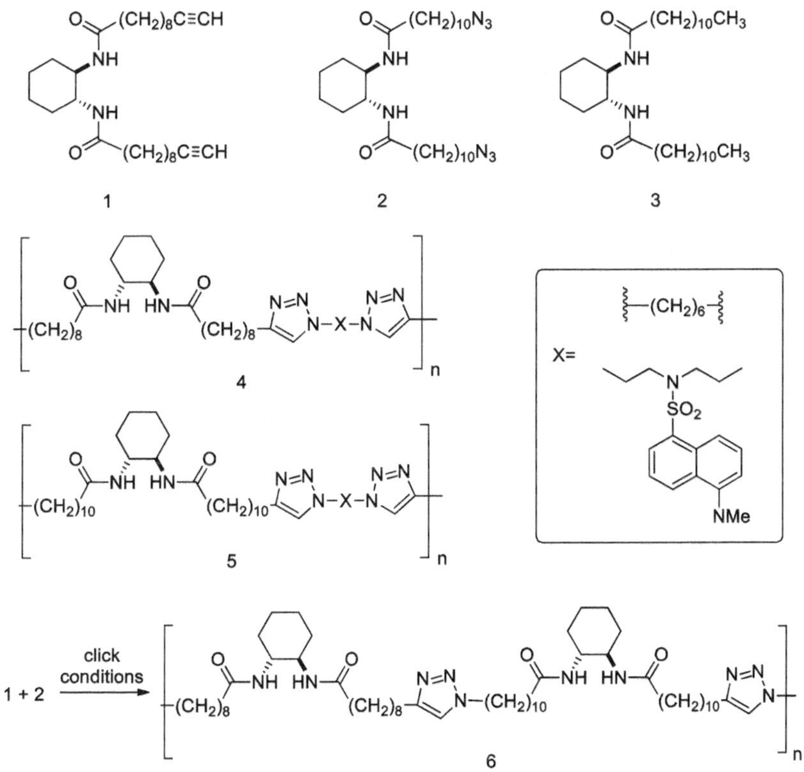

Scheme 38 Application of CuAAC for poly(amide-triazole) polymeric organogelator.

Lau and his colleagues first reported triazole-based dendronized polymer physical gel obtained from AB type hetero-bifunctional macromonomers under click polymerization conditions (Scheme 39).[52] Different physical and gelating properties were exhibited by three dendronized polymers due to the presence of amide functionality at regular intervals and different chain lengths of dendron appendages. Only G2 dendronized polymer exhibited gelating power in aromatic solvents, THF, ethanol, acetone, and ethyl acetate. The intramolecular hydrogen bonding interactions were displayed by all poly(amide-triazole)s however, only G2 dendronized gel possessed optimal dendron size and interaction range suitable for sustainable 3D structural arrangements required for gelation.

Quite recently, synthesis of a photo-responsive poly(amide-triazole) polymer has been reported under click chemistry conditions (Scheme 40).[53] The main polymer chain containing photo-responsive azobenzene units represented a rare example of such physical organogel with highly reversible sol-gel transformation without the requirement to convert *trans*-azobenzene to *cis*-conformers. No organogelating properties were displayed by previously reported materials such as azobenzene containing polytriazoles.[54,55]

Lee *et al.* disclosed the synthesis of triazole-containing hydrogels with time-dependent swelling behavior designed for sustained release of a model anti-cancer Rhodamine 6G drug. Initially, hydroxyethyl methacrylate (HEMA)-alkyne was prepared by coupling of *N*,*N*′-dicyclohexylcarbodiimide on HEMA, then 2-azido-1-ethyldimethylamine was synthesized (Scheme 41).[56] Furthermore, they were reacted under CuAAC conditions to generate a triazole ring containing monomer.

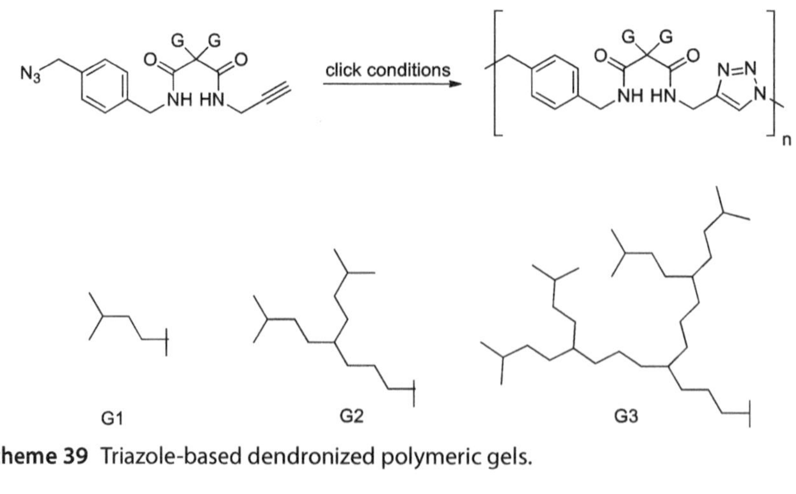

Scheme 39 Triazole-based dendronized polymeric gels.

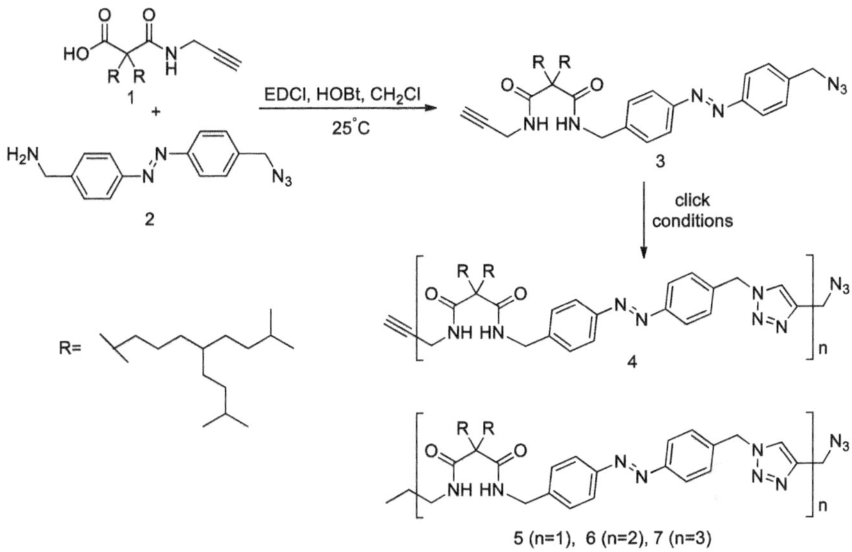

Scheme 40 Photoresponsive poly(amide-triazole) polymeric organogel.

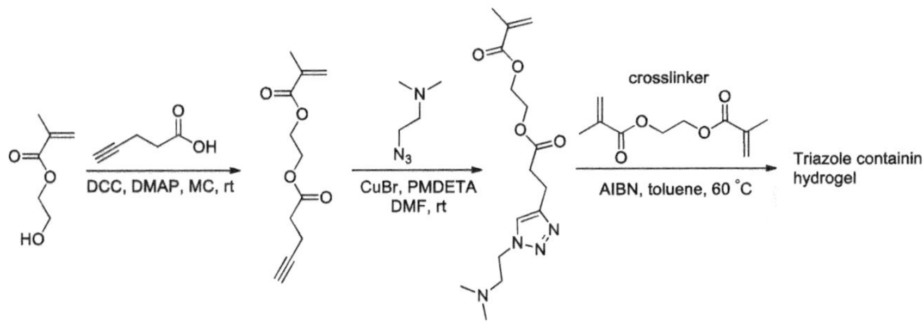

Scheme 41 Preparation of triazole containing hydrogel for sustain drug release.

Subsequently, the free radical polymerization of the monomer was accomplished using azobisisobutyronitrile (AIBN) initiator to achieve thermal and pH-responsive hydrogels.

These hydrogels exhibited a continuous swelling for 7 days maintaining a prolonged and sustained drug release caused by aggregation between triazole rings. As time elapses, the development of interactions with slowly penetrating water molecules disrupts the hydrophobic aggregations thus triggering the drug release. The drug release controllability of the hydrogels highlighted them as suitable drug carrier candidates.

Schmidt and his research group synthesized α-cyclodextrin/poly(ethylene glycol) (α-CD/PEG) hydrogel to study the additive effects of double hydrophilic block copolymer (DHBCs). For this purpose, reversible deactivation radical polymerization technique and click chemistry strategy were applied as a combined approach to synthesize poly(*N*-vinylpyrrolidone)-b-poly(oligo ethylene glycol methacrylate) (PVP-b-POEGMA) as a novel DHBC (Scheme 42).[57] These DHBC-based hydrogels showed thermoadaptive behavior when heated to different temperatures after α-CD addition. These DHBC-based α-CD hydrogels showed different mechanical properties when heated above Tcp or Tsp and cooled to ambient temperature. The thermoadaptibility of DHBC-based α-CD hydrogels has been considered as a stimulating feature for their possible applications in sensing.

Very recently, Nagashama and his colleagues used azide-modified mammalian cells and alkyne-modified biocompatible polymers under copper-free click reaction and generated living multifunctional hydrogels (Scheme 43).[58] Interestingly, these gels retained original cellular functionalities that could be exploited further as active cross-linking sites. This excellent methodology can be adopted to modify the virus and bacterial

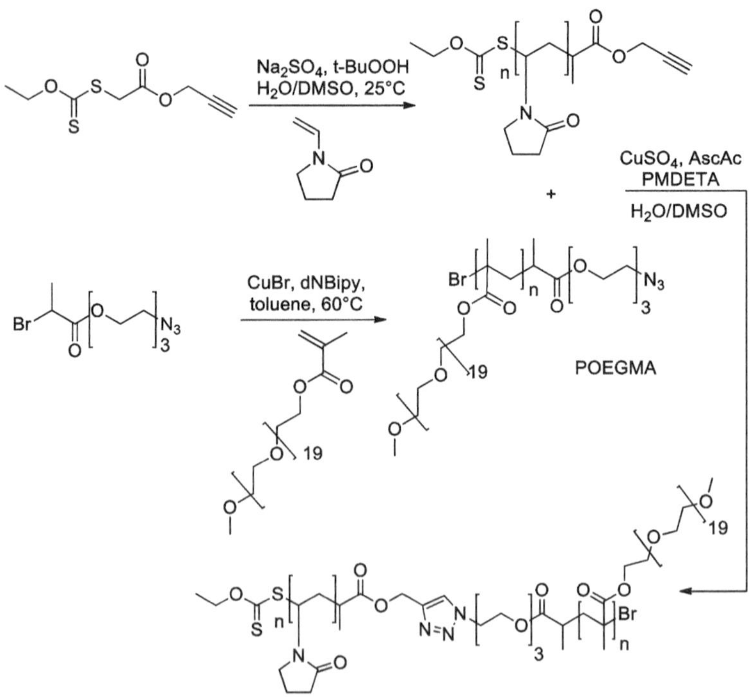

Scheme 42 Applications of CuAAC for the synthesis of PVP-b-POEGMA.

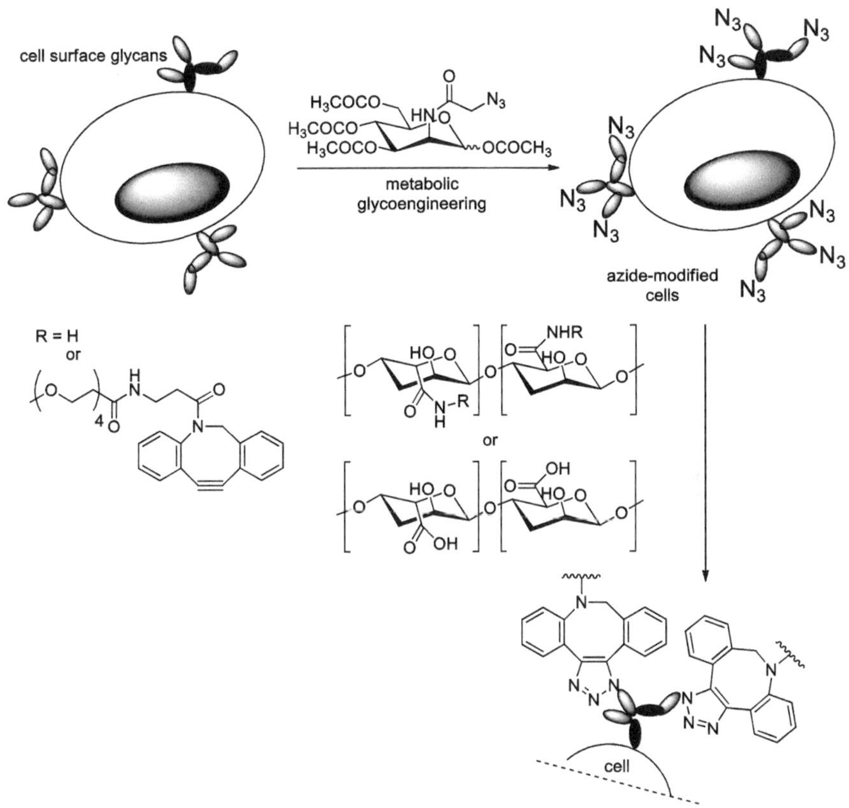

Scheme 43 Triazole for the preparation of living functional hydrogels.

cell surfaces through metabolic glycoengineering for generating living cell-based next generation hydrogels.

Conclusion

Over the last few years, both classes of triazoles have been well exploited for synthesis, functionalization, and/or modification of polymers of various types. However, the fidelity of CuAAC-based click synthetic strategies allows their applications in combination with polymerization techniques to synthesize and functionalize both natural and synthetic polymers. The examples represented in this chapter highlight the applications of triazole in polymer chemistry as an emerging research area with unlimited potential.

References

1. Martens S, Holloway JO, Du Prez FE. Click and click-inspired chemistry for the design of sequence-controlled polymers. *Macromol Rapid Commun* 2017;**38**(24):1700469.
2. Tunca U. Click and multicomponent reactions work together for polymer chemistry. *Macromol Chem Phys* 2018;**219**(16):1800163.
3. Diaz-Ortiz A, Prieto P, Carrillo J, Martin R, Torres I. Applications of metal-free 1, 2, 4-triazole derivatives in materials science. *Curr Org Chem* 2015;**19**(7):568–84.
4. Wang M, Wudl F. Solution-processable conjugated polymers containing alternating 1-alkyl-1, 2, 4-triazole and N–S–N links. *J Mater Chem* 2010;**20**(27):5659–63.
5. Lim E. Synthesis and characterization of carbazole-benzothiadiazole-based conjugated polymers for organic photovoltaic cells with triazole in the main chain. *Int J Photoenergy* 2013;**2013**.
6. Liu B, Wen M-T, Shen M-L, Miao W-N, He T-T, Xu L. A new 3D cadmium coordination polymer containing 3-amino-1H-1, 2, 4-triazole: synthesis, structure, and property. *Inorg Chem Commun* 2018;**88**:38–41.
7. Cetin A, Korkmaz A, Kaya E. Synthesis, characterization and optical studies of conjugated Schiff base polymer containing thieno [3, 2-b] thiophene and 1, 2, 4-triazole groups. *Opt Mater* 2018;**76**:75–80.
8. Sinirlioglu D, Aslan A, Muftuoglu AE, Bozkurt A. Synthesis and proton conductivity studies of methacrylate/methacrylamide-based azole functional novel polymer electrolytes. *J Appl Polym Sci* 2014;***131***(4).
9. Wang H, Zhu H, Fu W, Zhang Y, Xu B, Gao F, et al. A high strength self-healable antibacterial and anti-inflammatory supramolecular polymer hydrogel. *Macromol Rapid Commun* 2017;**38**(9):1600695.
10. Mahata D, Jana M, Jana A, Mukherjee A, Mondal N, Saha T, et al. Lignin-graft-polyoxazoline conjugated triazole a novel anti-infective ointment to control persistent inflammation. *Sci Rep* 2017;**7**:46412.
11. Díaz DD, Punna S, Holzer P, McPherson AK, Sharpless KB, Fokin VV, et al. Click chemistry in materials synthesis. 1. Adhesive polymers from copper-catalyzed azide-alkyne cycloaddition. *J Polym Sci A Polym Chem* 2004;**42**(17):4392–403.
12. van Steenis DJV, David OR, van Strijdonck GP, van Maarseveen JH, Reek JN. Click-chemistry as an efficient synthetic tool for the preparation of novel conjugated polymers. *Chem Commun* 2005;**34**:4333–5.
13. Diaz DD, Tellado JJM, Velazquez DG, Ravelo AG. Polymer thermoreversible gels from organogelators enabled by 'click' chemistry. *Tetrahedron Lett* 2008;**49**(8):1340–3.
14. Golas PL, Tsarevsky NV, Sumerlin BS, Matyjaszewski K. Catalyst performance in "click" coupling reactions of polymers prepared by ATRP: ligand and metal effects. *Macromolecules* 2006;**39**(19):6451–7.
15. Golas PL, Matyjaszewski K. Marrying click chemistry with polymerization: expanding the scope of polymeric materials. *Chem Soc Rev* 2010;**39**(4):1338–54.
16. Öztürk T, Meyvacı E. Synthesis and characterization poly (ϵ-caprolactone-b-ethylene glycol-b-ϵ-caprolactone) ABA type block copolymers via "click" chemistry and ring-opening polymerization. *J Macromol Sci A* 2017;**54**(9):575–81.
17. Uyar Z, Degirmenci M, Genli N, Yilmaz A. Synthesis of well-defined bisbenzoin end-functionalized poly (ε-caprolactone) macrophotoinitiator by combination of ROP and click chemistry and its use in the synthesis of star copolymers by photoinduced free radical promoted cationic polymerization. *Des Monomers Polym* 2017;**20**(1):42–53.
18. Kim KO, Kim J, Choi T-L. Controlled ring-opening metathesis polymerization of a monomer containing terminal alkyne and its versatile postpolymerization functionalization via click reaction. *Macromolecules* 2014;**47**(13):4525–9.

19. Fishman JM, Kiessling LL. Synthesis of functionalizable and degradable polymers by ring-opening metathesis polymerization. *Angew Chem Int Ed* 2013;**52**(19):5061–4.
20. Lutz J-F, Börner HG, Weichenhan K. Combining ATRP and "click" chemistry: a promising platform toward functional biocompatible polymers and polymer bioconjugates. *Macromolecules* 2006;**39**(19):6376–83.
21. Mantovani G, Ladmiral V, Tao L, Haddleton DM. One-pot tandem living radical polymerisation–Huisgens cycloaddition process ("click") catalysed by N-alkyl-2-pyridylmethanimine/Cu (I) Br complexes. *Chem Commun* 2005;**16**:2089–91.
22. Ranjan R, Brittain WJ. Tandem RAFT polymerization and click chemistry: an efficient approach to surface modification. *Macromol Rapid Commun* 2007;**28**(21):2084–9.
23. O'Reilly RK, Joralemon MJ, Hawker CJ, Wooley KL. Fluorogenic 1, 3-dipolar cycloaddition within the hydrophobic core of a shell cross-linked nanoparticle. *Chem A Eur J* 2006;**12**(26):6776–86.
24. Li M, De P, Gondi SR, Sumerlin BS. Responsive polymer-protein bioconjugates prepared by RAFT polymerization and copper-catalyzed azide-alkyne click chemistry. *Macromol Rapid Commun* 2008;**29**(12–13):1172–6.
25. Malkoch M, Thibault RJ, Drockenmuller E, Messerschmidt M, Voit B, Russell TP, et al. Orthogonal approaches to the simultaneous and cascade functionalization of macromolecules using click chemistry. *J Am Chem Soc* 2005;**127**(42):14942–9.
26. Ossipov DA, Hilborn J. Poly (vinyl alcohol)-based hydrogels formed by "click chemistry". *Macromolecules* 2006;**39**(5):1709–18.
27. Jiang X, Vogel EB, Smith MR, Baker GL. "Clickable" polyglycolides: tunable synthons for thermoresponsive, degradable polymers. *Macromolecules* 2008;**41**(6):1937–44.
28. Altintas O, Hizal G, Tunca U. ABCD 4-miktoarm star quarterpolymers using click [3 + 2] reaction strategy. *J Polym Sci A Polym Chem* 2008;**46**(4):1218–28.
29. Johnson JA, Finn M, Koberstein JT, Turro NJ. Synthesis of photocleavable linear macromonomers by ATRP and star macromonomers by a tandem ATRP-click reaction: precursors to photodegradable model networks. *Macromolecules* 2007;**40**(10):3589–98.
30. Lundberg P, Hawker CJ, Hult A, Malkoch M. Click assisted one-pot multi-step reactions in polymer science: accelerated synthetic protocols. *Macromol Rapid Commun* 2008;**29**(12–13):998–1015.
31. Altintas O, Yankul B, Hizal G, Tunca U. One-pot preparation of 3-miktoarm star terpolymers via click [3 + 2] reaction. *J Polym Sci A Polym Chem* 2007;**45**(16):3588–98.
32. Mespouille L, Vachaudez M, Suriano F, Gerbaux P, Coulembier O, Degée P, et al. One-pot synthesis of well-defined amphiphilic and adaptative block copolymers via versatile combination of "click" chemistry and ATRP. *Macromol Rapid Commun* 2007;**28**(22):2151–8.
33. Tsarevsky NV, Sumerlin BS, Matyjaszewski K. Step-growth "click" coupling of telechelic polymers prepared by atom transfer radical polymerization. *Macromolecules* 2005;**38**(9):3558–61.
34. Durmaz H, Dag A, Altintas O, Erdogan T, Hizal G, Tunca U. One-pot synthesis of ABC type triblock copolymers via in situ click [3 + 2] and Diels–Alder [4 + 2] reactions. *Macromolecules* 2007;**40**(2):191–8.
35. Srinivasachari S, Liu Y, Zhang G, Prevette L, Reineke TM. Trehalose click polymers inhibit nanoparticle aggregation and promote pDNA delivery in serum. *J Am Chem Soc* 2006;**128**(25):8176–84.
36. Hasegawa T, Umeda M, Numata M, Li C, Bae A-H, Fujisawa T, et al. 'Click chemistry' on polysaccharides: a convenient, general, and monitorable approach to develop (1 → 3)-β-D-glucans with various functional appendages. *Carbohydr Res* 2006;**341**(1):35–40.
37. Yoo EJ, Ahlquist M, Kim SH, Bae I, Fokin VV, Sharpless KB, et al. Copper-catalyzed synthesis of N-sulfonyl-1, 2, 3-triazoles: controlling selectivity. *Angew Chem Int Ed* 2007;**46**(10):1730–3.

38. Schatz C, Louguet S, Le Meins JF, Lecommandoux S. Polysaccharide-block-polypeptide copolymer vesicles: towards synthetic viral capsids. *Angew Chem Int Ed* 2009;**48**(14):2572–5.
39. Pujari SS, Seela F. Cross-linked DNA: propargylated ribonucleosides as "click" ligation sites for bifunctional azides. *J Org Chem* 2012;**77**(9):4460–5.
40. Weller RL, Rajski SR. DNA methyltransferase-moderated click chemistry. *Org Lett* 2005;**7**(11):2141–4.
41. El-Sagheer AH, Brown T. New strategy for the synthesis of chemically modified RNA constructs exemplified by hairpin and hammerhead ribozymes. *Proc Natl Acad Sci* 2010;**107**(35):15329–34.
42. Wu P, Feldman AK, Nugent AK, Hawker CJ, Scheel A, Voit B, et al. Efficiency and fidelity in a click-chemistry route to triazole dendrimers by the copper (I)-catalyzed ligation of azides and alkynes. *Angew Chem Int Ed* 2004;**43**(30):3928–32.
43. Moreno P, Quélèver G, Peng L. Synthesis of poly (aminoester) dendrimers via 'click' chemistry in combination with the divergent and convergent strategies. *Tetrahedron Lett* 2015;**56**(26):4043–6.
44. Guerra S, Nguyen TLA, Furrer J, Nierengarten J-FO, Barberá J, Deschenaux R. Liquid-crystalline dendrimers designed by click chemistry. *Macromolecules* 2016;**49**(9):3222–31.
45. Chen P-G, Huang Z-H, Sun Z-Y, Li Q-Q, Chen Y-X, Zhao Y-F, et al. Synthesis of an MUC1 glycopeptide dendrimer based on β-cyclodextrin by click chemistry. *Synlett* 2017;**28**(15):1961–5.
46. Niederhafner P, Šebestík J, Ježek J. Glycopeptide dendrimers. Part I. *J Pept Sci* 2008;**14**(1):2–43.
47. Chow H-F, Lau K-N, Ke Z, Liang Y, Lo C-M. Conformational and supramolecular properties of main chain and cyclic click oligotriazoles and polytriazoles. *Chem Commun* 2010;**46**(20):3437–53.
48. Noro A, Hayashi M, Matsushita Y. Design and properties of supramolecular polymer gels. *Soft Matter* 2012;**8**(24):6416–29.
49. Suzuki M, Hanabusa K. Polymer organogelators that make supramolecular organogels through physical cross-linking and self-assembly. *Chem Soc Rev* 2010;**39**(2):455–63.
50. Iha RK, Wooley KL, Nystrom AM, Burke DJ, Kade MJ, Hawker CJ. Applications of orthogonal "click" chemistries in the synthesis of functional soft materials. *Chem Rev* 2009;**109**(11):5620–86.
51. Díaz DD, Rajagopal K, Strable E, Schneider J, Finn M. "Click" chemistry in a supramolecular environment: stabilization of organogels by copper (I)-catalyzed azide–alkyne [3 + 2] cycloaddition. *J Am Chem Soc* 2006;**128**(18):6056–7.
52. Lau KN, Chow HF, Chan MC, Wong KW. Dendronized polymer organogels from click chemistry: a remarkable gelation property owing to synergistic functional-group binding and dendritic size effects. *Angew Chem Int Ed* 2008;**47**(36):6912–6.
53. Wang H-Z, Chow H-F. A photo-responsive poly (amide–triazole) physical organogel bearing azobenzene residues in the main chain. *Chem Commun* 2018;**54**(60):8391–4.
54. Xue X, Zhu J, Zhang Z, Zhou N, Tu Y, Zhu X. Soluble main-chain azobenzene polymers via thermal 1, 3-dipolar cycloaddition: preparation and photoresponsive behavior. *Macromolecules* 2010;**43**(6):2704–12.
55. Kang X, Zhao J, Li H, He S. Synthesis of a main-chain liquid crystalline azo-polymer via "click" chemistry. *Colloid Polym Sci* 2013;**291**(9):2245–51.
56. Mishra V, Jung SH, Park JM, Jeong HM, Lee HI. Triazole-containing hydrogels for time-dependent sustained drug release. *Macromol Rapid Commun* 2014;**35**(4):442–6.
57. Li T, Kumru B, Al Nakeeb N, Willersinn J, Schmidt B. Thermoadaptive supramolecular α-cyclodextrin crystallization-based hydrogels via double hydrophilic block copolymer templating. *Polymers* 2018;**10**(6):576.
58. Nagahama K, Kimura Y, Takemoto A. Living functional hydrogels generated by bioorthogonal cross-linking reactions of azide-modified cells with alkyne-modified polymers. *Nat Commun* 2018;**9**(1):2195.

CHAPTER 5

Triazole-Based Glycoconjugates

Arruje Hameed[a] and Tahir Farooq[b,*]
[a]Department of Biochemistry, Government College University, Faisalabad, Pakistan
[b]Department of Applied Chemistry, Government College University, Faisalabad, Pakistan
*Corresponding author. E-mail: tahirfarooqfsd@gmail.com

Introduction

Carbohydrates are a vital source of energy, an important part of the structural building block of genetic materials, and the most abundant class of biomolecules. They play a crucial part in many intracellular and cellular interactions in the form of cell surface receptors and signaling molecules.[1,2] The diversity of functional groups, linkage pattern, and presence of a number of rings make them valuable biomolecules for the construction of a variety of potent molecular architectures. Considering their valued potential to act as diagnostic and pharmaceutical agents, researchers have focused their attention to synthesize glycoconjugates of a diverse variety.[3] Furthermore, carbohydrates are considered as privileged moieties because of their hydrophilic properties, optimum pharmacokinetics, and low toxicity, thus they are preferred for the development of novel molecular scaffolds.[4,5]

Over the last two decades, the triazoles have become attractive pharmacophore motifs due to their unprecedented stability toward metabolic degradation, excellent capacity to act as the bioisostere of amide functionality, and broad spectrum of bioactivity. Triazole containing molecules show a number of potential therapeutic applications. Therefore, much research has been carried out to develop hybrid molecules with enhanced biological profiles by combining triazoles with other moieties. In this connection, the aforementioned fascinating structural aspects of carbohydrates have been well exploited using triazole-chemistry protocols to synthesize an array of biologically potent molecules (Figure 1).[6] The versatility of the triazole-chemistry protocols has allowed the construction of sugar-based simple, as well as complex, architectures like glycodendrimers, neoglycopolymers, and neoglycoconjugates with interesting pharmacological activities.[7] The subsequent sections of this chapter highlights the noteworthy developments made in the synthesis of triazole-based glycoconjugates.

Advances in Triazole Chemistry
https://doi.org/10.1016/B978-0-12-817113-4.00009-3

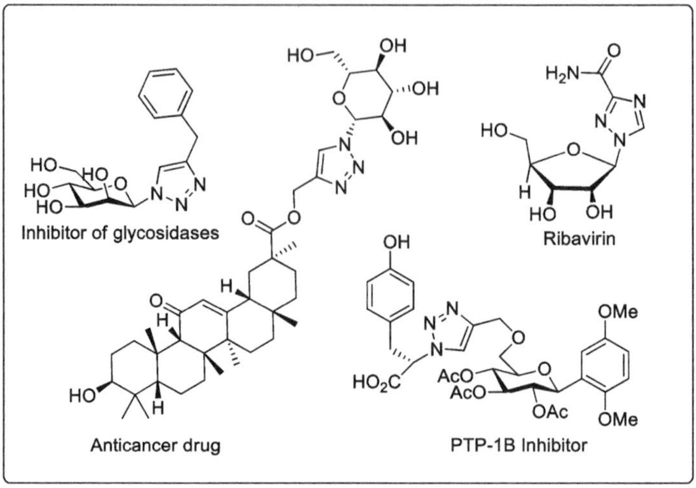

Figure 1 Known 1,2,4- and 1,2,3-triazole-based glycoconjugates.

1,2,4-Triazole and carbohydrates

Over the last few years, the 1,2,4-triazoles have attracted much attention due to their interesting biological activities, and much effort has been made to construct 1,2,4-triazole-tethered carbohydrates with desired therapeutic potentials. A few years ago, Kun and his co-workers presented an excellent methodology for the synthesis of a series of 3-(β-D-glucopyranosyl)-5-substituted-1,2,4-triazoles as glycogen phosphorylase inhibitors startingfrom O-perbenzoylated5-(β-D-glucopyranosyl)tetrazoles (Scheme 1). A library of triazolic-carbohydrates was generated using heterocyclic, mono-, and bicyclic aromatic and aliphatic substituents at 5-position. The C-glucopyranosyl-1,2,4-triazoles was a novel skeleton, and the derivatives with 5-(2-naphthyl)- and 5-(4-aminophenyl) substituents were found as the best inhibitors of glycogen phosphorylase.[8] In a recent approach, Kun

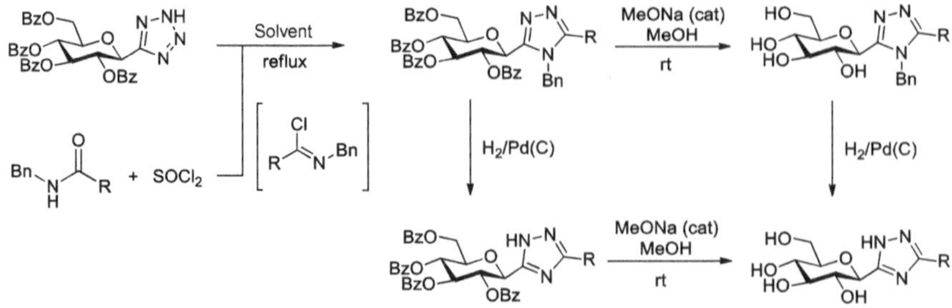

Scheme 1 Synthesis of 3-(β-D-glucopyranosyl)-5-substituted-1,2,4-triazoles.

and his colleagues revealed the synthesis of nine 3-(β-D-glucopyranosyl)-5-substituted-1,2,4-triazole derivatives as glycogen phosphorylase inhibitors with promising pharmacokinetic profiles. These derivatives were selected using quantum mechanics-polarized ligand docking (QM-PLD) and synthesized according to the given routes (Scheme 2).[9]

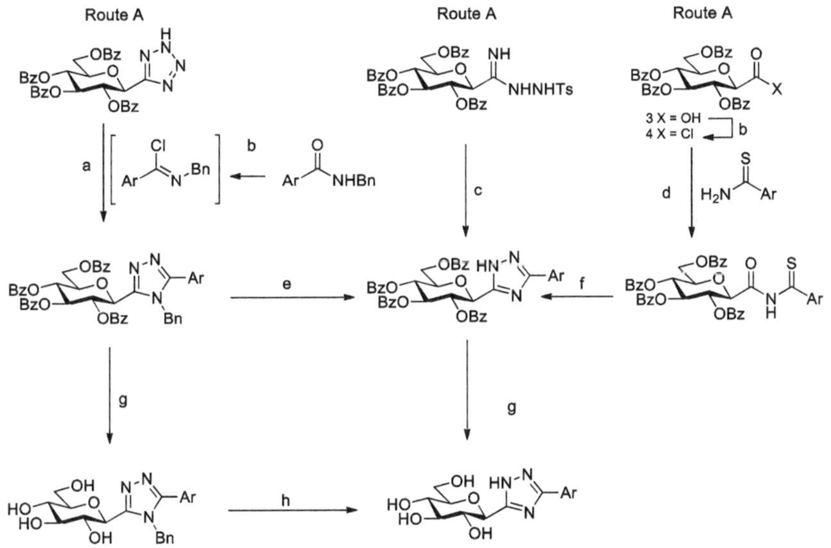

Conditions: a) anhydr. m-xylene at boiling temp, b) $SOCl_2$ at boiling temp. c) 1.ArCOCl. anhydr. $CHCl_3$, pyridine 0°C-rt, 2.Bu_4NF, THF at boiling temp, d) anhydr. CH_3CN, pyridine, rt, e) H_2 (1 atm), Pd(C), anhydr. EtOAcnor THF, both at boiling temp.; f) $NH_2NH_2.H_2O$, pyridine, rt, g) cat. NaOMe, anhydr. MeOH, $CHCl_3$, rt; h) H_2 (1 atm) , Pd(C), anhydr. Meoh, EtOAc at boiling temp.

Scheme 2 New routes for synthesis of 3-(β-D-glucopyranosyl)-5-substituted-1,2,4-triazoles.

In another study, various glycopyranosyl aminoguanidine nitrates were stereoselectively converted to 1-(-D- and -L-glycopyranosyl)-5-methyl-1*H*-1,2,4-triazoleswith*invitro*immune-andantitumoractivities(Scheme3).[10] D'Accorso and his co-workers described the synthesis of 1,2,4-triazole D-ribose derivatives with antitumor activities (Scheme 4).[11] Moderate antiproliferative activities were seen in compounds where sugar moiety was linked to 1,2,4-triazole via sulfur. No perceptible changes in bioactivities were observed even after the introduction of second heterocyclic substituents. A few sugar-containing triazolic compounds were found as promising leads for future antitumoral studies.

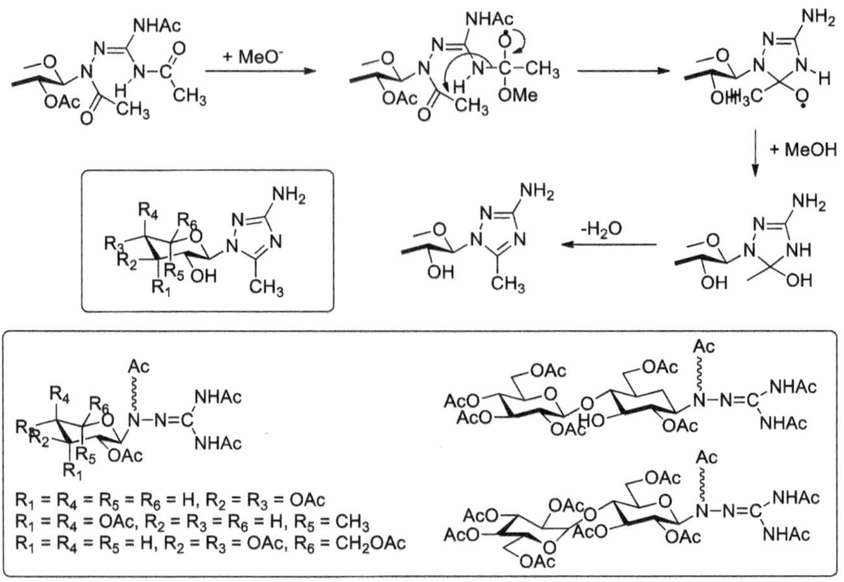

Scheme 3 Stereoselective synthesis of 1-(-D- and -L-glycopyranosyl)-5-methyl-1*H*-1,2,4-triazoles.

Using different glycosyl halides, the *S*-glycosides were synthesized regioselectively from 5-(3-chlorobenzo[b]thien-2-yl)-4*H*-1,2,4-triazole-3-thiol and its 3-methylsulfanyl and 3-benzylsulfanyl derivatives (Scheme 5). The microwave-assisted synthesis improved the reaction yields without affecting the regioselectivity of the reaction (Scheme 6).[12]

The *S*,*N*1- and *S*,*N*2-bis(glycosylated) derivatives with antibacterial and antifungal activates were synthesized regioselectively through a glycosylation process (Scheme 7).[13] Few of these glycosides demonstrated better antibacterial and antifungal potential compared with baneocin and chloramphenicol used as reference drugs. Various *N*- and *S*-β-D-glucosides with differently substituted 1,2,4-triazole were accessed through an efficient and facile procedure (Scheme 8). These triazolic glucosides displayed an appreciable antimicrobial potential. Furthermore, they were found as suitable candidates for further studies with possible applications in viral infection and chemotherapy of cancer.[14]

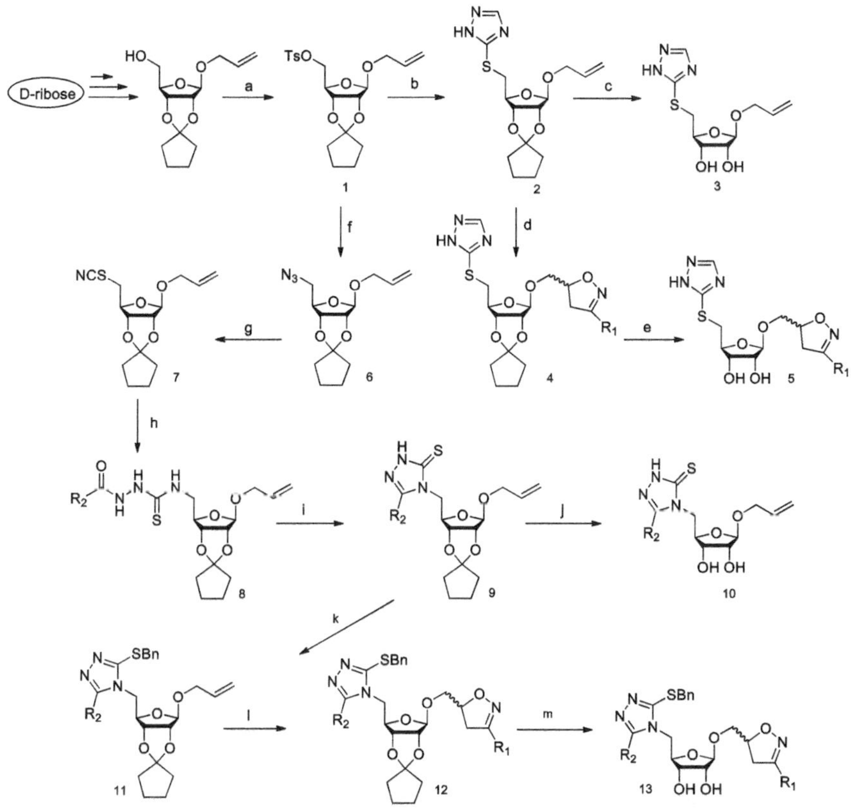

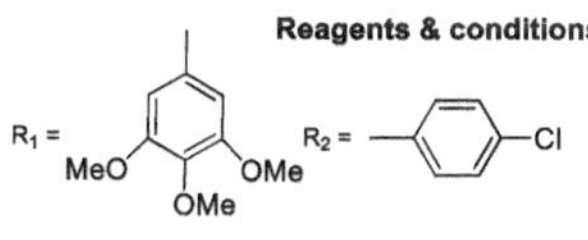

Reagents & conditions: a) TsCl pyridine; b) triazolyl sodium thiolate, DMF, 50 °C c) CH_3CO_2H, 55 °C d) R_1HCNOH, chloramine-T, EtOH, H_2O e) CH_3CO_2H, 40 °C f) NaN_3, DMF g) CS_2. $(Ph)_3P$, dioxane h) $R_1CONHNH_2$, dioxane, reflux i) NaEtO, EtOH, reflux, 3 h; j) CH_3CO_2H, 40 °C k) BnCl, K_2CO_3, acetone l) R_1HCNOH, chloramine-T, EtOH, H_2O m) CH_3CO_2H, 40 °C

Scheme 4 Synthesis of 1,2,4-triazole D-ribose derivatives with antitumor activities.

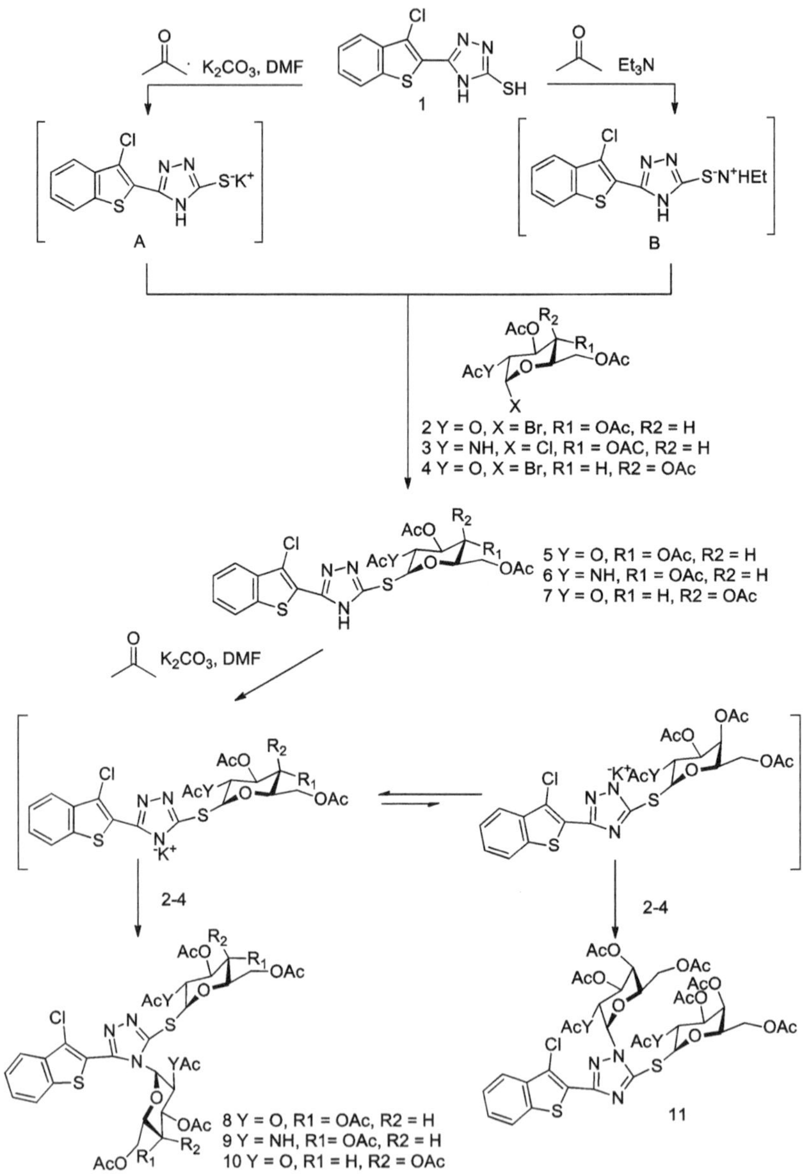

Scheme 5 Regioselective synthesis of 1,2,4-triazole containing *S*-glycosides.

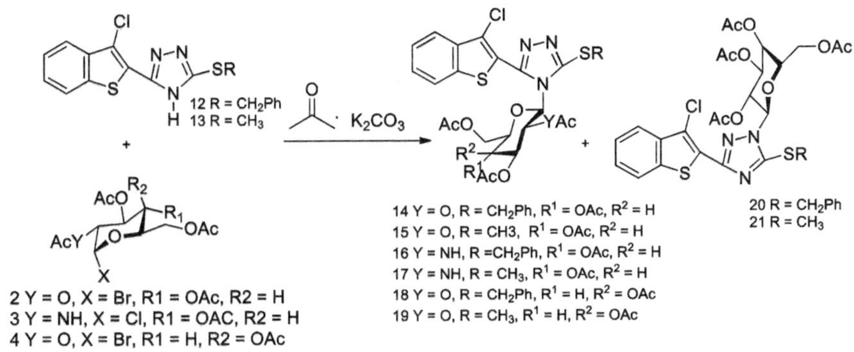

Scheme 6 Microwave-assisted synthesis of *S*-glycosides with 1,2,4-triazole.

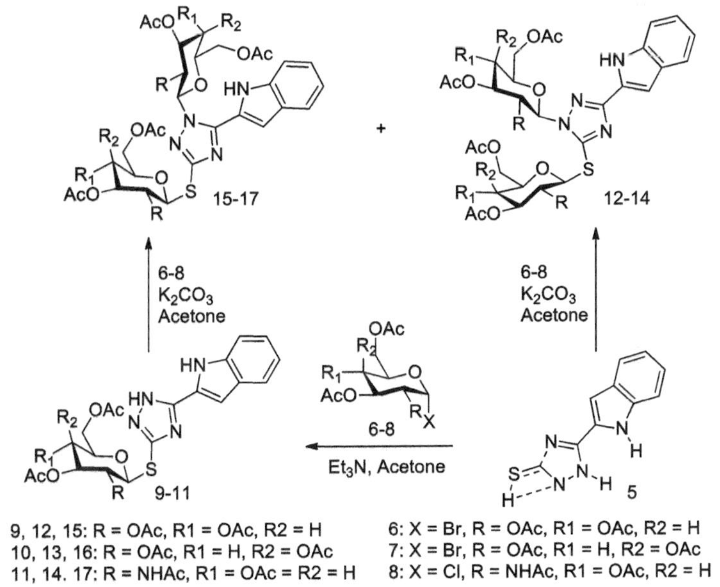

Scheme 7 *S*- and *S,N*-bis(glycosides) with 1,2,4-triazolic moiety.

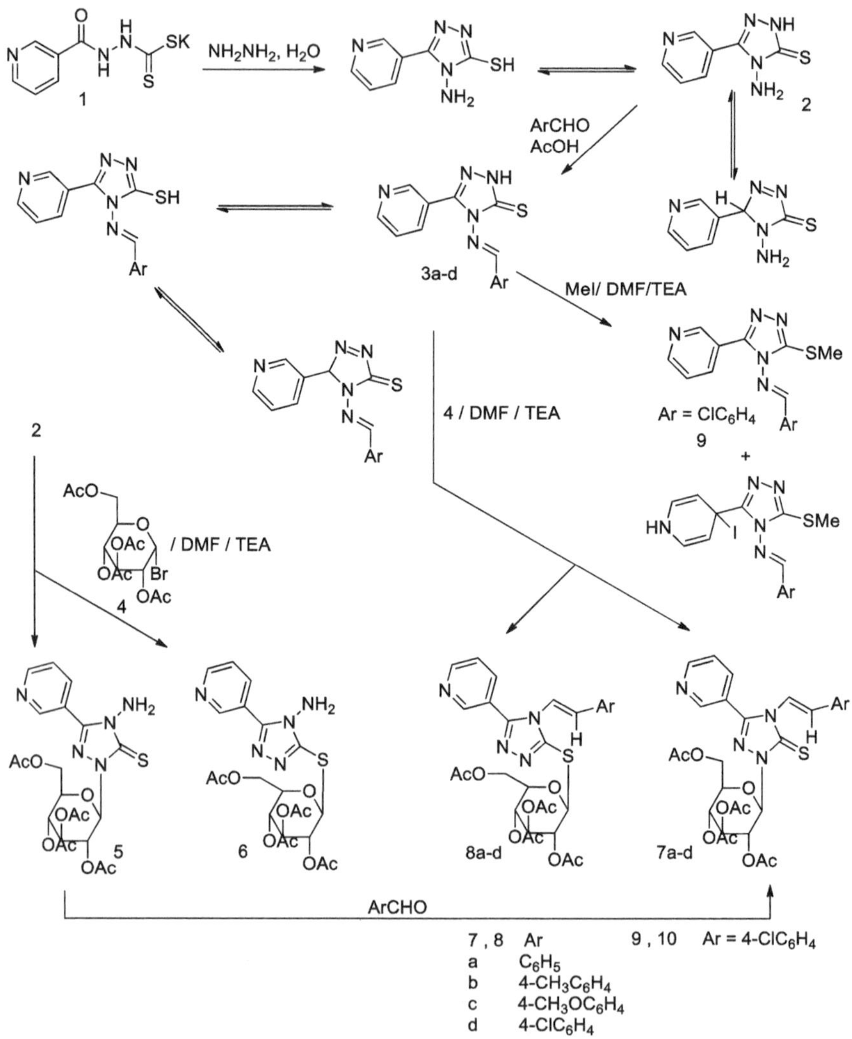

Scheme 8 Synthesis of *N*- and *S*-β-D-glucosides with differently substituted 1,2,4-triazole.

1,2,3-Triazoles and carbohydrates

The azide or alkyne functionality could easily be incorporated on sugar moieties through simple reactions. Furthermore, they undergo click reactions to produce triazole containing carbohydrate conjugates. In fact, the facile applications of Cu-catalyzed azide alkyne cycloadditions (CuAAC) in carbohydrate chemistry has helped the production of a variety of glycoconjugates including simpler to complex architectures like oligomers, glycol dendrimers, glycoclusters, glycopolymers, sugar-based macrocycles, glycoproteins, glycopeptide conjugates, and glycolipid conjugates, etc.[15, 16]

Incorporation of azide functionality on carbohydrates

The hazardous nature of azides especially of low molecular weight often make them dangerous to handle, thus they have found limited applications in synthetic reactions. However, under a range of reaction conditions, the remarkable inert behavior and stability of glycosyl azides has been well explored in the field of click-carbohydrates. The alkyne and azide functionalities are introduced on the anomeric position of a sugar ring for click reactions, and the rest of the ring positions are usually used for standard protection-deprotection reactions.[17] The carbohydrates are equipped with azides and alkynes for facetious synthesis of mono-, di-, or polyfunctionalized molecular frameworks that react with complementary functionalized molecules to produce triazole containing conjugates.[18] In general, glycosyl halides undergo nucleophilic substitution reaction with sodium azide to generate corresponding glycosyl azides in high yield.[19] In another route, the (TMS)N_3 is made to react with acetylated sugars to furnish respective azides in excellent yields.[20, 21] The 1,2-*cis*-glycosyl halides could be converted to 1,2-*trans*-azido sugars.[22, 23] The good leaving groups like triflate, mesyl, tosyl, and halides have also been used to incorporate azide substituent selectively on primary carbon of sugar moieties.[24] The fate of SN2 reactions at secondary carbons of carbohydrates are influenced by the anomeric configuration and stereochemistry of the neighboring groups. An inversion of configuration is observed in glycosyl azide products.[25] The sugar epoxides undergo ring opening reactions when attacked by azide ions to furnish diaxial products.[26] Azide functionality is also introduced to carbohydrates by following the methodology of radical addition to glycals. In one such attempt, 2-azido sugars were prepared using azidonitration method.[27, 28] The diazo transfer, azidophenylselenation, and azidochlorination reactions could be used to achieve 2-azoido sugars.[29, 30] In an indirect insertion of an azido group using triflyl azide, the N_2 is converted into amine with no change in configuration in diazo transfer reactions. The amino sugars were also made to undergo diazo transfer reactions.[31] High yields of azido sugars were also received from unprotected amino-sugars in short reaction time using diazo transfer as a protecting strategy for amines (Scheme 9).[32, 33]

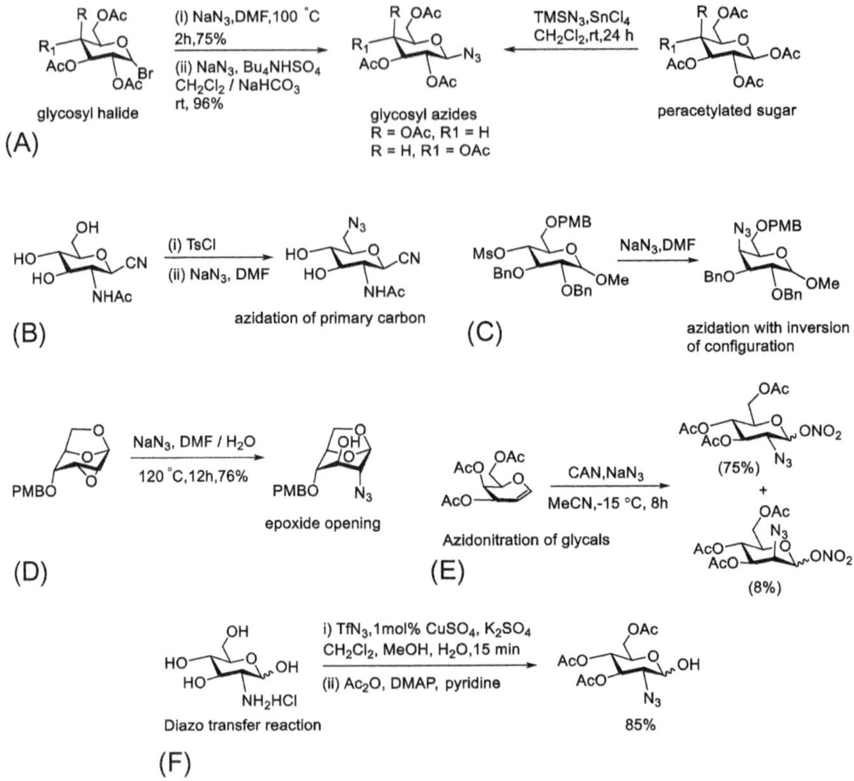

Scheme 9 Synthetic routes for glycosyl azides.

The sialic acid derivatives did not undergo direct azidation like other carbohydrates; therefore, chemoenzymatic methodology has been developed to achieve sialic acid-containing carbohydrates. A variety of sialosides with α-2,6 or α-2,3 linkages have been generated through this powerful emerging approach.[34, 35] There is no need to isolate intermediates in this one-pot, three-enzyme-coupled reaction process (Scheme 10).[36] This methodology has shown successful applications for synthesis of various desired products.[37] Recently, a series of disialyl glycans and GD3 ganglioside oligosaccharides of chemotherapeutic potential were prepared using two-step multi-enzyme strategy. The monosialylated oligosaccharides with α-2,6- or α-2,3-linkage were synthesized initially through one-pot, three-enzyme approach. Subsequently, oligosaccharides containing two sialyl groups were received as a result of α-2,8-sialyltransferase activity (Scheme 11).[38] Furthermore, the potentials of azide functionality at sialosides could be explored by using the counterpart at lipids, peptides, and sugars under CuAAC conditions.[39] The

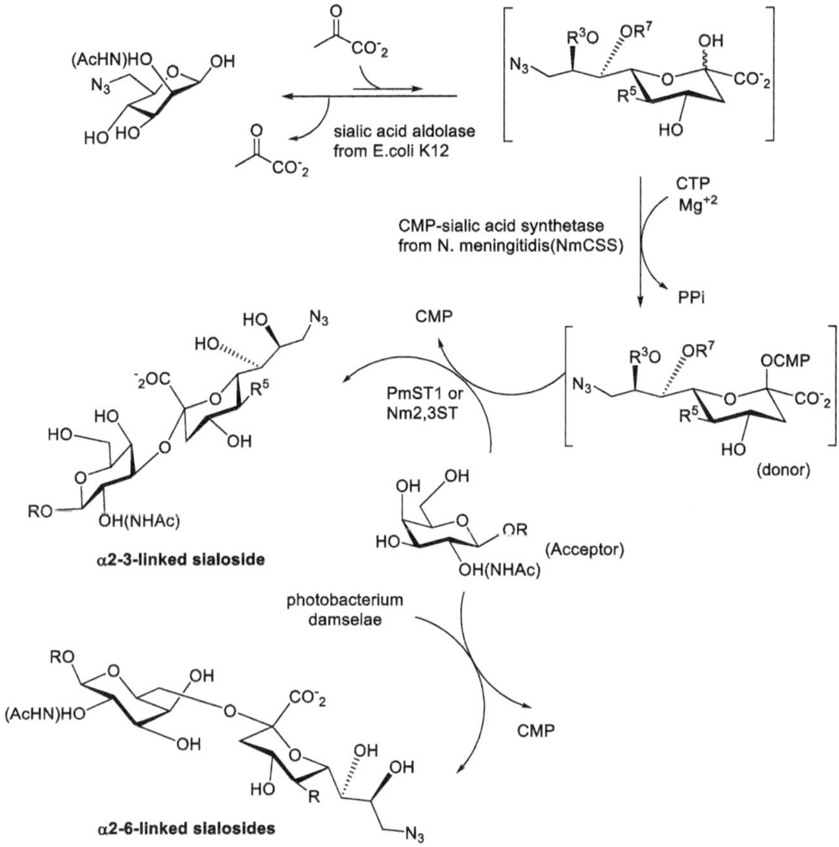

Scheme 10 Chemoenzymatic azidation of α-2,6 or α-2,3 link sialosides.

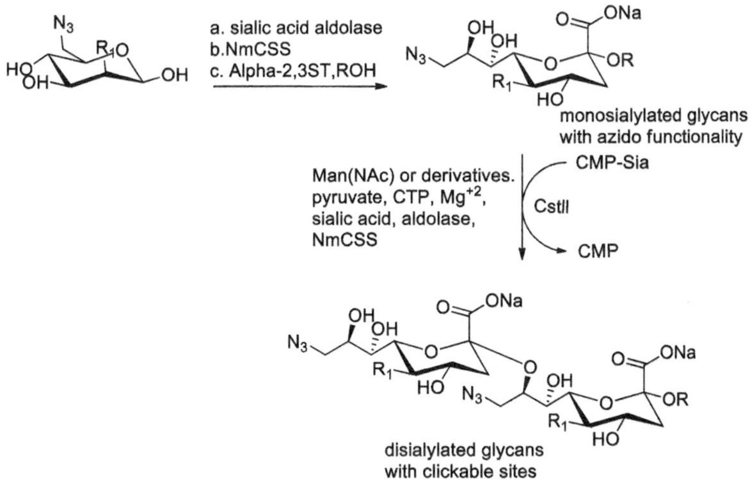

Scheme 11 Two-step multi-enzymatic synthesis of azide-functionalized GD3 oligosaccharides.

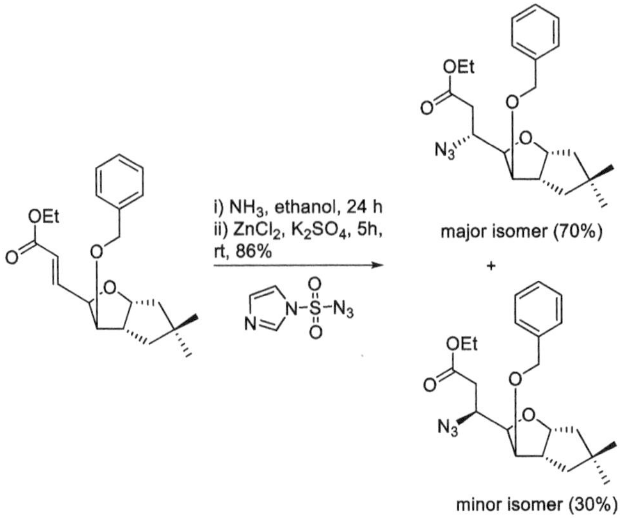

Scheme 12 One-pot synthesis of glycosylated azido ester.

α,β-unsaturated olefinic ester undergoes diazo transfer reaction to furnish glycosyl β-azido ester through a different route. This one-pot reaction proceeds under mild conditions and involves metal-catalyzed 1,4-conjugate addition of ammonia (Scheme 12).[40] Such incorporation of the azido group at β-position of an α,β-unsaturated olefinic ester is not successful through straightforward diazo transfer reaction.[41, 42]

Incorporation of alkyne functionality on carbohydrates

The aforementioned methods highlighted the well-established routes to incorporate azide functionality at specific positions on carbohydrates for click chemistry reactions. Similarly, alkyne functionality is also incorporated on carbohydrates through propargylation reaction. Furthermore, they are used to prepare glycoconjugates through click reactions.

Triazolyl glycoconjugate synthesis using click protocol

Recently, the preparation of a conjugate molecule with collective properties of two coupling components with distinctive properties has become an attractive research area for synthetic chemists.[43, 44] In this connection, the CuAAC strategy has emerged as a powerful, reliable, and practically simple approach for the construction of glycoconjugates with unusual biological potential distinctive from starting segments (Scheme 13). In a high-yielding and expedient CuAAC reaction, the mono/dipeptide alkynes were made

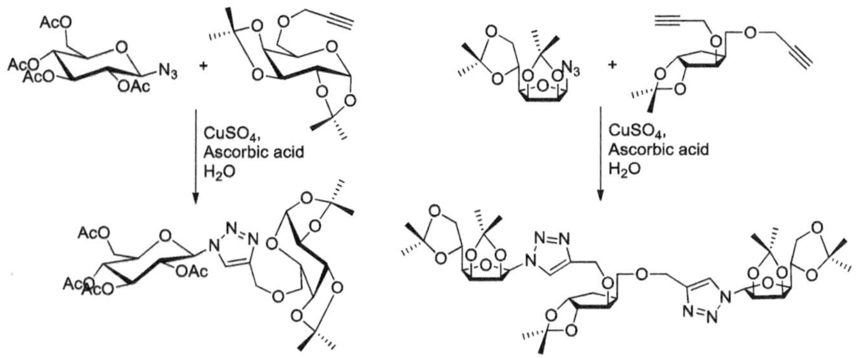

Scheme 13 Synthesis of glycoconjugates using click chemistry.

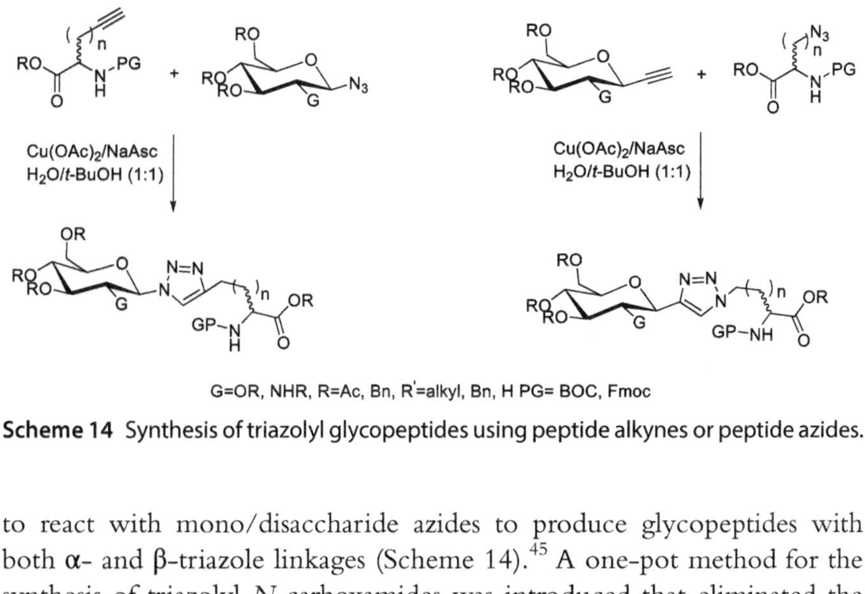

Scheme 14 Synthesis of triazolyl glycopeptides using peptide alkynes or peptide azides.

to react with mono/disaccharide azides to produce glycopeptides with both α- and β-triazole linkages (Scheme 14).[45] A one-pot method for the synthesis of triazolyl *N*-carboxamides was introduced that eliminated the involvement of bases, ligands, and solvents. This facile approach was a welcome advance for the synthesis of potent molecules (Scheme 15).[46]

In general, molecules with multiple azide functionalities are known for stability issues. In this connection, a one-pot strategy was disclosed for *in situ* conversion of amines to azide functionalities and subsequently to corresponding 1,2,3-triazoles under click conditions. This one-pot microwave-assisted reaction completes in about 20 min to furnish multivalent carbohydrate-based entities (Scheme 16).[47] According to another report, the ethisterone reacted with azido sugars under click conditions to generate a library of corresponding triazolylethisterone glycoconjugates in high yield

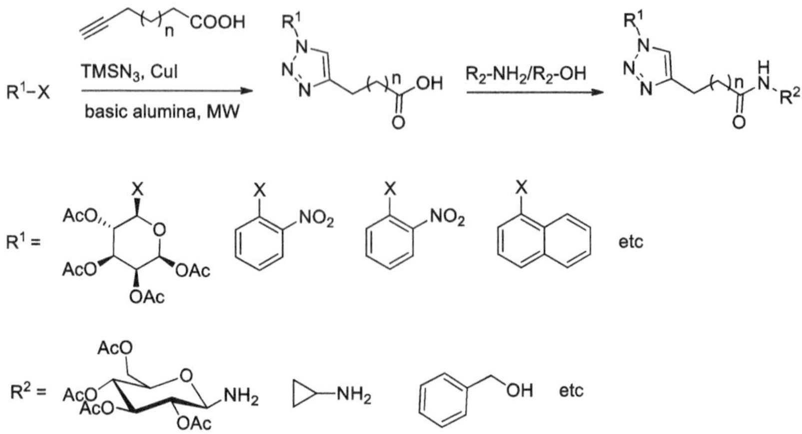

Scheme 15 Preparation of triazolyl *N*-carboxamides by one-pot four-component reaction.

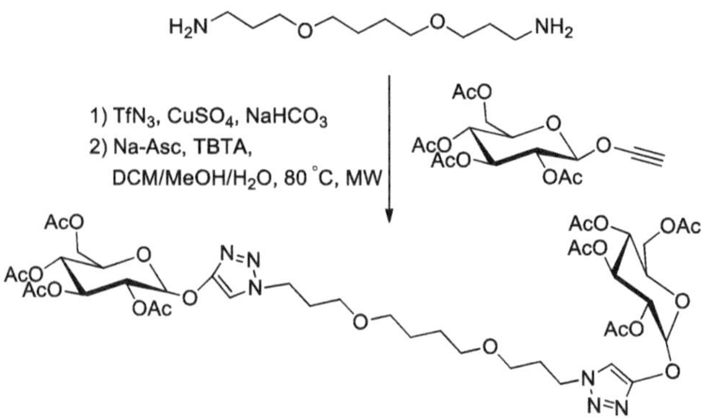

Scheme 16 Glycoconjugates by one-pot strategy.

(Scheme 17).[48] The same research group further explored the potential of such transformations. They reacted glycosylalkynes obtained from triazolyl azido alcohols with ethisterone to produce bis(triazolyl)ethisterone glycoconjugates using click protocol (Scheme 18).[49] The triazolyl-glycoconjugates of α-tocopherol were prepared from glycosylalkynes and α-tocopherol azide using standard click reactions. These glycoconjugates displayed better water solubility and also their radical scavenging potential was found analogous to parent α-tocopherol. However, unstable glycoconjugates were obtained when sugars with azide functionality were reacted with α-tocopherol alkyne under the same conditions (Scheme 19).[50] The noscapine are natural alkaloids with well-known antitussive potential. The click triazole carrying noscapine-based glycoconjugates displayed much better therapeutic profiles compared with parent alkaloid (Scheme 20).[51]

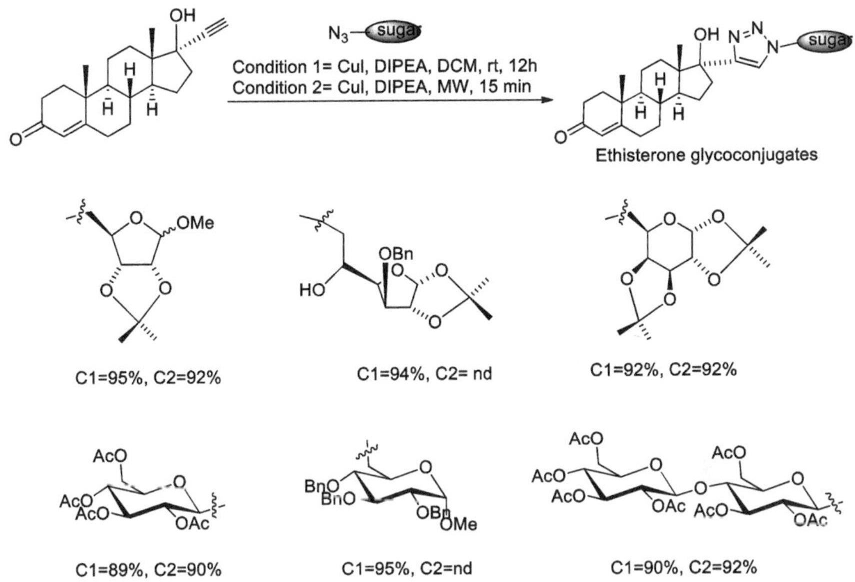

Scheme 17 Click-inspired synthesis of triazolyl ethisterone glycoconjugates.

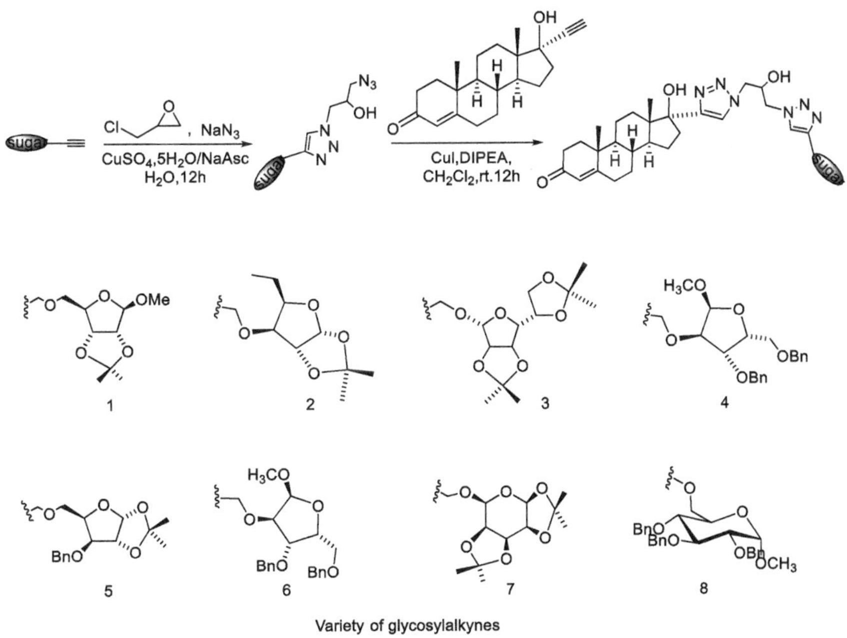

Scheme 18 Click protocol for preparation of bis(triazolyl)ethisterone glycoconjugates.

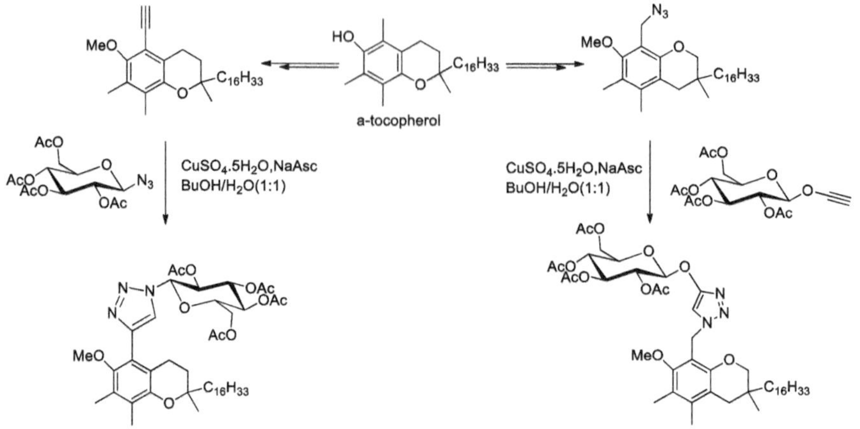

Scheme 19 Preparation of glycoconjugates using α-tocopherol.

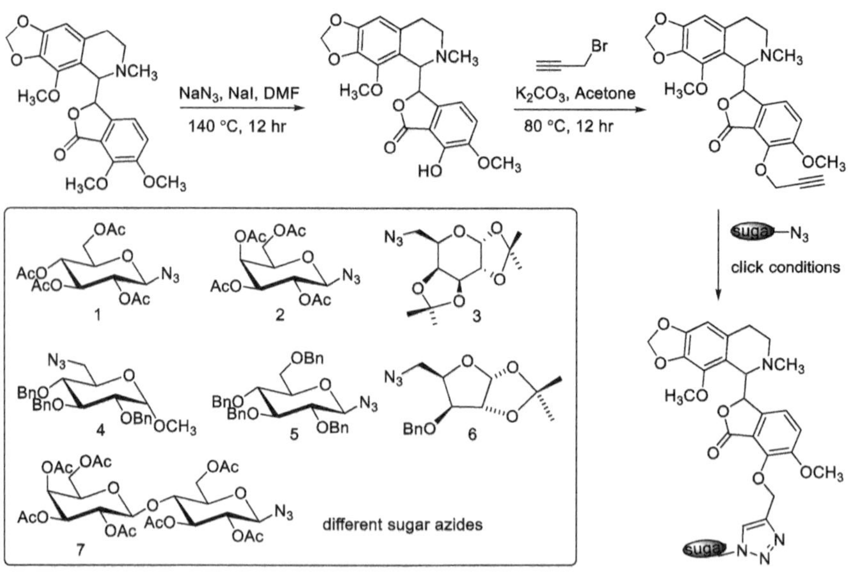

Scheme 20 Preparation of noscapine-based glycoconjugates.

Double or multi-clicking and glycoconjugates

A one-pot two-step method was developed for the synthesis of morpholine-fused triazolyl-glycoconjugates from sugars containing alkyne functionality. Initially, the glycosylalkynes were converted into triazolyl azido alcohols, which subsequently underwent click reaction to furnish morpholine-fused

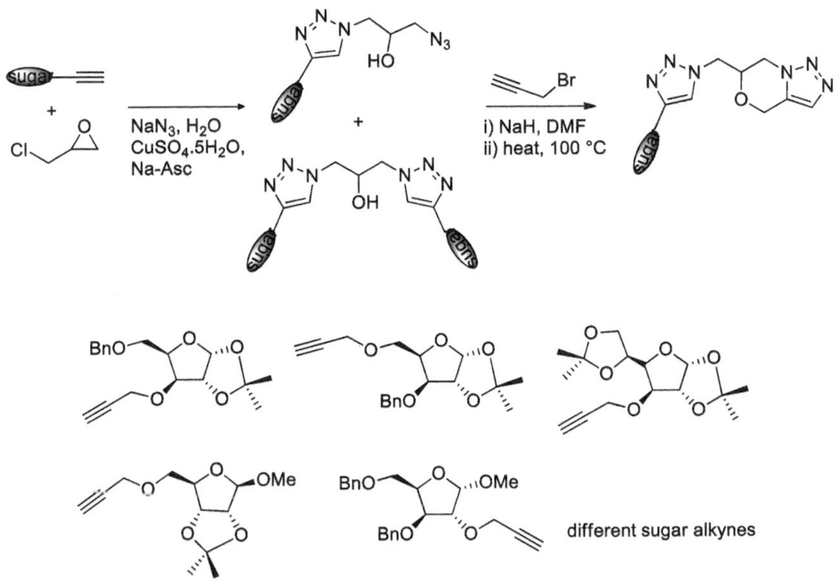

Scheme 21 Preparation of morpholine-fused glycoconjugates.

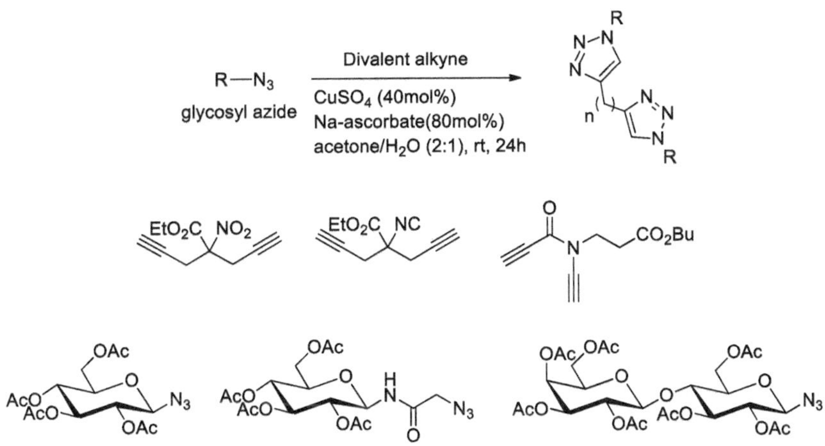

Scheme 22 Cycloaddition reactions for triazole containing glycoamino acid mimics.

triazolyl-glycoconjugates in high yields (Scheme 21).[52] In another report, the azidoacetamide and glycosyl azides underwent CuAAC reaction with alkyne containing divalent molecules and produced triazolyl-divalent glycoconjugates (Scheme 22). Furthermore, they were used for the synthesis of divalent-glycopeptide mimics.[53] According to another method, the polyfunctionalized adamantane were made to react with sugar azides under click conditions to produce glycoclusters having four triazoles (Scheme 23).[54]

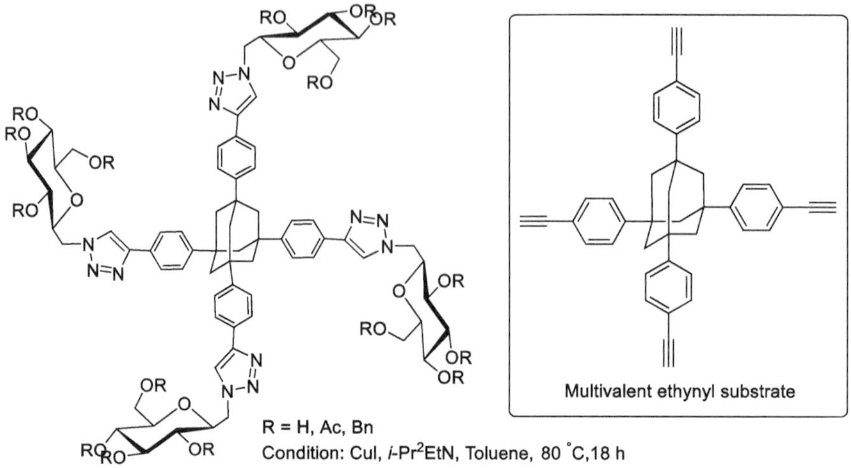

Scheme 23 Synthesis of triazole *C*-glycoclusters from polyfunctionalized adamantine.

Synthesis of macrocyclic glycoconjugates using click chemistry

The carbohydrate-based building blocks have been well exploited to synthesize glycomacrocycles using click chemistry reactions. The macrocycles with great structural diversity are accessible by linking multi-functionalized carbohydrates through different positions. Resultantly, the physiochemical properties like lipophilicity, hydrophilicity, biodisponibility, and solubility could be modulated. The chirality of carbohydrates helps to prepare chiral cavities with applications in asymmetric catalysis, chiral recognition, and host-guest chemistry.[55] The triazolyl-macrocycles find potential applications in molecular reactors, supramolecular structures, artificial receptors, drug discovery, and drug carrier systems, etc. due to the chemically stable nature of triazoles and their potential to be involved in various noncovalent interactions like dipole-dipole interactions and hydrogen bonding.[7, 56] The C2- and C3-symmetric carbohydrate-based macrocycles were synthesized using alkyne and azide functionalities at the two terminals of glycoconjugates under CuAAC reaction conditions (Scheme 24).[57, 58] Different furanosides containing alkyne or azido functionality were reacted under click conditions to achieve macrocycles comprising sugar residues and triazole units (Scheme 25).

According to another report, the D-manno-, D-gluco-, and D-galacto-derived azido compounds containing alkyne substituent at 2- or 3-position

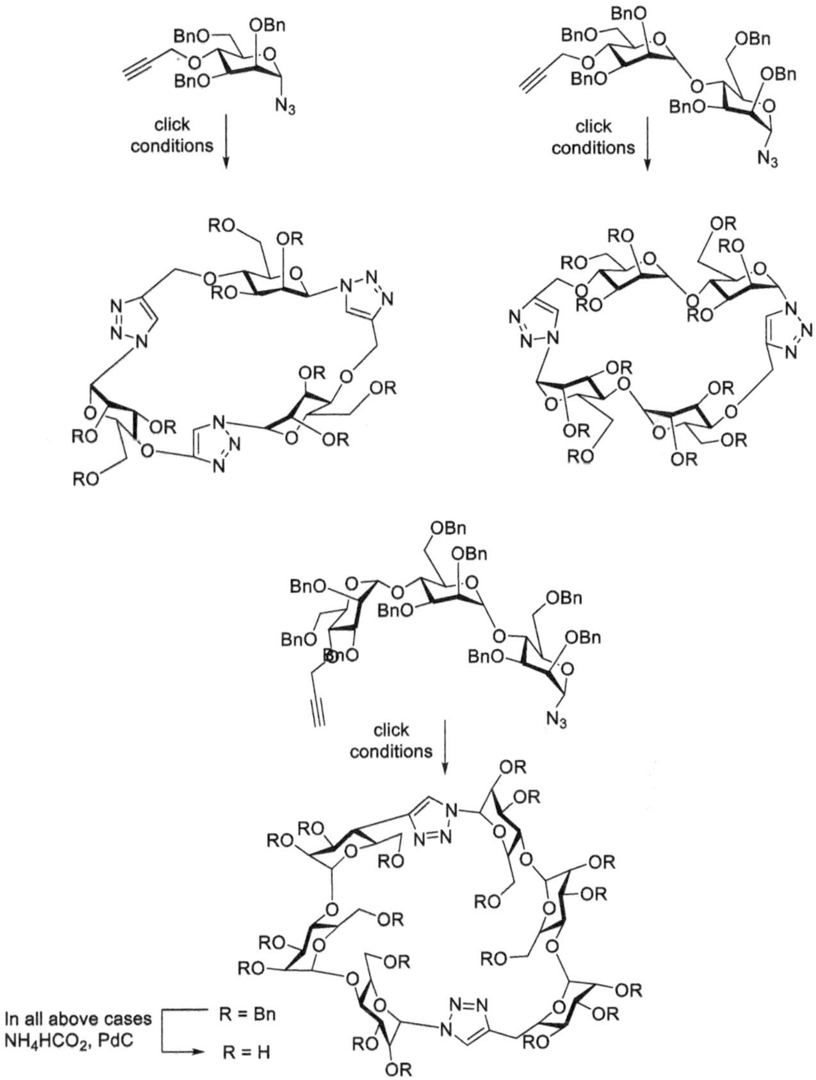

Scheme 24 Macrocycles from glycoconjugates containing alkyne and azide functionalities.

of the sugar ring were exploited to synthesize macrocycles encompassing two sugar units connected through two triazoles (Scheme 26).[59] Factors like anomeric configuration, nature of sugar units, and size of the ring influenced the product formation. In another reaction, the macrocycle precursor (glycohybrid) prepared through simple standard reactions was subjected to click

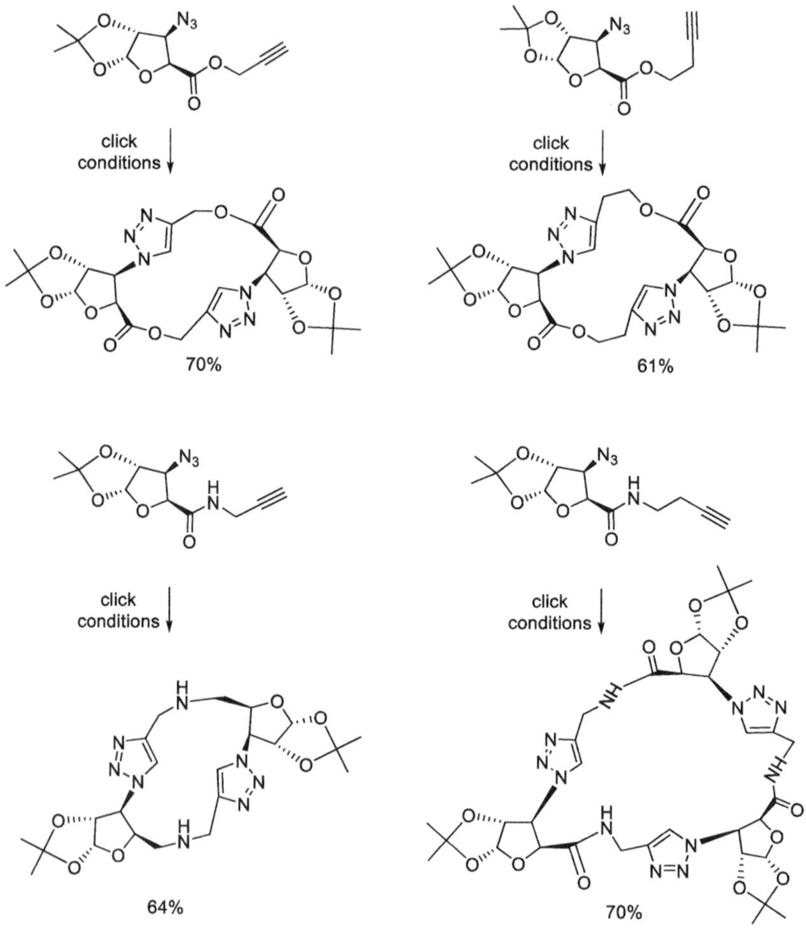

Scheme 25 Macrocycles from functionalized furanosides.

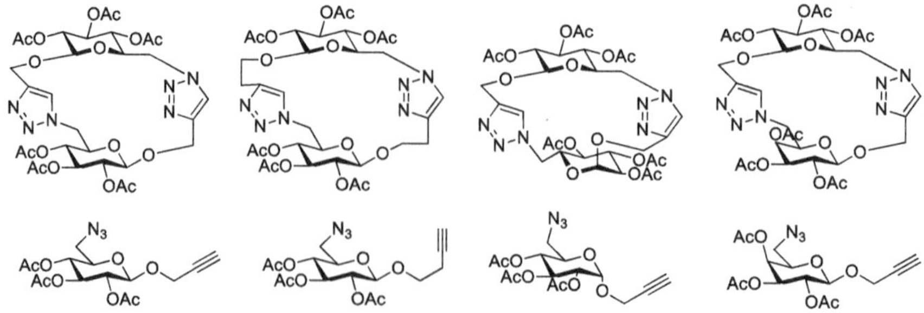

Scheme 26 Macrocycles synthesized from functionalized sugars.

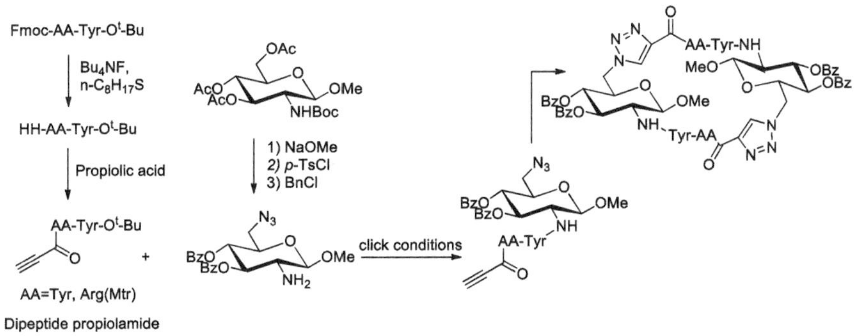

Scheme 27 Preparation of hybrid macrocycles containing carbohydrate and amino acid.

condition to achieve C2-symmetric macromolecules with diverse applications (Scheme 27).[59] In another approach, the 1,7-octadiyne and 6-azido glycosides were made to react under combined CuAAC and ring-closure metathesis (RCM) methodologies to achieve macrocycles with interesting properties. The Grubbs 1-mediated RCM followed the click reaction to furnish inseparable macrocycles that subsequently underwent reduction of olefin to give pure *E* and *Z* isomers (Scheme 28).[60]

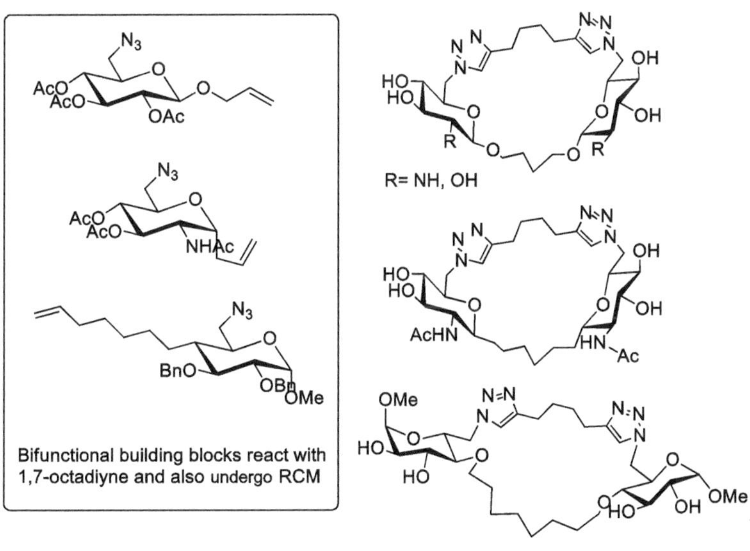

Scheme 28 Synthesis of macrocycles using combined click chemistry and RCM approach.

Triazole-fused bicyclic macrocycles

The triazole-linked macrocycles with 2 to 6C-glucopyranoside residues were prepared when bifunctionalized oligomers underwent intramolecular CuAAC reaction (Scheme 29).[61]

The one-pot two-step reaction involving intramolecular CuAAC and sonogashira reaction described the formation of triazole-fused tetracyclic glycosides in good to high yields (Scheme 30). This environmental

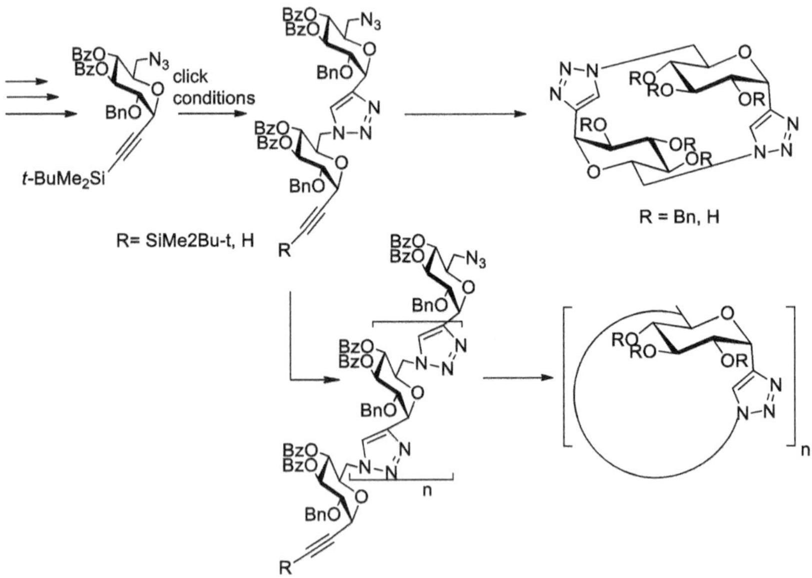

Scheme 29 Intramolecular CuAAC for triazole containing cycloglucopyranosides.

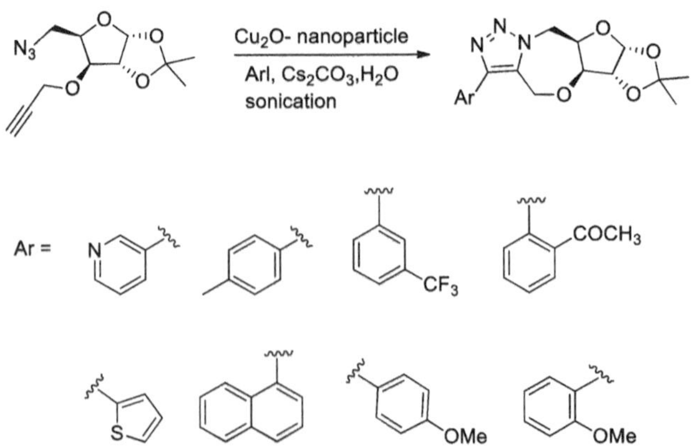

Scheme 30 One-pot preparation of triazole-fused tetracyclic glycosides.

friendly reaction utilized recyclable cuprous oxide nanoparticles in aqueous media.[62] Likewise when bifunctionalized sucrose molecules were subjected to click reactions they underwent both intermolecular and intramolecular cycloadditions to furnish corresponding macrocycles. The intramolecular cyclization was favored when the reaction was performed with lower concentration of starting material and catalyst, but unexpectedly macrocycles with 1,5-triazole regioisomer were also produced. However, the use of elongated azido-acetylenes exclusively produced macrocycles with 1,4-triazole regioisomer suggesting steric repulsion in the small cavity of macrocycles containing 1,4-triazoles as the major factor for the production of macrocycles with 1,5-triazoles (Scheme 31).[63] The galactose-based bifunctional precursors containing alkyne and azido functionalities at anomeric positions were developed for the synthesis of cyclic pseudo-oligosaccharides. They were made to undergo microwave-assisted CuAAC reactions. The dimers and trimers were produced with low concentration of the solution. The reactions run with concentrated solutions furnished low yields of cyclic tetramers and pentamers. The dimers and trimers were further transformed to sialosides (Scheme 32).[64]

Another report described the synthesis of macrocycles prepared from a galacturonic acid precursor. The outcome of the click reaction was found to be influenced by the reaction conditions (Scheme 33).[65] The complexes of macrocycles with Cu(II) showed a metal:ligand ratio of 1:1. Muthana *et al.* described a high-yielding chemoenzymatic intramolecular CuAAC reaction of trisaccharides for the synthesis of macrocyclic systems. Furthermore, the methodology was used for the production of interesting macrocycles containing sialic acid functionality (Scheme 34).[66]

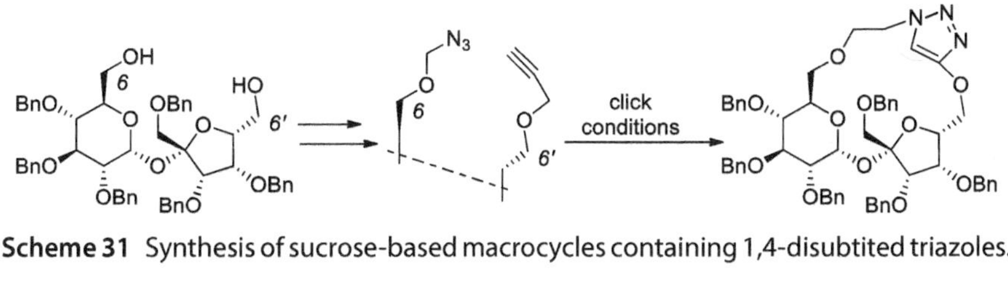

Scheme 31 Synthesis of sucrose-based macrocycles containing 1,4-disubtited triazoles.

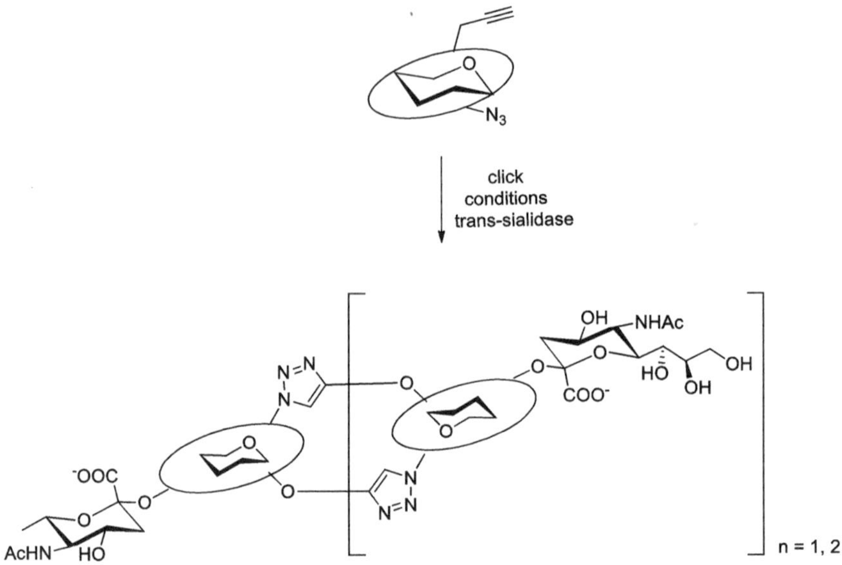

Scheme 32 Synthesis of cyclic pseudo-oligosaccharides via click reactions.

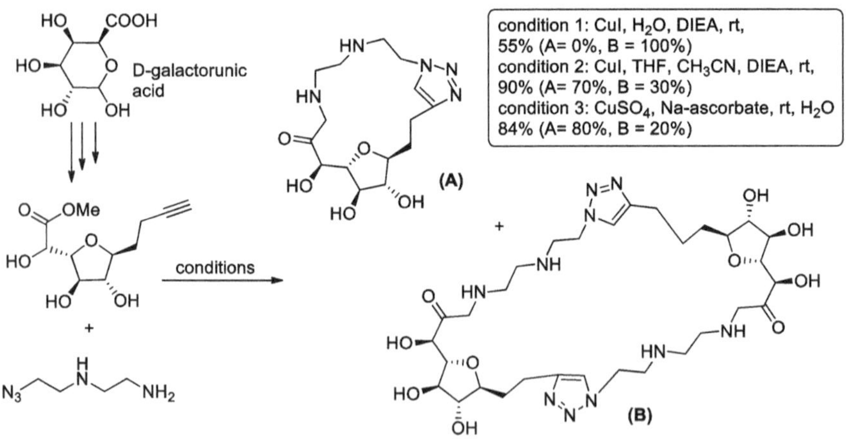

Scheme 33 Click reactions for macrocycles of galactorunic acid derivatives.

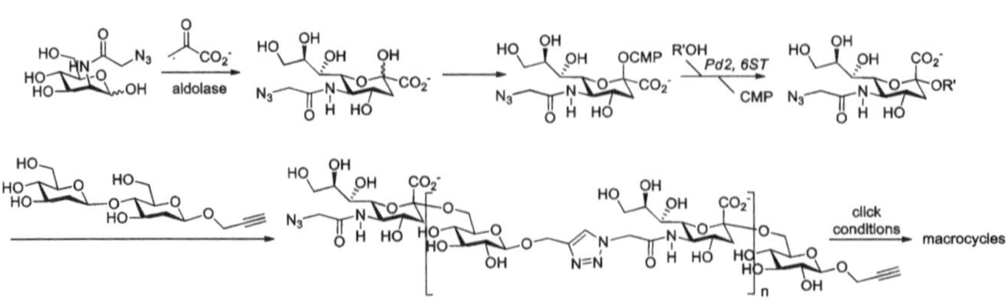

Scheme 34 Click synthesis of macrocycles with sialic acid functionality.

Click triazoles and glycopolymers

In recent years, the synthesis of multi-functionalized glycopolymers with interesting properties has become an attractive research area. The CuAAC approach has been found to be a very efficient methodology for polymerization and functionalization of polymers.[67, 68] The click protocol has been well explored in combination with well-known classical polymerization methods like ring opening polymerization, reversible addition fragmentation chain transfer, and atom transfer radical polymerization for the synthesis of glycopolymer frameworks from sugar-based precursors.[69] The prepolymerization and postpolymerization functionalizations could easily be managed by highly efficient click reactions. However, the postpolymerization functionalization is highly preferred to avoid undesirable side reactions of functional groups.

A series of narrow molecular weight glycopolymers were obtained from terminal alkynes bearing poly(methacrylate)s through click reactions with sugar azides. Using this powerful strategy, both unprotected and protected C-6 and α- or β-anomeric sugar azides could be used to generate a library of polymer scaffolds (Scheme 35).[70] The propargylated fluorine reacted with lactosyl azide under click conditions to furnish triazole containing monomer. Afterward, this monomer underwent Suzuki-Miyaura coupling polymerization subsequently affording the polyfluorene-based glycopolymer, which shows fluorescence quenching with calcium ions (Scheme 36).[71] The glycopolymers with controlled molecular weight and desired chain end fidelity were achieved using cobalt catalyzed chain transfer

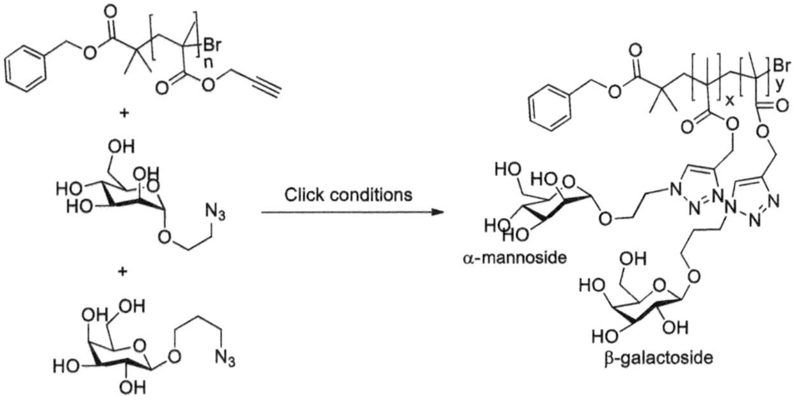

Scheme 35 Combination of living radical polymerization and click approaches.

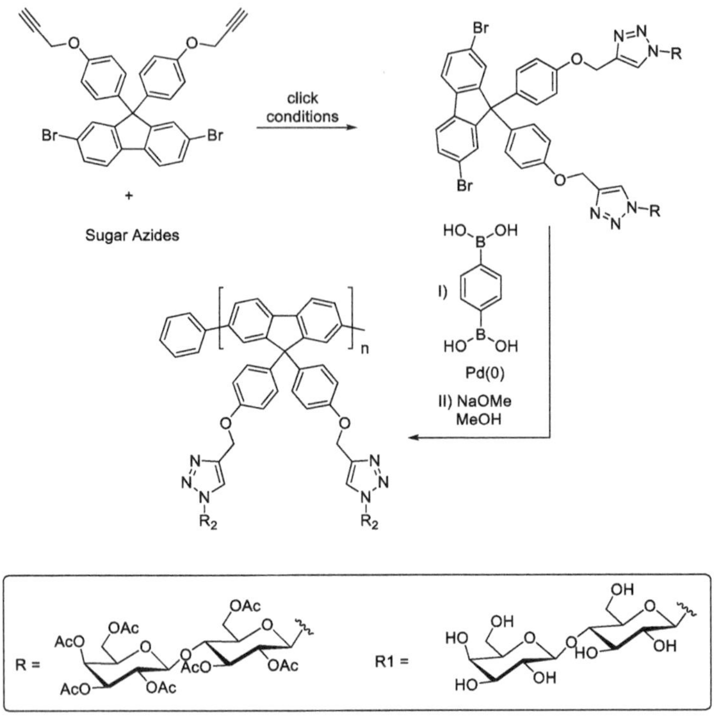

Scheme 36 Combined application of click reaction, Suzuki coupling, and polymerization.

(CCCT) polymerization and CuAAC reactions. The monomer with alkyne functionality underwent homopolymerization to furnish a polymer with ω-terminal vinyl group that was subsequently subjected to 1,4-conjugate thiol addition. Afterward, this polymer gave click reactions with various glycosyl azides (Scheme 37).[69]

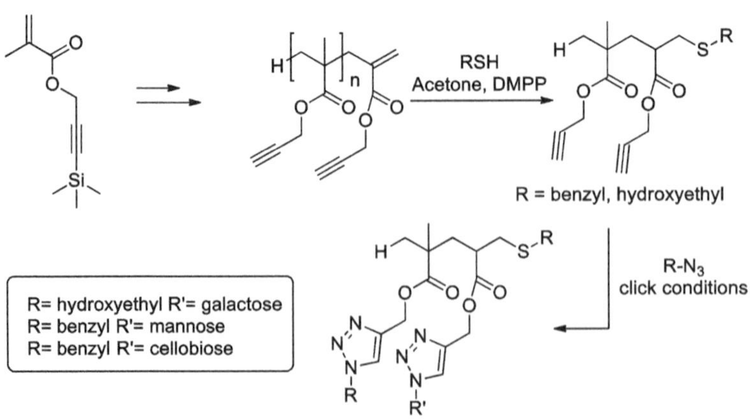

Scheme 37 Preparation of end-functionalized glycopolymers.

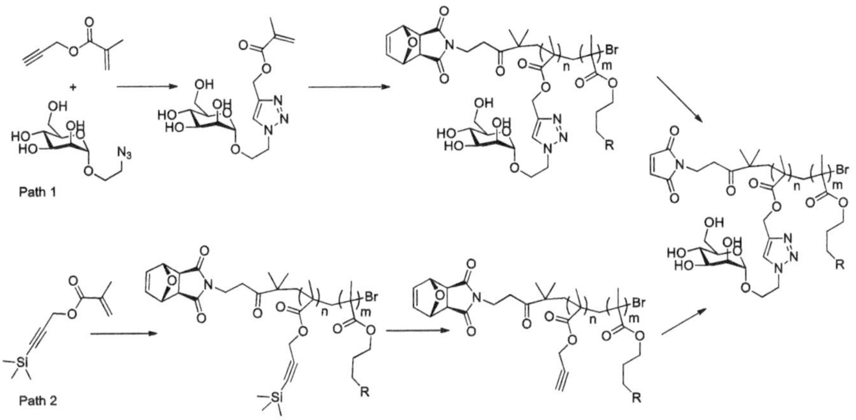

Scheme 38 Synthesis of maleimide containing glycopolymers using ATRP technique and CuAAC methodology.

In one such study, the ATRP technique and CuAAC methodology were employed to synthesize glycopolymers containing maleimide functionality. The azido sugar reacted with an alkyne and afforded monomer that underwent polymerization. The TMS-protected alkyne was converted to a corresponding polymer that was made to react with azido sugar to furnish the glycopolymer. Furthermore, maleimide was made to react through retro-Diels-Alder to achieve a corresponding functionalized glycopolymer (Scheme 38).[68, 72]

Conclusion

In recent years, triazole-chemistry has widely been used for the preparation of simple to complex architectures with interesting analytical (receptors for ionic and organic compounds), chemical (asymmetric synthesis, chiral catalysis, and chiral recognition), and biological (DNA binders, enzyme inhibitors, protein ligands, glyocolipid analogs) applications. Due to the natural abundance of carbohydrates with well-defined configuration and ease of functionalization, the triazolyl glycoconjugates have inspired the designing and development of novel chemotherapeutic drugs, new biomaterials, and organometallic catalysts over the last few decades. The emergence of click chemistry has really revolutionized the synthesis of triazole containing carbohydrates with an ever-broadening spectrum of applications in almost all fields of chemical and biological sciences.

References

1. Bertozzi CR, Kiessling LL. Chemical glycobiology. *Science* 2001;**291**(5512):2357–64.
2. Angata T, Varki A. Chemical diversity in the sialic acids and related α-keto acids: an evolutionary perspective. *Chem Rev* 2002;**102**(2):439–70.
3. Schmidt RR, Kinzy W. Anomeric-oxygen activation for glycoside synthesis: the trichloroacetimidate method. In: *Advances in carbohydrate chemistry and biochemistry.* vol. 50. Academic Press; 1994. p. 21–123.
4. Tiwari VK, Mishra RC, Sharma A, Tripathi RP. Carbohydrate based potential chemotherapeutic agents: recent developments and their scope in future drug discovery. *Mini Rev Med Chem* 2012;***12***(14):1497–519.
5. Varki A. Biological roles of oligosaccharides: all of the theories are correct. *Glycobiology* 1993;**3**(2):97–130.
6. Somesh S, Madhuri K. Triazolyl glycoconjugates and their impact in medicinal chemistry. *Org Chem* 2016;**12**(4):104–26.
7. Dondoni A. Triazole: the keystone in glycosylated molecular architectures constructed by a click reaction. *Chem Asian J* 2007;**2**(6):700–8.
8. Kun S, Bokor É, Varga G, Szőcs B, Páhi A, Czifrák K, et al. New synthesis of 3-(β-D-glucopyranosyl)-5-substituted-1, 2, 4-triazoles, nanomolar inhibitors of glycogen phosphorylase. *Eur J Med Chem* 2014;**76**:567–79.
9. Kun S, Begum J, Kyriakis E, Stamati EC, Barkas TA, Szennyes E, et al. A multidisciplinary study of 3-(β-d-glucopyranosyl)-5-substituted-1, 2, 4-triazole derivatives as glycogen phosphorylase inhibitors: computation, synthesis, crystallography and kinetics reveal new potent inhibitors. *Eur J Med Chem* 2018;**147**:266–78.
10. Yu J, Li Z, Lu W, Zhang S, Cai M. Novel stereoselective synthesis of 1-(β-d-and α-l-glycopyranosyl)-5-methyl-1H-1, 2, 4-triazoles. *Tetrahedron Asymmetry* 2003;**14**(17):2507–13.
11. Avanzo RE, Anesini C, Fascio ML, Errea MI, D'Accorso NB. 1, 2, 4-Triazole D-ribose derivatives: design, synthesis and antitumoral evaluation. *Eur J Med Chem* 2012;**47**:104–10.
12. El Sayed H, Kassem AA, Abdel-Hamid HM, Louis F, Khattab SA, Aouad MR. Regioselectivity in the glycosylation of 5-(3-chlorobenzo [b] thien-2-yl)-4H-1, 2, 4-triazole-3-thiol. *Carbohydr Res* 2009;**344**(6):725–33.
13. El Sayed H, El Sayed H, El Fattah MEDA, Boraei AT, El-Nabi HMA. Regioselective synthesis, characterization and antimicrobial evaluation of S-glycosides and S, N-diglycosides of 1, 2-dihydro-5-(1H-indol-2-yl)-1, 2, 4-triazole-3-thione. *Eur J Med Chem* 2013;**66**:106–13.
14. Khalil NS. Efficient synthesis, structure, and antimicrobial activity of some novel N-and S-β-D-glucosides of 5-pyridin-3-yl-1, 2, 4-triazoles. *Carbohydr Res* 2006;**341**(13):2187–99.
15. Aragão-Leoneti V, Campo VL, Gomes AS, Field RA, Carvalho I. Application of copper (I)-catalysed azide/alkyne cycloaddition (CuAAC)'click chemistry'in carbohydrate drug and neoglycopolymer synthesis. *Tetrahedron* 2010;**66**(49):9475–92.
16. He X-P, Zeng Y-L, Zang Y, Li J, Field RA, Chen G-R. Carbohydrate CuAAC click chemistry for therapy and diagnosis. *Carbohydr Res* 2016;**429**:1–22.
17. Bräse S, Banert K. *Organic azides: syntheses and applications.* John Wiley & Sons; 2010.
18. Santoyo-González F, Hernández-Mateo F. Azide–alkyne 1, 3-dipolar cycloadditions: a valuable tool in carbohydrate chemistry. In: *Heterocycles from carbohydrate precursors.* Springer; 2007. p. 133–77.
19. Pfleiderer W, Bühler E. Pteridine, XXXII. Ein neuer Weg zur Synthese von Pteridin-N-8-, Purin-N-9-und Triazolo [4.5-d] pyrimidin-N-3-glykosiden. *Chem Ber* 1966;**99**(9):3022–39.
20. Paulsen H, Györgydeák Z, Friedmann M. Konformationsanalyse, III. exo-Anomerer Effekt und Circulardichroismus von Glycopyranosylaziden. *Chem Ber* 1974;**107**(5):1568–78.

21. Bianchi A, Bernardi A. Traceless Staudinger ligation of glycosyl azides with triaryl phosphines: stereoselective synthesis of glycosyl amides. *J Org Chem* 2006;**71**(12):4565–77.
22. Györgydeák Z, Szilágyi L, Paulsen H. Synthesis, structure and reactions of glycosyl azides. *J Carbohydr Chem* 1993;**12**(2):139–63.
23. Györgydeák Z, Thiem J. Synthesis and transformation of glycosyl azides. *Adv Carbohydr Chem Biochem* 2006;**60**:103–82.
24. Sicherl F, Wittmann V. Orthogonally protected sugar diamino acids as building blocks for linear and branched oligosaccharide mimetics. *Angew Chem Int Ed* 2005;**44**(14):2096–9.
25. Vogel C, Gries P. Synthesis of a 4-amino-4-deoxy-D-galacturonic acid derivative. *J Carbohydr Chem* 1994;**13**(1):37–46.
26. Arndt S, Hsieh-Wilson LC. Use of cerny epoxides for the accelerated synthesis of glycosaminoglycans. *Org Lett* 2003;**5**(22):4179–82.
27. Lemieux R, Ratcliffe R. The azidonitration of tri-O-acetyl-D-galactal. *Can J Chem* 1979;**57**(10):1244–51.
28. Bongat AF, Demchenko AV. Recent trends in the synthesis of O-glycosides of 2-amino-2-deoxysugars. *Carbohydr Res* 2007;**342**(3–4):374–406.
29. Bovin NV, Zurabyan SÉ, Khorlin AY. Addition of halogenoazides to glycals. *Carbohydr Res* 1981;**98**(1):25–35.
30. Czernecki S, Ayadi E, Randriamandimby D. Seleno glycosides. 2. Synthesis of phenyl 2-(N-acetylamino)-and 2-azido-2-deoxy-1-seleno-. alpha.-D-glycopyranosides via azido-phenylselenylation of diversely protected glycals. *J Org Chem* 1994;**59**(26):8256–60.
31. Vasella A, Witzig C, Chiara JL, Martin-Lomas M. Convenient synthesis of 2-azido-2-deoxy-aldoses by diazo transfer. *Helv Chim Acta* 1991;**74**(8):2073–7.
32. Alper PB, Hung S-C, Wong C-H. Metal catalyzed diazo transfer for the synthesis of azides from amines. *Tetrahedron Lett* 1996;**37**(34):6029–32.
33. Nyffeler PT, Liang C-H, Koeller KM, Wong C-H. The chemistry of amine – azide interconversion: catalytic diazotransfer and regioselective azide reduction. *J Am Chem Soc* 2002;**124**(36):10773–8.
34. Yu H, Chokhawala HA, Huang S, Chen X. One-pot three-enzyme chemoenzymatic approach to the synthesis of sialosides containing natural and non-natural functionalities. *Nat Protoc* 2006;**1**(5):2485.
35. Meng X, Yao W, Cheng J, Zhang X, Jin L, Yu H, et al. Regioselective chemoenzymatic synthesis of ganglioside disialyl tetrasaccharide epitopes. *J Am Chem Soc* 2014;**136**(14):5205–8.
36. Yu H, Huang S, Chokhawala H, Sun M, Zheng H, Chen X. Highly efficient chemoenzymatic synthesis of naturally occurring and non-natural α-2, 6-linked sialosides: a P. damsela α-2, 6-sialyltransferase with extremely flexible donor–substrate specificity. *Angew Chem* 2006;**118**(24):4042–8.
37. Yu H, Cao H, Tiwari VK, Li Y, Chen X. Chemoenzymatic synthesis of C8-modified sialic acids and related α2–3-and α2–6-linked sialosides. *Bioorg Med Chem Lett* 2011;**21**(17):5037–40.
38. Yu H, Cheng J, Ding L, Khedri Z, Chen Y, Chin S, et al. Chemoenzymatic synthesis of GD3 oligosaccharides and other disialyl glycans containing natural and non-natural sialic acids. *J Am Chem Soc* 2009;**131**(51):18467–77.
39. Song X, Yu H, Chen X, Lasanajak Y, Tappert MM, Air GM, et al. A sialylated glycan microarray reveals novel interactions of modified sialic acids with proteins and viruses. *J Biol Chem* 2011. https://doi.org/10.1074/jbc.M111.274217.
40. Mishra A, Tiwari VK. One-pot synthesis of glycosyl-β-azido ester via diazotransfer reaction toward access of glycosyl-β-triazolyl ester. *J Org Chem* 2015;**80**(10):4869–81.
41. Lakshmipathi P, Rao AR. Efficient conjugate addition of hydrogen azide to enoates. *Tetrahedron Lett* 1997;**38**(14):2551–2.
42. Xu L-W, Li L, Xia C-G, Zhou S-L, Li J-W. The first ionic liquids promoted conjugate addition of azide ion to α, β-unsaturated carbonyl compounds. *Tetrahedron Lett* 2004;**45**(6):1219–21.

43. Singh Y, Spinelli N, Defrancq E, Dumy P. A novel heterobifunctional linker for facile access to bioconjugates. *Org Biomol Chem* 2006;**4**(7):1413–9.
44. Virta P, Katajisto J, Niittymäki T, Lönnberg H. Solid-supported synthesis of oligomeric bioconjugates. *Tetrahedron* 2003;**59**(28):5137–74.
45. Kuijpers BH, Groothuys S, Keereweer AR, Quaedflieg PJ, Blaauw RH, van Delft FL, et al. Expedient synthesis of triazole-linked glycosyl amino acids and peptides. *Org Lett* 2004;**6**(18):3123–6.
46. Pal R, Sarkar S, Chatterjee N, Sen AK. Efficient synthesis of 1, 4-disubstituted triazolyl N-carboxamides via a simple and convenient MCR using basic alumina as solid support. *Tetrahedron Lett* 2013;**54**(41):5642–6.
47. Beckmann HSG, Wittmann V. One-pot procedure for diazo transfer and azide – alkyne cycloaddition: triazole linkages from amines. *Org Lett* 2007;**9**(1):1–4.
48. Kumar D, Mishra KB, Mishra BB, Mondal S, Tiwari VK. Click chemistry inspired highly facile synthesis of triazolyl ethisterone glycoconjugates. *Steroids* 2014;**80**:71–9.
49. Mishra KB, Mishra BB, Tiwari VK. Efficient synthesis of ethisterone glycoconjugate via bis-triazole linkage. *Carbohydr Res* 2014;**399**:2–7.
50. Singh AK, Gopu K. Synthesis and antioxidant properties of novel α-tocopherol glycoconjugates. *Tetrahedron Lett* 2010;**51**(8):1180–4.
51. Mishra KB, Mishra RC, Tiwari VK. First noscapine glycoconjugates inspired by click chemistry. *RSC Adv* 2015;**5**(64):51779–89.
52. Mishra KB, Tiwari VK. Click chemistry inspired synthesis of morpholine-fused triazoles. *J Org Chem* 2014;**79**(12):5752–62.
53. Sahoo L, Singhamahapatra A, Kumar K, Loganathan D. Synthesis of divalent glycoamino acids with bis-triazole linkage. *Carbohydr Res* 2013;**381**:51–8.
54. Dondoni A, Marra A. C-glycoside clustering on calix [4] arene, adamantane, and benzene scaffolds through 1, 2, 3-triazole linkers. *J Org Chem* 2006;**71**(20):7546–57.
55. Xie J, Bogliotti N. Synthesis and applications of carbohydrate-derived macrocyclic compounds. *Chem Rev* 2014;**114**(15):7678–739.
56. Tiwari VK, Mishra BB, Mishra KB, Mishra N, Singh AS, Chen X. Cu-catalyzed click reaction in carbohydrate chemistry. *Chem Rev* 2016;**116**(5):3086–240.
57. Bodine KD, Gin DY, Gin MS. Highly convergent synthesis of C 3-or C 2-symmetric carbohydrate macrocycles. *Org Lett* 2005;**7**(20):4479–82.
58. Bodine KD, Gin DY, Gin MS. Synthesis of readily modifiable cyclodextrin analogues via cyclodimerization of an alkynyl-azido trisaccharide. *J Am Chem Soc* 2004;**126**(6):1638–9.
59. Billing JF, Nilsson UJ. C 2-symmetric macrocyclic carbohydrate/amino acid hybrids through copper (I)-catalyzed formation of 1, 2, 3-triazoles. *J Org Chem* 2005;**70**(12):4847–50.
60. Dörner S, Westermann B. A short route for the synthesis of "sweet" macrocycles via a click-dimerization–ring-closing metathesis approach. *Chem Commun* 2005;**22**:2852–4.
61. Conte ML, Grotto D, Chambery A, Dondoni A, Marra A. Convergent synthesis and inclusion properties of novel C n-symmetric triazole-linked cycloglucopyranosides. *Chem Commun* 2011;**47**(4):1240–2.
62. Chatterjee N, Pal R, Sarkar S, Sen AK. Synthesis of triazole-fused tetracyclic glycosides in aqueous medium: application of nanodomain cubic cuprous oxide as reusable catalyst in one-pot domino Sonogashira-cyclization. *Tetrahedron Lett* 2015;**56**(25):3886–9.
63. Lewandowski B, Jarosz S. Application of 1′, 2, 3, 3′, 4, 4′-hexa-O-benzylsucrose in the preparation of sucrose macrocycles via a click chemistry route: regioselectivity study. *Synth Commun* 2011;**41**(14):2161–8.
64. Campo VL, Carvalho I, Da Silva CH, Schenkman S, Hill L, Nepogodiev SA, et al. Cyclooligomerisation of azido-alkyne-functionalised sugars: synthesis of 1, 6-linked cyclic pseudo-galactooligosaccharides and assessment of their sialylation by Trypanosoma cruzi trans-sialidase. *Chem Sci* 2010;**1**(4):507–14.

65. Allam A, Dupont L, Behr JB, Plantier-Royon R. Convenient synthesis of a galacturonic acid based macrocycle with potential copper-complexation ability. *Eur J Org Chem* 2012;**2012**(4):817–23.
66. Muthana S, Yu H, Cao H, Cheng J, Chen X. Chemoenzymatic synthesis of a new class of macrocyclic oligosaccharides. *J Org Chem* 2009;**74**(8):2928–36.
67. Kempe K, Krieg A, Becer CR, Schubert US. "Clicking" on/with polymers: a rapidly expanding field for the straightforward preparation of novel macromolecular architectures. *Chem Soc Rev* 2012;**41**(1):176–91.
68. Geng J, Mantovani G, Tao L, Nicolas J, Chen G, Wallis R, et al. Site-directed conjugation of "clicked" glycopolymers to form glycoprotein mimics: binding to mammalian lectin and induction of immunological function. *J Am Chem Soc* 2007;**129**(49):15156–63.
69. Nurmi L, Lindqvist J, Randev R, Syrett J, Haddleton DM. Glycopolymers via catalytic chain transfer polymerisation (CCTP), Huisgens cycloaddition and thiol–ene double click reactions. *Chem Commun* 2009;**19**:2727–9.
70. Ladmiral V, Mantovani G, Clarkson GJ, Cauet S, Irwin JL, Haddleton DM. Synthesis of neoglycopolymers by a combination of "click chemistry" and living radical polymerization. *J Am Chem Soc* 2006;**128**(14):4823–30.
71. Chen Q, Cui Y, Zhang T-L, Cao J, Han B-H. Fluorescent conjugated polyfluorene with pendant lactopyranosyl ligands for studies of Ca2+-mediated carbohydrate–carbohydrate interaction. *Biomacromolecules* 2009;**11**(1):13–9.
72. Geng J, Lindqvist J, Mantovani G, Haddleton DM. Simultaneous copper (I)-catalyzed azide–alkyne cycloaddition (CuAAC) and living radical polymerization. *Angew Chem Int Ed* 2008;**47**(22):4180–3.

CHAPTER 6

Triazoles in Nanotechnology

Arruje Hameed[a] and Tahir Farooq[b,*]
[a]Department of Biochemistry, Government College University, Faisalabad, Pakistan
[b]Department of Applied Chemistry, Government College University, Faisalabad, Pakistan
*Corresponding author. E-mail: tahirfarooqfsd@gmail.com

Introduction

In the last few decades, the exceptional advancements in nanotechnology have revolutionized human daily life to new standards. It has become the science of future due to the unprecedented applications of nanomaterials in drug delivery, bioimaging, catalysis, agriculture, pollution remediation, energy production, and storage, etc. The nanomaterials showing a variety of applications in the aforesaid domains are generally classified as nanoparticles (NPs), nanocomposites (NCs), and carbon nanotubes (CNTs) (Figure 1). These nanosized-materials exhibit size-dependent electrical, magnetic, and optical properties that differ from bulk equivalents. Considering the ever-increasing scope of nanomaterials, a number of approaches have been developed for their preparation and surface functionalization.[1] The nanomaterials have high surface area that can be functionalized with a variety of organic, inorganic, and biomolecules to induce desired properties in accordance to their applications. Over the years, a number of physical, chemical, and biological methods have been developed for their surface functionalization.[2]

Over the years, triazoles (both 1,2,4-triazole and 1,2,3-triazole) have shown a wide spectrum of applications in medicines, electronics, catalysis, and pollution remediation. Thus, there has been an increasing trend to functionalize nanomaterials with triazoles for an array of applications (Figure 2). Especially, the recent advances in click chemistry have made it a leading approach for surface functionalization of any type of nanomaterials for any kind of application.[3, 4] Herein, we present some pioneering examples of surface functionalization/modification of NPs, NCs, and CNTs with both types of triazoles for a range of applications.

Advances in Triazole Chemistry
https://doi.org/10.1016/B978-0-12-817113-4.00004-4

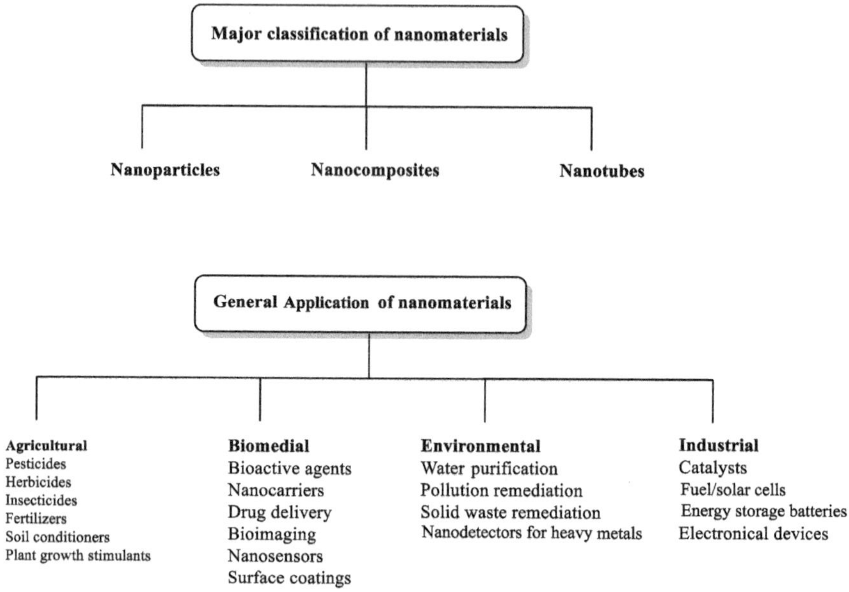

Figure 1 General classification and applications of nanomaterials.

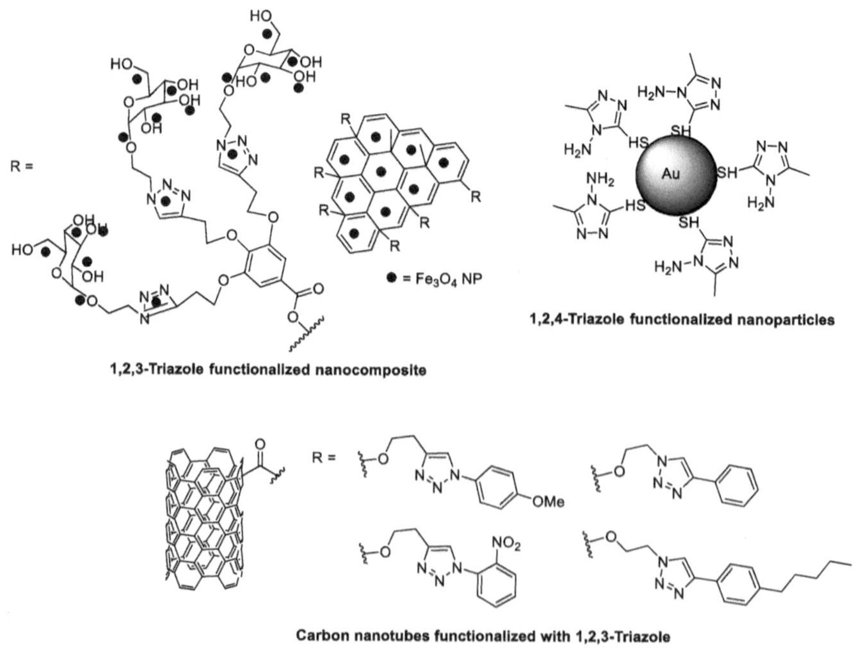

Figure 2 Triazole functionalized known nanomaterials.

Nanoparticles functionalized with 1,2,4-triazole

A NP is supposed to be a microscopic particle of less than 100 nm in size that behaves as a complete unit in terms of its properties. The noble metal NPs of Au, Ag, Cu, and SiO_2 have commonly been employed for a range of applications including drug delivery, catalysis, nanosensors, bioimaging, and wastewater treatment.

Over the last decade, researchers have widely exploited the colorimetric properties of metal NPs like Au and Ag for colorimetric detections. Such colorimetric NPs with their potential to undergo surface functionalization have found wide applications in bioimaging and chemical sensing.[5] AuNPs exhibit strong surface plasmon resonance (SPR) absorption and high extinction coefficient at a visible region, and these properties are greatly influenced by the size, shape, and dielectric constant.[6] Especially, the presence of other species in solution cause aggregation of AuNPs that change the color of an AuNP solution due to change in SPR. This sensitivity of AuNPs makes them useful as nanoprobes for colorimetric detection of biomolecules, drugs, and metal ions.

Chromium plays vital roles in the activation of an array of enzymes and influences the metabolism of carbohydrates, proteins, and fats in humans. Its deficiency causes insulin resistance while excessive amounts disturbs the cellular structures.[7] Therefore, a close and precise monitoring of Cr^{+3} is of great importance for human health. For that very purpose, a number of techniques and methods have been developed. However, they find limited applications as they are complicated and expensive. Thus, development of cost-effective and reliable methods has been desired for the real-time detection of Cr^{+3}. In this connection, Shahrivari *et al.* functionalized the surface of AuNPs with 4-amino-5-methyl-4*H*-1,2,4-triazole-3-thiol (AMTT) for selective colorimetric detection of Cr^{+3}. The AMTT-Au NPs exhibited an instant aggregation in the presence of Cr^{+3}, and the color change could easily be observed by the naked eye. These triazole-functionalized nanoprobes showed sensitivity at pH 3–5 and detected Cr^{+3} selectively even in plasma samples. Thus, the attachment of 1,2,4-triazole at the surface of AuNPs converted them into nanosensors for Cr^{+3} avoiding the use of sophisticated and expensive instrumentation (Figure 3).[4]

Similarly, the 1,2,4-triazole functionalized AuNPs have found recent applications in food safety-measures and diagnosis of diseases. In our nervous system, dopamine is an important neurotransmitter, and its controlled level is highly important for normal coordinated functioning of central and peripheral nervous system. Usually, dopamine is determined quantitatively

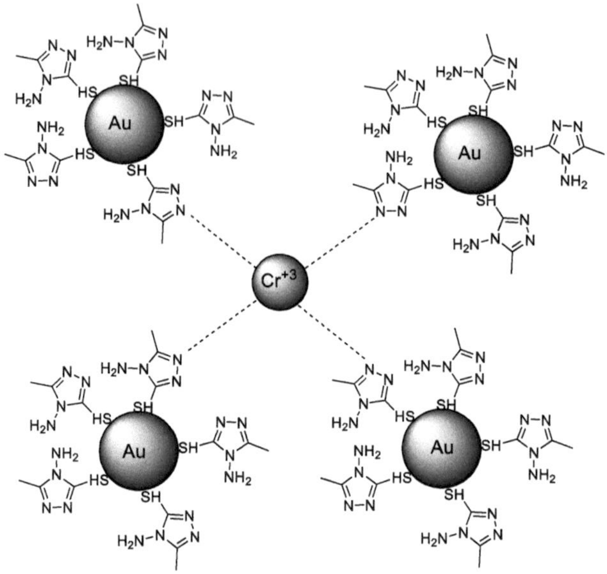

Figure 3 1,2,4-Triazole functionalized AuNPs for selective colorimetric detection of Cr^{+3}.

using expensive spectroscopic or chromatographic approaches. However, the surface functionalization of AuNPs with 4-amino-3-hydrazino-5-mercapto-1,2,4-triazole (AHMT) presented a cost-effective and reliable method for colorimetric detection of dopamine. Feng *et al.* employed AHMT-AuNPs for colorimetric determination of dopamine in the presence of different analytes.[8] Their results established AHMT-AuNPs as highly selective and sensitive nanosensors for dopamine. Very recently, the selective sensitivity of AHMT-AuNPs toward dopamine was studied by employing quantum chemistry calculations and molecular dynamic simulations. The AHMT-AuNPs underwent aggregations as a result of H-bonding between dopamine and N of triazole. The structural interactions of dopamine with triazole on the surface of AuNPs ensured its selective colorimetric detection (Figure 4).[9]

Silver is widely used in the photography, medical, and battery industries owing to its excellent thermal and electrical activities and antibacterial potential. Resultantly, industrial wastewater contains a heavy amount of Ag, which acts as an environmental contaminant in various ways. To remove the Ag from wastewater, various methods have been developed using different approaches. Over the years, adsorption has become the major practical method for the removal of Ag from wastewater.[10] Furthermore, the

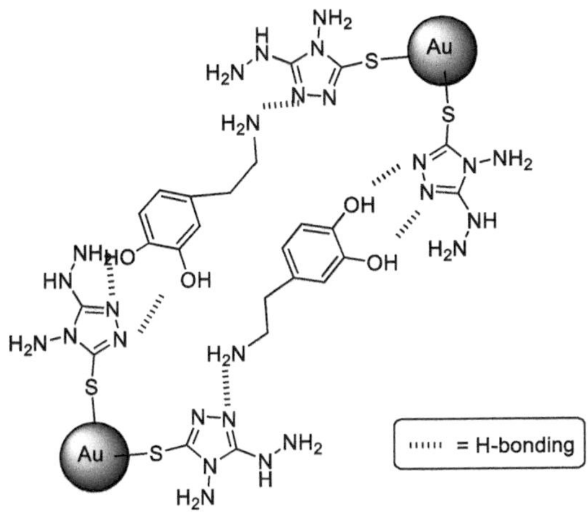

Figure 4 1,2,4-Triazole functionalized AuNPs for colorimetric detection of dopamine.

adsorption-based methods have been revolutionized by the involvement of NPs. Silica has widely been used for the adsorption of Ag from wastewater; however, it shows low reusability and poor selectivity for Ag. In this connection, Fu *et al.* enhanced the selectivity of silica-NPs (SNPs) for Ag using 3-amino-5-mercapto-1,2,4-triazole for surface functionalization of SNPs. The functionalized SNPs displayed a selective uptake of Ag due to its complexation with thiol group of 1,2,4-triazole, and the adsorption equilibrium was established in just 20 min at room temperature. Furthermore, the ATT-SNPs showed excellent reusability up to five cycles.[11] In a very recent effort, Veena *et al.* prepared a promising next generation of antimicrobial and anticancer agents by functionalization of the surface of AuNPs with 4-amino-3,5-dimercapto-1,2,4-triazole (ADMT). The prepared ADMT-AuNPs exhibited anticancer activity against breast adenocarcinoma (MCF-7) cell lines and antibacterial potential against Gram-positive and Gram-negative bacteria.[12]

In another very recent report, Khorramabadi *et al.* described the preparation of 3-mercapto-1,2,4-triazole functionalized Fe_3O_4-based CuNPs as heterogeneous nanocatalysts. This nanocatalyst was further employed for the synthesis of a range of tetrazoles from amino acids (Figure 5). This triazole functionalized magnetic nanocatalyst exhibited excellent reusability even after five cycles. Tetrazoles are N-rich heterocycles with diverse applications as herbicides, diuretics, antivirals, antidiabetic, and anticancerous agents. This cheaper nanocatalyst provided a greener method for tetrazole preparation.

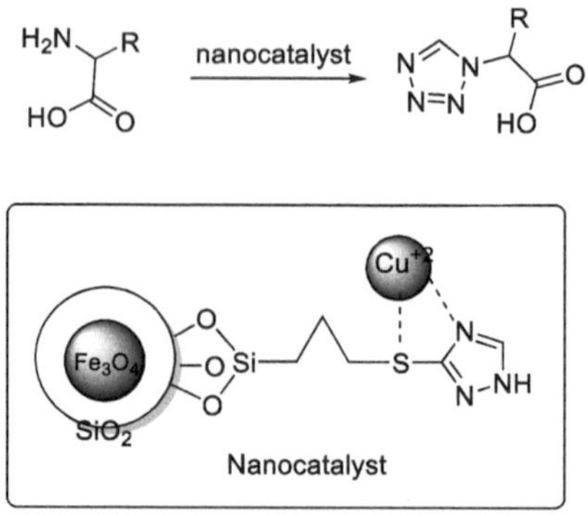

Figure 5 1,2,4-Triazole functionalized nanoparticle-based nanocatalyst.

Furthermore, the catalyst was recommended for the preparation of other heterocyclic compounds.[13]

Nanoparticles functionalized with 1,2,3-triazole

Chen *et al.* functionalized the AuNPs using click chemistry for colorimetric detection of Cr^{+3} over a pH range of 4–7. The AuNPs functionalized with 1,2,3-triazoles were employed as nanoprobes for selective visual detection of Cr^{+3} in lake water samples. In fact, the specific interaction of Cr^{+3} with triazolic functionality caused aggregation of the NPs resulting in a color change from red to purple. This colorimetric method proved to be an efficient and practical method for selective detection of Cr^{+3} in wastewater for environmental remediation.[14]

The high photoactivity, electrical properties, and chemical stability of TiO_2 NPs have made them ideal candidates for solar cells and photocatalysis. However, the TiO_2-based nanostructures are highly desired for integrated applications in dye-sensitized solar cells, photoreduction of CO_2, and photocatalytic degradation. The TiO_2 nanochains exhibited high dye loading for efficient light harvesting due to the large surface area. The well-crystallized anatase phase made nanochains highly photoactive under UV light. Until recently, the known methods for the preparation of TiO_2 nanochains include sol-gel, solvothermal, hydrothermal, and template-assisted methods.[15] These methods were considered impractical due to kinetic and thermodynamic factors. Thus, a facile, simple, and inexpensive approach has really been desirous. Accordingly, Xia *et al.* functionalized TiO_2 NPs with clickable functionalities, which subsequently underwent click reaction to produce one-dimensional TiO_2 nanochains (Figure 6).[16] This click

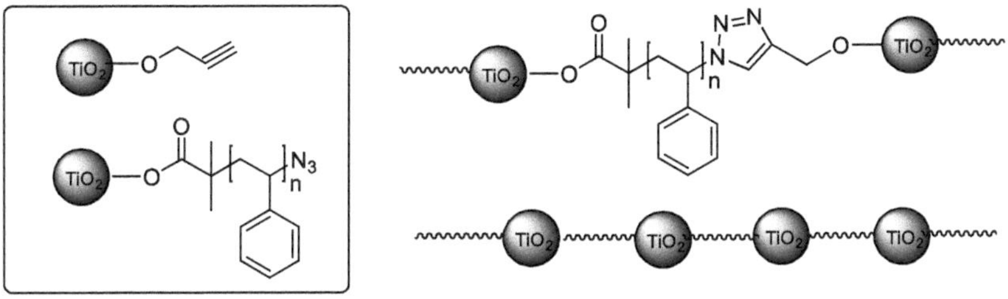

Figure 6 Click reaction for the preparation of one-dimensional TiO_2 nanochains.

reaction-based polymerization methodology was suggested as a simple approach for the preparation of TiO_2 nanochains in high yield.

As described earlier, the functionalized NPs have successfully been employed for colorimetric detection of heavy metal ions including Cr^{+3}. Very recently, Mondal *et al.* also developed 1,2,3-triazole functionalized AuNPs as dual colorimetric sensors for selective and simultaneous detection of Cr^{+3} and Eu^{+3} in lake water samples (Figure 7). The citrate was employed as a reducing as well as a capping agent for the preparation of AuNPs from chloroauric acid. The as-prepared AuNPs were subsequently functionalized with 1,2,3-triazole-4,5-dicarboxylic acid (TADA), which served as nanoprobes with low interference. The TADA@AuNPs underwent aggregation selectively with Cr^{+3} and Eu^{+3} compared with other metal ions and lanthanide systems. The aggregation-induced color change was used as a visual detection parameter in pH range 4–10. The TADA@AuNPs were suggested as suitable candidates for the development of highly sensitive colorimetric assays for simultaneous dual detection of Cr^{+3} and Eu^{+3} in wastewater.[17]

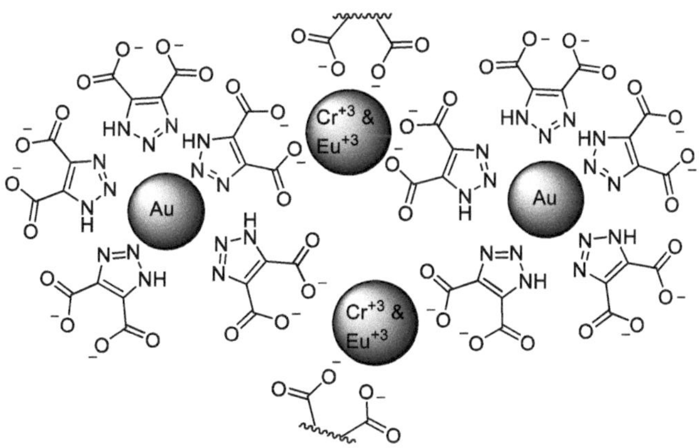

Figure 7 1*H*-1,2,3-Triazole functionalized AuNPs for colorimetric detection of heavy metals.

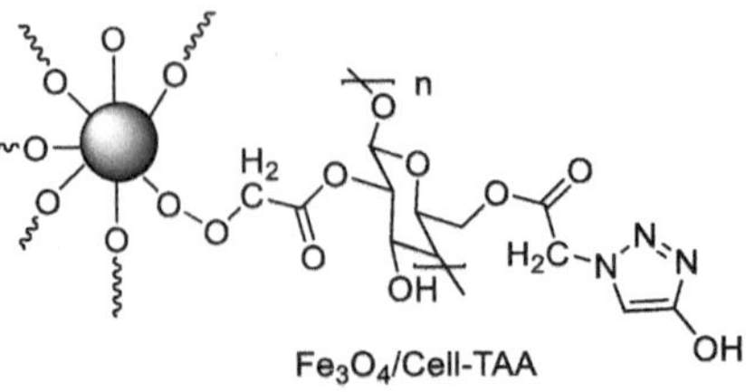

Figure 8 1,2,3-Triazole functionalized Fe_3O_4 as nanocarriers for cancer therapy.

The field of targeted and controlled drug delivery has really exploited the active surface sites and large surface area of NPs. A number of NPs have been used as intracellular carriers for a variety of therapeutics. Among them, the Fe_3O_4 magnetic nanoparticles (MNPs) have become ideal candidates due to their biodegradability, superparamagnetisim, and large magnetic moments. However, they find limited applications as nanocarriers of drugs because they undergo oxidation and agglomeration in biological media due to magnetic interactions and high surface area.[18] Generally, the stability of MNPs is improved by chemical modifications, especially coating them with polymeric material, which has gained considerable attention because of good dispersibility and easy magnetic separation. Recently, Movagharnegad *et al.* used modified cellulose to functionalize superparamagnetic Fe_3O_4 NPs, which were finally used as carriers of doxorubicin (a model drug). Initially, the bromoacetylation of cellulose was performed, and the product was further reacted with NaN_3. The prepared Fe_3O_4/Cell-N_3 NPs were reacted with propargyl alcohol under click conditions to achieve Fe_3O_4/Cell/Triazole (TAA). The model drug exhibited a pH-dependent controlled-release profile from these triazole containing MNPs (Figure 8).[19] The prepared Fe_3O_4/Cell/TAA were suggested as suitable candidates for cancer therapy. They have also been suggested as contrasting agents for the magnetic resonance imaging of cancerous cells.

Nanocomposites functionalized with 1,2,4-triazole

The incorporation of NPs into the standard matrix of materials produces NCs. The incorporation of NPs improves thermal, electrical, and mechanical properties of component materials. A number of NCs with a wide variety of interesting applications have been developed over the last few decades.

Over the last few decades, the NCs based on metal NPs and functional polymers have shown promising applications as biosensors, medicinal agents, targeted delivery systems, and antibacterial agents. The NCs for such applications are required to show biocompatibility, biological activity, hydrophilicity, and structural ability to bind with other compounds. These promising characteristics are shown by the least explored copolymers of 1-vinyl-1,2,4-triazoles.[20] Also, the silver NPs show a wide spectrum of antimicrobial properties and are frequently used in medical instruments, surgical masks, and wound coverings. In 2017, Pozdnyakov *et al.* used copolymers of 1-vinyl-1,2,4-triazole with *N*-vinylpyrrolidine and different ratios of silver NPs to prepare novel functional polymer NCs (Figure 9).[21] The functional lactam and triazole rings in stabilizing matrix stabilize the silver NPs through coordination bonds. This interaction avoids the aggregation of SNPs for a considerably longer time. The prepared 1,2,4-triazole functionalized NC was suggested as an ideal for antimicrobial applications due to its non-toxic and hydrophilic nature.

Considering the promising features and biocompatibility of polymers of 1-vinyl-1,2,4-triazole, Sosedova *et al.* prepared nanobiocomposites using poly(1-vinyl-1,2,4-triazole) (PVT) as encapsulating matrix for silver NPs (Figure 10).[22] The prepared hydrophilic Ag-PVT exhibited excellent antifungal and antibacterial potential. As a part of toxicity studies, the subacute intragastric administration of the nanobiocomposite did not alter the structural morphology of brain tissues in animal models. In a very recent report, Pozdnyakov *et al.* used copolymers of 1-vinyl-1,2,4-triazole and acrylonitrile as a polymeric matrix for the preparation of silver containing NCs (Figure 11).[23] The thermal stability of prepared insoluble NCs with a porous structure made it a suitable candidate for hi-tech catalytic materials.

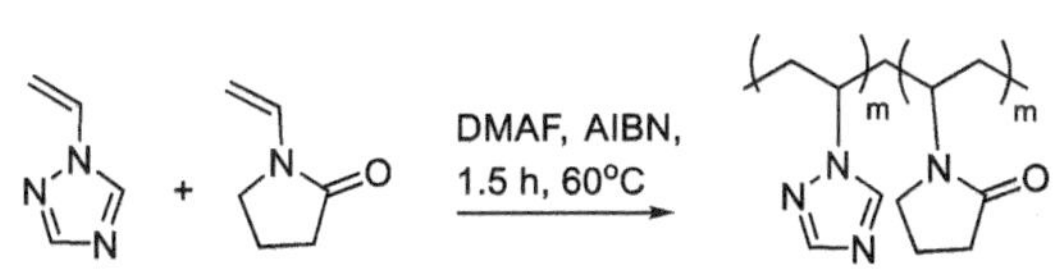

Figure 9 1,2,4-Triazole containing copolymer for functional nanocomposite.

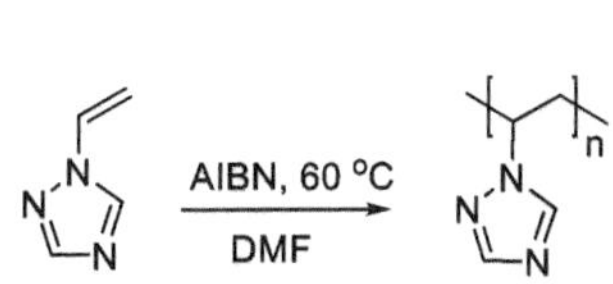

Figure 10 Poly(1-vinyl-1,2,4-triazole) as matrix for nanobiocomposite.

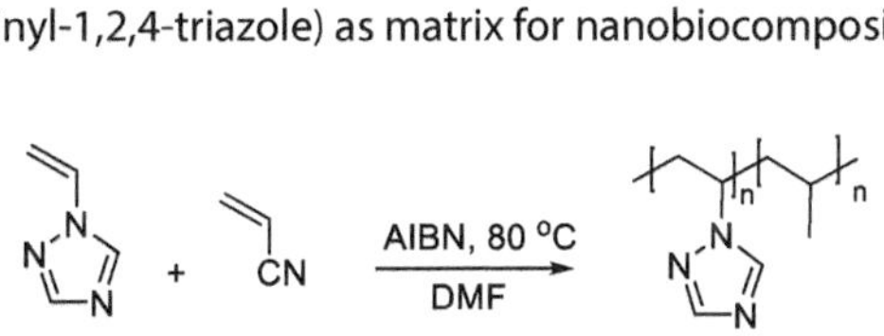

Figure 11 Copolymeric matrix for nanocomposite.

Pozdnyakov *et al.* also employed copolymers of 1-vinyl-1,2,4-triazole and acrylonitrile as a stabilizing matrix for CuNPs to prepare insoluble porous NCs (Figure 12).[24] The prepared NC efficiently catalyzed the one-pot multicomponent click reaction for regioselective synthesis of di- and trisubstituted 1,2,3-triazoles. The NC did not show any phase change and exhibited excellent catalytic efficiency even after five cycles.

The polymer-based resins have really been promising candidates for clinical dentistry. However, over the years, they have shown a number of issues like high volumetric shrinkage, poor water resistance, and mechanical strength. Such problems could be solved by modifying the materials. The modifications could be induced by developing hybrid additives, filler content optimization, and alternative monomer production. The NPs offer higher surface area for functionalization, which could be used to improve wear resistance and mechanical properties and reduce polymerization shrinkage. The surface-functionalized NP could be used as filler in dental applications.[25, 26] Recently, Yushau *et al.* prepared and modified the surface of SiO_2 NPs with 3-amino-1,2,4-triazoles. The blend of bis-phenol-A glycidyl methacrylate/triethylene glycol dimethylacrylate was prepared, and subsequently the triazole-functionalized SiO_2 NPs were added as filler in different weight percentage. The final product was prepared as dental NC B1-B6. The prepared NCs with low toxicity and high solubility were suggested as promising candidates for dentistry applications. The NCs showed improvement in sorption and mechanical properties due to the surface functionalization of SiO_2 NPs with 1,2,4-triazoles.[27]

The successful applications of chemotherapy are anticipated to selectively target cancer cells without interfering with normal cells in the

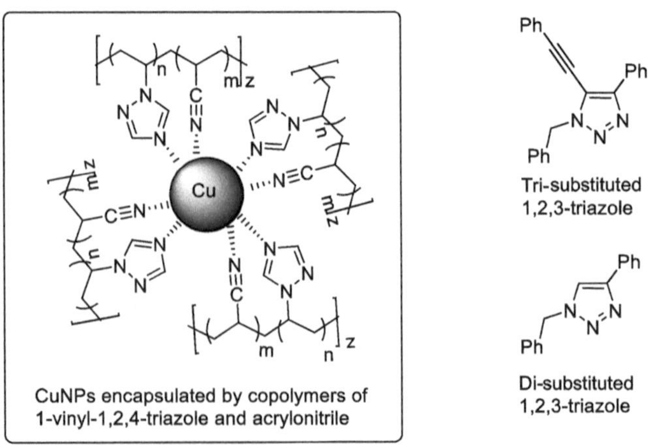

Figure 12 1,2,4-Triazole functionalized nanocomposite as catalyst.

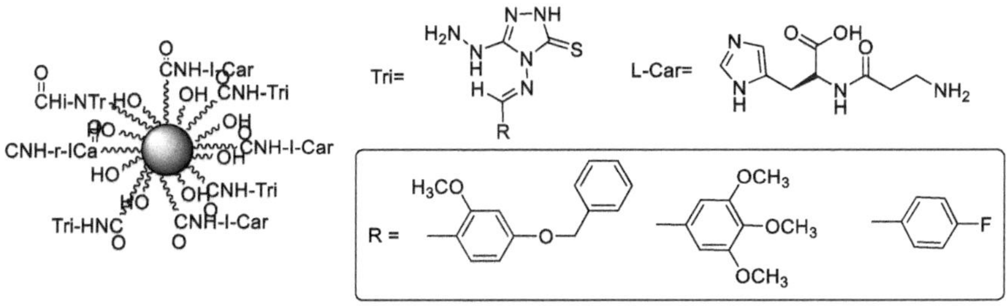

Figure 13 1,2,4-Triazole functionalized nanoconjugates.

human body. Currently, it has been a real challenge to develop an anticancer drug targeting cancerous cells selectively with minimal side effects. A number of efforts have been made to enhance the efficacy and selectivity of known anti-cancerous drugs.[28] These efforts include the use of nanocarriers, growth factors, and antibodies, etc. The ease of functionalization, chemical inertness, and excellent fluorescent properties of carbon NPs have made them attractive candidates for applications in drug delivery, bioimaging, and nanocarriers. The selectivity of chemotherapy could be improved using peptides as carrier vectors because of their excellent physiochemical properties for targeting the tumor cells.[29] Very recently, Ajmal *et al.* prepared double conjugates of CNPs using L-carnosine (a dipeptide) and different 1,2,4-triazole derivatives (Figure 13).[28] Both the dipeptide and triazoles were selected after molecular docking studies. The pediatric brain tumor cell lines (SJGBM2 and CHLA-200) were used to study the selective anticancer activity of prepared conjugates compared with a control drug. They were highly biocompatible and did not show any toxicity against embryonic kidney cell lines. The docking studies, biophysical assays, and *in vivo* evaluation proved the selective anticancer efficacy of titled conjugates and paved the way for tailored synthesis of novel drug delivery systems.

Nanocomposites functionalized with 1,2,3-triazoles

Hasantabar *et al.* prepared a PXTE-TiO_2 NC and used it for the removal of dye molecules and Cd ions from water samples (Figure 14). The NC was prepared in two steps starting with the preparation of poly(xanthoneamide-triazole-ethercalix) (PXTE) through click reaction. Then it was immobilized on TiO_2 NPs to prepare title NC. The prepared NC exhibited thermal stability and crystalline nature due to TiO_2 NPs. The PXTE-TiO_2 NC was employed to remove methylene blue and cadmium ions from aqueous samples. Thus, the prepared NCs were suggested as promising candidates for water refinement applications.

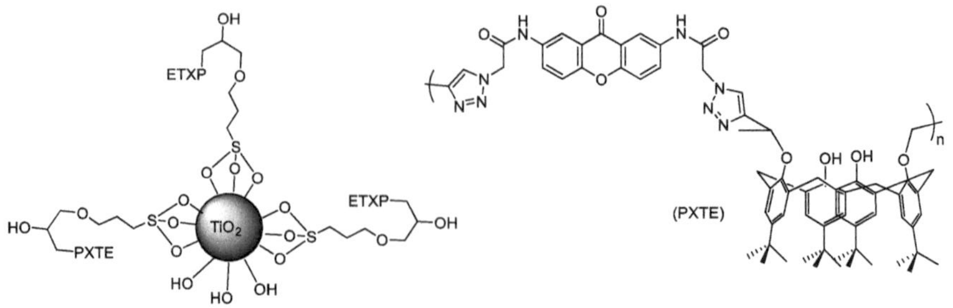

Figure 14 PXTE-TiO_2 nanocomposite for the removal of dyes and Cd ions.

The exceptional mechanical, thermal, and electrical properties of graphene have received special attraction in biotechnology, electronics, and material sciences. Its two-dimensional carbon nanostructure has made it a promising candidate in nanotechnology especially for applications in environmental pollution remediation. For various applications, grapheme and its derivatives are chemically modified by functionalizing the surface with inorganic and organic molecules. In recent years, grapheme oxide (GO) and reduced GO have differently been functionalized for various applications.[30] In 2015, Namvari *et al* used click chemistry approach to functionalize GO for the preparation of NCs as biosorbents. GO was first functionalized with two types of alkyne moieties and subsequently reacted with azide-modified glucose under click conditions before reduction with sodium ascorbate. The triazole-functionalized grapheme nanosheets were decorated with Fe_3O_4 NPs for the preparation of magnetic NCs (Figure 15). The prepared NCs exhibited excellent efficiency as adsorbents for the removal of dye from wastewater. Thus, the NCs were suggested as promising candidates for pollution remediation.[31]

The grafting of polymers on graphene surface improves its electrical properties, interfacial interactions, and solubility. Generally, polymers are grafted on GO using radical coupling, atom-transfer radical polymerization, and *in situ* polymerization approaches.[32] However, most of these grafting techniques often offer some serious limitations. Thus the development of more practical methods are highly desired for efficient functionalization of GO with polymers. Mahapatra *et al.* employed click reaction to graft hyperbranched polyurethane (HBPU) on GO for the preparation of NCs with good thermo- and mechanical properties. The GO and HBPU were functionalized with clickable alkynyl and azido functionalities, respectively (Figure 16). The prepared NC displayed enhanced thermal and mechanical properties because of excellent dispersion of GO in HBPU matrix.

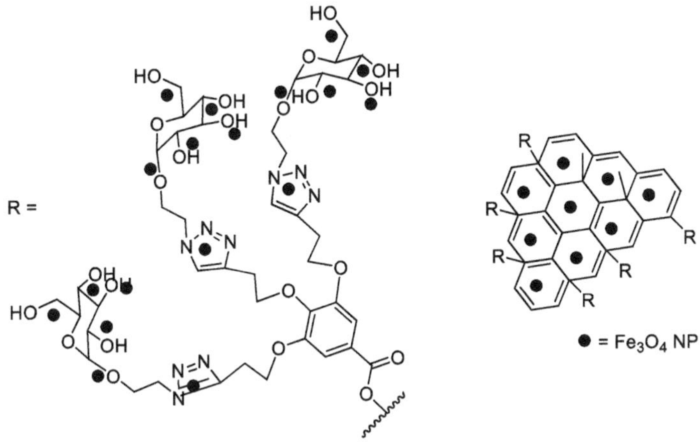

Figure 15 Nanocomposite (rGO-gallic-Glc-Fe_3O_4) as sorbent.

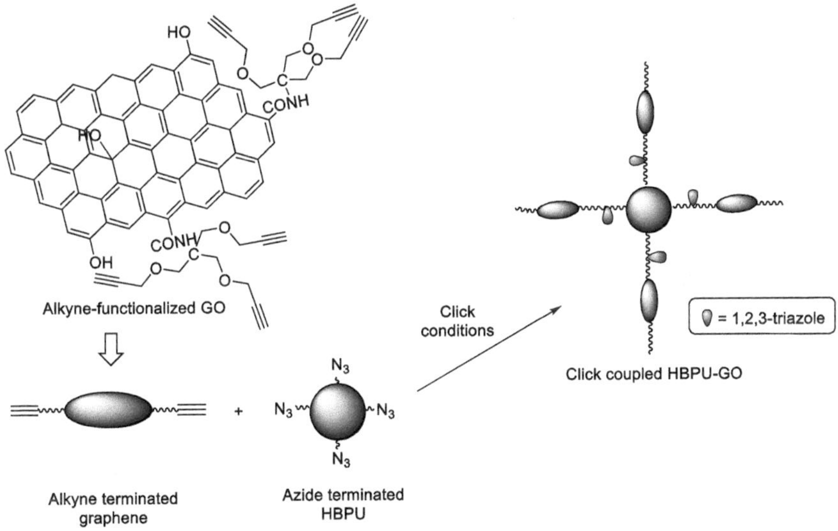

Figure 16 Triazole functionalized nanocomposites of HBPU and GO.

Furthermore, it was suggested as the promising choice for the preparation of shape-memory materials.[3] Very recently, Borthakur *et al.* also used a click chemistry approach to functionalize GO sheets with glucose molecules. Initially, the GO and glucose were functionalized with azide and alkyne, which subsequently underwent click reaction (Figure 17).[33] The functionalized sheets were decorated with CuNPs to achieve Cu-frGO NC, which showed activity like peroxidase enzyme. The prepared NC was used as colorimetric nanoprobe to measure chromium ions in water samples.

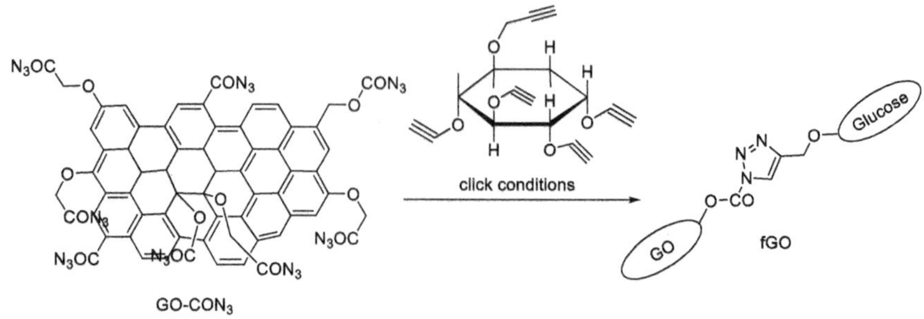

Figure 17 Click triazoles for GO functionalization with glucose.

Furthermore, they were suggested as highly reliable peroxidase mimics and also promising candidates for wastewater treatment.

Over the last many decades, biofouling has become a major ecotoxicological issue for the marine industry. It has raised operational costs of the shipping industry. A number of chemicals including complexing agents and acid pickling are frequently employed antifouling approaches. Some of the available antifouling paints involve Cu and organotin as the main component. All these available options pose serious environmental threats due to their toxic profiles.[34] Over the last few years, nanotechnology has presented a number of facile, economical, and non-hazardous solutions to a number of environmental problems. Furthermore, the efficiency and environmental friendliness has also been enhanced by click chemistry approach for functionalization of nanomaterials. In this connection, Iannazzo *et al.* presented 1,2,3-triazole/MWCNT polyester-based gelcoat NCs as an environmental friendly, antibiofouling coating (Figure 18).[35] At first, the triazole functionalized multi-walled carbon nanotubes (MWCNTs) were prepared as biocide agents and used as nanofillers of gelcoat for the preparation of title NC. It inhibited the replication of micro-organisms to exert antibiofouling properties. The NC was suggested as a promising candidate for future hi-tech antibiofouling coating.

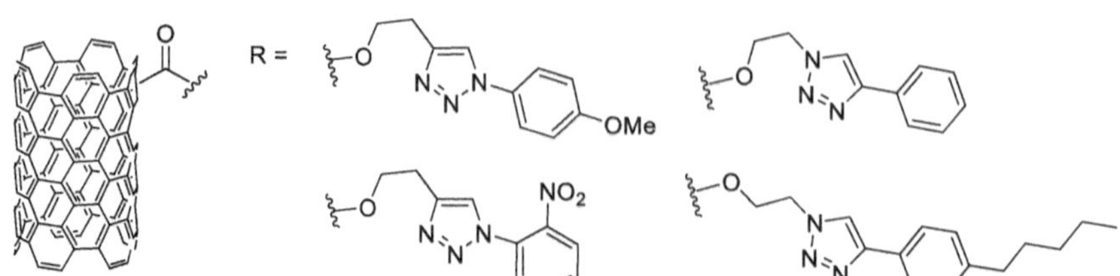

Figure 18 1,2,3-Triazole functionalized MWCNTs as nanofiller of nanocomposite.

Carbon nanotubes functionalized with 1,2,4-triazoles

The CNTs include single-walled carbon nanotubes (SWNTs) and MWCNTs, and have become promising candidates for a variety of applications in nanomaterials. Generally, two strategies are adopted to improve their solubility and incorporation into functional nanomaterials. The strategies include non-covalent functionalization by superamolecular interactions and covalent functionalization with different organic or inorganic compounds. The functionalized CNTs find a range of applications including energy storage, electronics, drug delivery, and environmental pollution remediation.[36]

Organic dyes are major industrial pollutants posing serious threats to the environment and human health. Mostly, they have non-degradable and stable organic structures thus contaminate wastewater entering the main water sources. Over the years, a number of methodologies have been developed for the adsorption, removal, and detoxification of synthetic dyes. The methods mainly involve biological treatment, chemical coagulation, precipitation, membrane filtration, and adsorption, etc. However, most of these methods require high cost; hence they are not adopted by many industries. Over the last few decades, simple operational procedures, high efficiency, and low cost have made adsorption as the leading approach for the removal of synthetic dyes from wastewater. The efficiency of the adsorption process depends mainly on the nature and structure of the adsorbent.[37] As described earlier, nanotechnology offers many economical and environmental friendly solutions to a wide range of pollution-related issues. Chemical inertness, thermal stability, and high stability have made CNTs attractive adsorbent for wastewater treatment. They have widely been used for the adsorption of a variety of synthetic dyes from industrial wastes. In recent years, a number of approaches have been adopted for their surface modification to enhance their sorption capacity toward environmental pollutants including dyes.[38]

Very recently, El-sharkawy *et al.* used MWCNTs for the preparation of a highly selective and efficient adsorbent for the removal of dyes from a water sample. At first, they prepared *N*,*N*-bis(4 hydroxysalicylidene)-4-*H*-1,2,4-triazole-3,5-diamine-diamine-Cu (II) as Schiff base and anchored it on MWCNT through covalent linkage (Figure 19).[39] The prepared ([Cu_2-L]@MWCNT) was evaluated for its potential to remove different organic dyes from water samples. The adsorption process followed second order kinetics, and the data fully followed Langmuir adsorption isotherm. The title adsorbent showed high sorption capacity for high molecular weight red dye compared to others. Thus, they proposed the prepared MWCNT-based

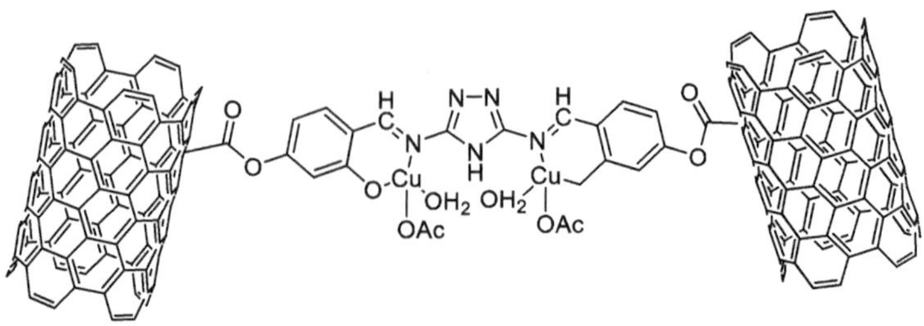

Figure 19 MWCNTs functionalized with triazole-containing Schiff base.

adsorbent as a promising advancement for the removal of dyes from industrial waste.

In additions to dyes, there are many other inorganic and organic compounds that have raised ecological concerns due to their ubiquitous presence and highly toxic nature. Among inorganics, the nitrite (NO_2^-), a widely used corrosion inhibitor and additive, has become highly toxic for the environment and human health. It reduces the oxygen transport capacity of blood by irreversibly oxidizing the hemoglobin. Similarly some of the organics like isomers of dihydroxybenzene (resorcinol (RC), catechol (CC), and hydroquinone (HQ)) are highly toxic but widely used in pharmaceutical, chemical, and cosmetic industry.[40] Therefore, their constant quantitative monitoring is highly desirous for an environmental point of view. In this connection, Yang *et al.* prepared a 1,2,4-triazole functionalized MWCNT-based nanohybrid for simultaneous monitoring of nitrite and isomers of dihydroxybenzene. They functionalized MWCNT with 3-amino-5-mercapto-1,2,4-triazole as MWCNT-SH and prepared AuNPs-graphene nanohybrids (Au-GR). Furthermore, the SH group of MWCNT-SH and AuNPs from Au-GR were made to interact synergistically to produce a novel MWCNT-SH@Au-GR nanohybrid (Figure 20).[41] This nanohybrid was used for the fabrication of electrodes for simultaneous monitoring of nitrite and phenolic compounds. Both differential pulse voltammetry and cyclic voltammetry techniques were adopted to evaluate the detection behavior of the novel nanosensor. The development of 1,2,4-triazole functionalized MWCNT-based nanohybrid was a welcome advance for the simultaneous monitoring of organic and inorganic toxicants.

As discussed earlier, the CNTs have variously been functionalized for different applications. Among different chemical modifications, the CNT functionalization with polymers has received considerable attention over the years. A number of polymers have been utilized for CNT functionalization and subsequent preparation of CNT/polymer NCs. Such composites

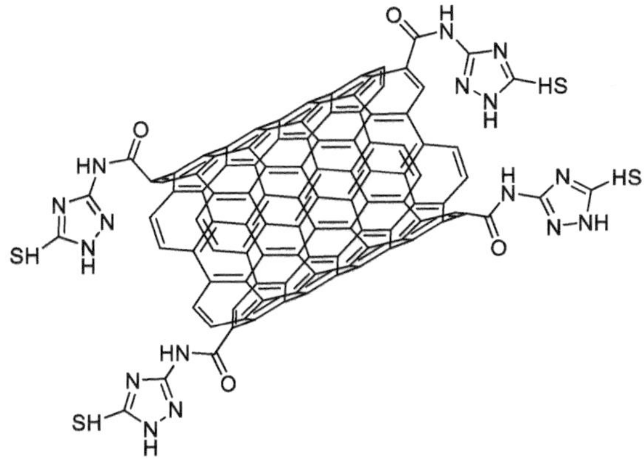

Figure 20 1,2,4-Triazole functionalized-MWCNT for the preparation of nanohybrid.

exhibit superior mechanical, electrical, and thermal properties compared with pristine polymer. The pure CNTs generally form aggregates in organic solvents due to their insoluble nature. This solubility issue causes negative effects on properties of CNT/polymer composite due to the heterogeneous dispersion of CNTs in polymeric matrix. The dispersion behavior of CNTs could be manipulated by modify their surfaces, processing method, and nature of the matrix. Over the last few years, a number of efforts have been made successfully to achieve efficient dispersion of CNTs in polymer matrix.[42, 43] In this connection, Takassi *et al.* used different ratios of MWCNTs (5, 10, and 15%) to prepare a series of MWCNT/polyimide composites (Figure 21).[44] Initially, they synthesized 3,5-diamino-*N*-(1*H*-[1,2,4-triazol-3-yl]-benzamide and reacted it *in situ* with a dianhydride in the presence of MWCNT and furnish MWCNT/poly(amic acid). Subsequently, it was further heated with different amounts of MWCNTs to produce a series of MWCNT/PI composites. The modified MWCNTs exhibited better dispersion in PI matrix, and the composite was thermally stable. The introduction of 1,2,4-triazole functionality increased the interaction of PI chains with MWCNTs leading to their increased dispersion in a polymeric matrix. Later, Takassi *et al.* used 3,5-diamino-*N*-(1*H*-[1,2,4] triazol-3-yl)-benzamide (DTB) to functionalize MWCNTs to improve their dispersion again in PI matrix (Figure 22).[45] They prepared PI-g-DTB-MWCNT composites to study their electrical, mechanical, and thermal properties compared with pure PI. The NCs exhibited better properties than PI due to the stronger interaction between PI chains and DTB-MWCNTs. The homogenous dispersion of 1,2,4-triaole functionalized MWCNTs also improved aforementioned properties of the prepared NCs.

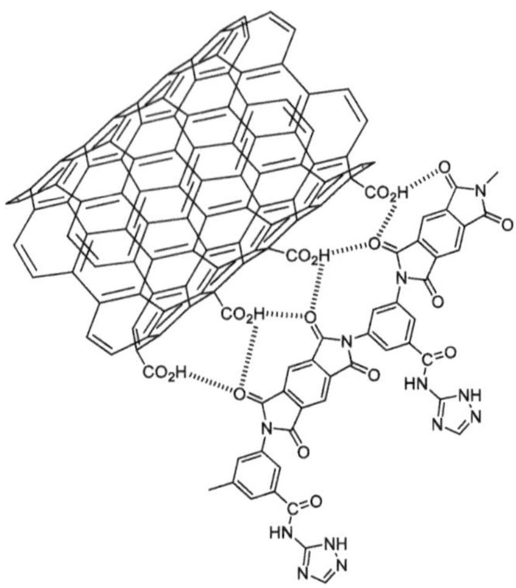

Figure 21 Functionalized MWCNT for polyimide-based nanocomposites.

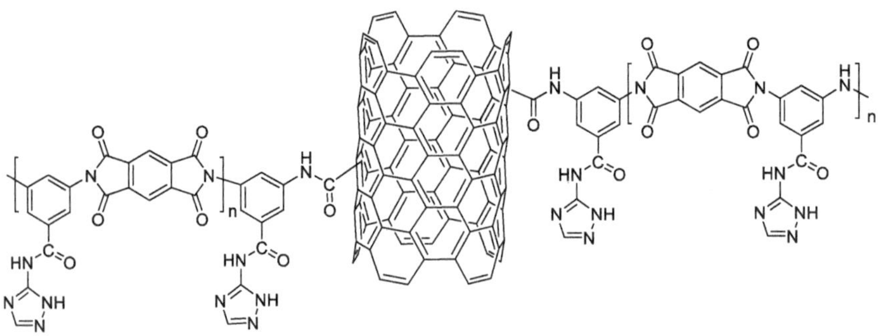

Figure 22 1,2,4-triaole functionalized MWCNTs for better dispersion in PI matrix.

Carbon nanotubes functionalized with 1,2,3-triazoles

As described earlier, the CNTs have been functionalized with a range of chemicals to induce desired properties as per applications. Over the years, click chemistry has emerged as a leading methodology for the construction of a variety of molecules for surface functionalization of CNTs. Zheng *et al.* functionalized SWCNTs with β-cyclodextrins (β-CD) using a click chemistry approach (Figure 23).[46] The β-CDs are cyclic oligosaccharides that have the ability to encapsulate a variety of molecular species in their hydrophobic cavities. The β-CD functionalized SWCNTs are expected to be promising for nanodevices, biomedicines, and supramolecular chemistry. As a pioneering example, He *et al.* prepared water-soluble Fe_3O_4-COOH and subsequently reacted them with propargyl alcohol or 3-azidopropan-1-amine

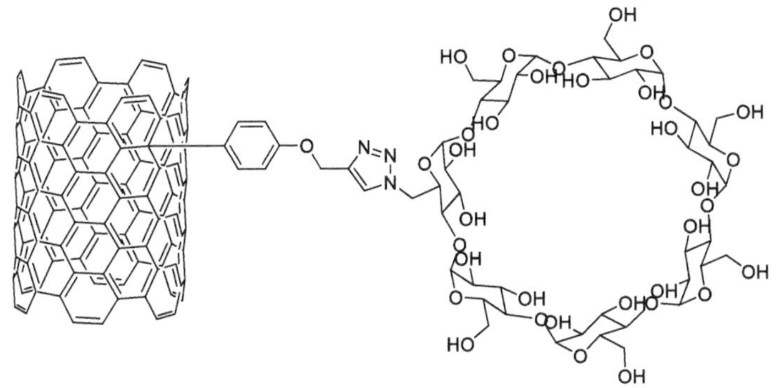

Figure 23 SWCNTs functionalized with β-cyclodextrins using 1,2,3-triazole.

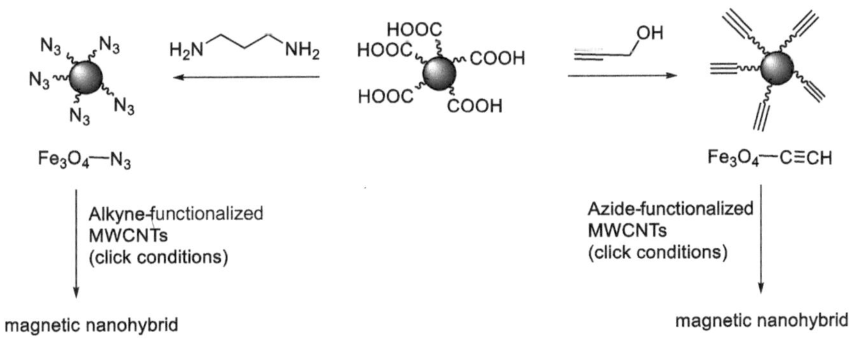

Figure 24 Click functionalized MWCNTs for the preparation of magnetic nanohybrids.

to introduce clickable functionalities. The as-prepared clickable magnetic NPs (Fe_3O_4-azide or Fe_3O_4-alkyne) were reacted with polymer functionalized MWCNTs to prepare magnetic nanohybrids (Figure 24).[47] This was a pioneering report for the preparation of magnetic nanohybrids using click chemistry and opened up new avenues for the construction of novel nanomaterials.

Attention was also diverted for the functionalization of CNT surfaces with environmental sensitive shells. Such functionalization has really been desirous for the development of CNT-based probes and nanosensors for environmental monitoring and pollution remediation, because they should be water-soluble and responsive to ionic strength, temperature, pH, and environmental stimuli. Su *et al.* functionalized MWCNTs with a temperature-responsive homopolymer using a two-step methodology. In the first step, the RAFT polymerization was used to produce

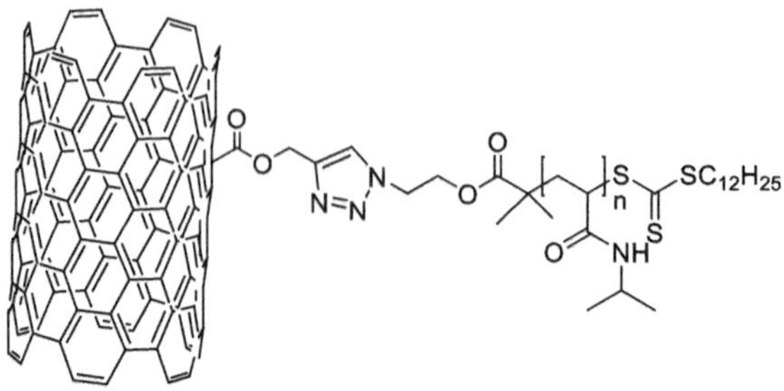

Figure 25 MWCNTs functionalized with temperature-responsive homopolymer.

azide-capped poly(*N*-isopropylacrylamide) N_3-PNIPAM. The alkyne containing MWCNTs were made to react with N_3-PNIPAM under click conditions. The product (MWCNTs-PNIPAM) exhibited good stability and solubility in water (Figure 25).[48]

The CNTs have become promising candidates for the preparation of nanohybrids due to their chemical inertness and large surface area. They have become attractive building blocks for the preparation of high performance materials like biofuel cells and biosensors. The recent advances have enabled researchers to functionalize CNTs with a range of enzymes for the preparation of novel hybrid materials.[49] The click chemistry has established its worth as well for the installation of enzymes on CNTs surface. The sensing, catalysis, and electron transfer properties of redox-active hemoproteins have made them attractive enzymes for the preparation of enzyme-immobilized electrodes. A few years ago, Onoda *et al.* employed a click chemistry approach to prepare hemoprotein-MWCNT hybrid materials by reacting alkyne functionality on MWCNT with cytochrome b_{562} with a tethered N_3-group on the heme (Figure 26).[50] This method was suggested

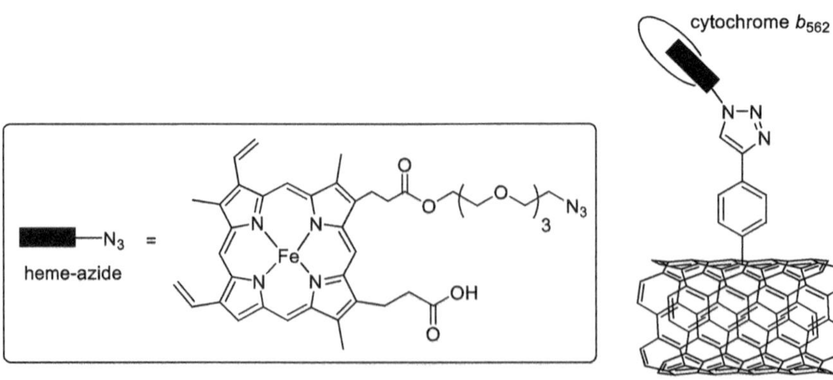

Figure 26 Hemoprotein-MWCNT hybrid nanomaterial.

as a pioneering effort for the development of bioelectrode interfaces using hemoprotein-carbon nanomaterials. The enzymes are expected to exhibit rigid orientation on SWCNT due to their covalent linkage. The hemoprotein matrices were anticipated as attractive materials for catalysis, sensors, and electron transfer mediators.

In recent years, the bioactive potential of glycofullerenes and glycodendrons has made them promising research candidates. However, they suffer from water solubility issues with 1–2 sugars and undergo aggregation in polar media.[51] They show a drastic increase in water solubility with increase in sugar units. The water solubility issues have restricted their potent applications in biological media. A number of efforts have been made to enhance their aqueous solubility for potent biological applications, and the recent advancements in click chemistry have revolutionized this domain. In this connection, Rodríguez-Pérez *et al.* prepared an artificial model assay for Ebola virus infection using CNTs as virus-mimicking platforms for multivalent presentation of carbohydrates (Figure 27).[52] They employed click conditions to attach glycofullerenes and glycodendrons on nanocarbon scaffolds (single- and multi-walled CNTs). The surface functionalization enhanced the hydrophilicity of resulting nanohybrids. The prepared nanoglycoconjugates were found to be potent inhibitors of viral infection in artificial cellular assay, and the maximum efficiency was shown by nanohybrids of glycofullerenes. The biocompatibility of these CNTs-based scaffolds highlighted them as promising candidates for a wider scope of applications in biomedics.

Conclusion

The recent advances in triazole chemistry have inspired almost all scientific disciplines including nanotechnology. The surface functionalization of nanomaterials has been used as a tool to tune their properties as per applications. However, over the last few years, triazoles have become promising candidates for tailoring the properties of functional nanomaterials. The triazole-functionalized nanomaterials have shown promising applications in drug delivery, wastewater treatment, environmental toxicant remediation, colorimetric nanosensors, nanocatalysts, and biomedical implants. Hence, triazole chemistry has a promising future in tailoring the functional properties of nanomaterials.

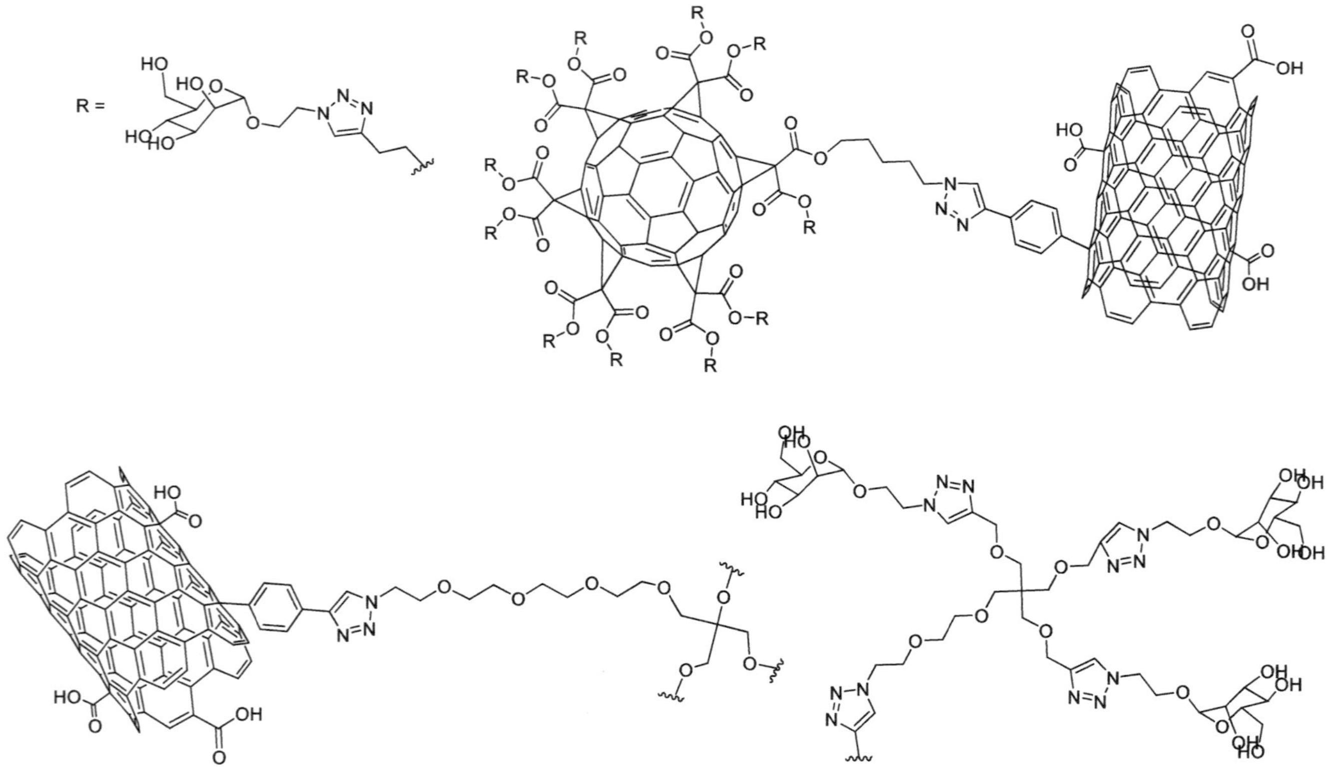

Figure 27 Click attachment of glycofullerenes and glycodendrons on MWCNTs.

References

1. Wongkaew N, Simsek M, Griesche C, Baeumner AJ. Functional nanomaterials and nanostructures enhancing electrochemical biosensors and lab-on-a-chip performances: recent progress, applications, and future perspective. *Chem Rev* 2018;**119**(1):120–94.
2. Wang ZM. *Toward functional nanomaterials.* vol. 5. Springer Science & Business Media; 2010.
3. Mahapatra SS, Yadav SK, Cho JW. Synthesis of click-coupled graphene sheets with hyperbranched polyurethane: effective exfoliation and enhancement of nanocomposite properties. *J Appl Polym Sci* 2017;**134**(12):44631–7.
4. Shahrivari S, Faridbod F, Ganjali MR. Highly selective and sensitive colorimetric determination of Cr3+ ion by 4-amino-5-methyl-4H-1, 2, 4-triazole-3-thiol functionalized Au nanoparticles. *Spectrochim Acta A Mol Biomol Spectrosc* 2018;**191**:189–94.
5. Oliveira E, Núñez C, Santos HM, Fernández-Lodeiro J, Fernández-Lodeiro A, Capelo JL, et al. Revisiting the use of gold and silver functionalised nanoparticles as colorimetric and fluorometric chemosensors for metal ions. *Sens Actuators B* 2015;**212**:297–328.
6. Daniel M-C, Astruc D. Gold nanoparticles: assembly, supramolecular chemistry, quantum-size-related properties, and applications toward biology, catalysis, and nanotechnology. *Chem Rev* 2004;**104**(1):293–346.
7. Cefalu WT, Hu FB. Role of chromium in human health and in diabetes. *Diabetes Care* 2004;**27**(11):2741–51.
8. Feng J-J, Guo H, Li Y-F, Wang Y-H, Chen W-Y, Wang A-J. Single molecular functionalized gold nanoparticles for hydrogen-bonding recognition and colorimetric detection of dopamine with high sensitivity and selectivity. *ACS Appl Mater Interfaces* 2013;**5**(4):1226–31.
9. Khavani M, Izadyar M, Housaindokht MR. Modeling of the functionalized gold nanoparticle aggregation in the presence of dopamine: a joint MD/QM study. *J Phys Chem C* 2018;**122**(45):26130–41.
10. Zhang L, Yang S, Han T, Zhong L, Ma C, Zhou Y, et al. Improvement of Ag (I) adsorption onto chitosan/triethanolamine composite sorbent by an ion-imprinted technology. *Appl Surf Sci* 2012;**263**:696–703.
11. Fu L, Zhang L, Wang S, Peng J, Zhang G. Selective adsorption of Ag+ by silica nanoparticles modified with 3-amino-5-mercapto-1, 2, 4-triazole from aqueous solutions. *J Mol Liq* 2017;**241**:292–300.
12. Veena V, Shivaprasad KH, Lokesh KS, Sharanagouda H, Ramakrishna D. Design, synthesis, computational and biological evaluation of 4-amino-3,5-dimercapto-1,2,4-triazole surface functionalized gold nanoparticles. *Asian J Chem* 2019;**31**:2875–84.
13. Khorramabadi V, Habibi D, Heydari S. 3-Mercapto-1, 2, 4-triazole functionalized Fe3O4 based Cu nanoparticles: a capable catalyst for the synthesis of diverse tetrazoles from amino acids. *Org Prep Proced Int* 2020;**52**:1–8.
14. Chen Y-C, Lee I-L, Sung Y-M, Wu S-P. Triazole functionalized gold nanoparticles for colorimetric Cr3+ sensing. *Sens Actuators B* 2013;**188**:354–9.
15. Lee H, Lee J, Lee Y-H, Lee J, Yoon M. Formation of Ag+–N–TiO_2 nanochains and their HPA-composites as highly visible light-sensitive photocatalysts toward two-color solar cells. *J Mater Chem* 2011;**21**(34):12829–35.
16. Xia R, Ruan Z, Zhang Y, Zhu H, Cao M, Chen P, et al. Click polymerization and characterization of TiO_2 nanoparticles to one-dimensional nanochains. *Chem Phys Lett* 2017;**687**:312–6.
17. Mondal P, Yarger JL. Colorimetric dual sensors of metal ions based on 1, 2, 3-triazole-4, 5-dicarboxylic acid-functionalized gold nanoparticles. *J Phys Chem C* 2019;**123**(33):20459–67.

18. Ulbrich K, Hola K, Subr V, Bakandritsos A, Tucek J, Zboril R. Targeted drug delivery with polymers and magnetic nanoparticles: covalent and noncovalent approaches, release control, and clinical studies. *Chem Rev* 2016;**116**(9):5338–431.
19. Movagharnegad N, Najafi Moghadam P, Nikoo A, Shokri Z. Modification of magnetite cellulose nanoparticles via click reaction for use in controlled drug delivery. *Polym-Plast Technol Eng* 2018;**57**(18):1915–22.
20. Pozdnyakov A, Kuznetsova N, Korzhova S, Ermakova T, Fadeeva T, Vetohina A, et al. Antimicrobial activity of Ag 0-nanocomposite copolymer of 1-vinyl-1, 2, 4-triazole with sodium acrylate. *Russ Chem Bull* 2015;**64**(6):1440–4.
21. Pozdnyakov A, Ivanova A, Emelyanov A, Ermakova T, Prozorova G. Nanocomposites with silver nanoparticles based on copolymer of 1-vinyl-1, 2, 4-triazole with N-vinylpyrrolidone. *Russ Chem Bull* 2017;**66**(6):1099–103.
22. Sosedova L, Novikov M, Titov E, Pozdnyakov A, Korzhova S, Ermakova T, et al. Synthesis, antimicrobial properties, and toxicity of a nanobiocomposite based on ag (0) particles and poly (1-vinyl-1, 2, 4-triazole). *Pharm Chem J* 2019;**52**(11):912–6.
23. Pozdnyakov AS, Emel'yanov AI, Kuznetsova NP, Ermakova TG, Korzhova SA, Khutsishvili SS, et al. Synthesis and characterization of silver-containing nanocomposites based on 1-vinyl-1, 2, 4-triazole and acrylonitrile copolymer. *J Nanomater* 2019;**2019**:1–7.
24. Pozdnyakov AS, Emel'yanov AI, Kuznetsova NP, Ermakova TG, Bolgova YI, Trofimova OM, et al. A polymer nanocomposite with CuNP stabilized by 1-vinyl-1, 2, 4-triazole and acrylonitrile copolymer. *Synlett* 2016;**27**(06):900–4.
25. Ruddell D, Maloney M, Thompson J. Effect of novel filler particles on the mechanical and wear properties of dental composites. *Dent Mater* 2002;**18**(1):72–80.
26. Wang H, Zhu M, Li Y, Zhang Q, Wang H. Mechanical properties of dental resin composites by co-filling diatomite and nanosized silica particles. *Mater Sci Eng C* 2011;**31**(3):600–5.
27. Yushau US, Almofeez L, Bozkurt A. Aminotriazole functional silica incorporated BisGMA/TEGDMA resins as dental nanocomposites. *Polym Polym Compos* 2019;**27**(8):488–95.
28. Ajmal M, Yunus U, Graham RM, Leblanc RM. Design, synthesis, and targeted delivery of fluorescent 1, 2, 4-triazole–peptide conjugates to pediatric brain tumor cells. *ACS Omega* 2019;**4**:22280–91.
29. da Silva JCE, Gonçalves HM. Analytical and bioanalytical applications of carbon dots. *TrAC Trends Anal Chem* 2011;**30**(8):1327–36.
30. Georgakilas V, Otyepka M, Bourlinos AB, Chandra V, Kim N, Kemp KC, et al. Functionalization of graphene: covalent and non-covalent approaches, derivatives and applications. *Chem Rev* 2012;**112**(11):6156–214.
31. Namvari M, Namazi H. Preparation of efficient magnetic biosorbents by clicking carbohydrates onto graphene oxide. *J Mater Sci* 2015;**50**(15):5348–61.
32. Ramezanzadeh B, Ghasemi E, Mahdavian M, Changizi E, Moghadam MM. Covalently-grafted graphene oxide nanosheets to improve barrier and corrosion protection properties of polyurethane coatings. *Carbon* 2015;**93**:555–73.
33. Borthakur P, Boruah PK, Das MR, Szunerits S, Boukherroub R. Cu (0) nanoparticle-decorated functionalized reduced graphene oxide sheets as artificial peroxidase enzymes: application for colorimetric detection of Cr (VI) ions. *New J Chem* 2019;**43**(3):1404–14.
34. Vladkova T, Akuzov D, Klöppel A, Brümmer F. Current approaches to reduction of marine biofilm formation. *J Chem Technol Metall* 2014;**49**(4):345–55.
35. Iannazzo D, Pistone A, Visco A, Galtieri G, Giofrè SV, Romeo R, et al. 1, 2, 3-triazole/MWCNT conjugates as filler for gelcoat nanocomposites: new active antibiofouling coatings for marine application. *Mater Res Express* 2015;**2**(11):115001.

36. Schnorr JM, Swager TM. Emerging applications of carbon nanotubes. *Chem Mater* 2011;**23**(3):646–57.
37. Zhan Y, Wan X, He S, Yang Q, He Y. Design of durable and efficient poly (arylene ether nitrile)/bioinspired polydopamine coated graphene oxide nanofibrous composite membrane for anionic dyes separation. *Chem Eng J* 2018;**333**:132–45.
38. Rajabi M, Mahanpoor K, Moradi O. Removal of dye molecules from aqueous solution by carbon nanotubes and carbon nanotube functional groups: critical review. *RSC Adv* 2017;**7**(74):47083–90.
39. El-Sharkawy R, El-Ghamry HA. Multi-walled carbon nanotubes decorated with Cu (II) triazole Schiff base complex for adsorptive removal of synthetic dyes. *J Mol Liq* 2019;**282**:515–26.
40. Wang C, Yuan R, Chai Y, Hu F. Simultaneous determination of hydroquinone, catechol, resorcinol and nitrite using gold nanoparticles loaded on poly-3-amino-5-mercapto-1, 2, 4-triazole-MWNTs film modified electrode. *Anal Methods* 2012;**4**(6):1626–8.
41. Yang C, Chai Y, Yuan R, Xu W, Chen S. Gold nanoparticle–graphene nanohybrid bridged 3-amino-5-mercapto-1, 2, 4-triazole-functionalized multiwall carbon nanotubes for the simultaneous determination of hydroquinone, catechol, resorcinol and nitrite. *Anal Methods* 2013;**5**(3):666–72.
42. Kobashi K, Ata S, Yamada T, Futaba DN, Yumura M, Hata K. A dispersion strategy: dendritic carbon nanotube network dispersion for advanced composites. *Chem Sci* 2013;**4**(2):727–33.
43. Sheikholeslam M, Pritzker M, Chen P. Hybrid peptide–carbon nanotube dispersions and hydrogels. *Carbon* 2014;**71**:284–93.
44. Takassi MA, Zadehnazari A, Farhadi A, Mallakpour S. Highly stable polyimide composite films based on 1, 2, 4-triazole ring reinforced with multi-walled carbon nanotubes: study on thermal, mechanical, and morphological properties. *Prog Org Coat* 2015;**80**:142–9.
45. Takassi M, Zadehnazari A. Nanocomposites of triazole functionalized multi-walled carbon nanotube with chemically grafted polyimide: preparation, characterization, and properties. *Fullerenes, Nanotubes, Carbon Nanostruct* 2016;**24**(2):128–38.
46. Guo Z, Liang L, Liang J-J, Ma Y-F, Yang X-Y, Ren D-M, et al. Covalently β-cyclodextrin modified single-walled carbon nanotubes: a novel artificial receptor synthesized by 'click' chemistry. *J Nanopart Res* 2008;**10**(6):1077–83.
47. He H, Zhang Y, Gao C, Wu J. 'Clicked' magnetic nanohybrids with a soft polymer interlayer. *Chem Commun* 2009;**13**:1655–7.
48. Su X, Shuai Y, Guo Z, Feng Y. Functionalization of multi-walled carbon nanotubes with thermo-responsive azide-terminated poly (N-isopropylacrylamide) via click reactions. *Molecules* 2013;**18**(4):4599–612.
49. Kobayashi K, Shimizu M, Nagamune T, Sasabe H, Fang Y, Knoll W. Monolayer formation of cytochrome b 562 on gold surfaces and its reconstitution reaction, studied by surface plasmon resonance spectroscopy. *Bull Chem Soc Jpn* 2002;**75**(8):1707–13.
50. Onoda A, Inoue N, Campidelli S, Hayashi T. Cofactor-specific covalent anchoring of cytochrome b 562 on a single-walled carbon nanotube by click chemistry. *RSC Adv* 2016;**6**(70):65936–40.
51. Muñoz A, Illescas BM, Luczkowiak J, Lasala F, Ribeiro-Viana R, Rojo J, et al. Antiviral activity of self-assembled glycodendro [60] fullerene monoadducts. *J Mater Chem B* 2017;**5**(32):6566–71.
52. Rodríguez-Pérez L, Ramos-Soriano J, Pérez-Sánchez A, Illescas BM, Muñoz A, Luczkowiak J, et al. Nanocarbon-based glycoconjugates as multivalent inhibitors of Ebola virus infection. *J Am Chem Soc* 2018;**140**(31):9891–8.

CHAPTER 7

Triazole-Based Plant Growth-Regulating Agents: A Recent Update

Arruje Hameed[a] **and Tahir Farooq**[b,*]
[a]Department of Biochemistry, Government College University, Faisalabad, Pakistan
[b]Department of Applied Chemistry, Government College University, Faisalabad, Pakistan
*Corresponding author. E-mail: tahirfarooqfsd@gmail.com

Introduction

Several commercially available plant growth regulators (PGRs) are frequently used for growth regulation, yield enhancement, and disease management in crop plants. The plant growth-regulating agents control the endogenous hormonal level and also interfere with the hormone-mediated signalings in plants. The PGRs are also termed as plant bioregulators or plant biostimulants, and they generally influence the production of gibberellins, cytokinins, auxins, abscisic acid (ABA), and ethylene. The plant height and organ size is controlled by brassinosteroids, auxins, and gibberellins.[1,2]

The triazole-based PGRs directly interfere with the biosynthesis of brassinosteroid, ABA, and gibberellic acid (GA) in plants (Figures 1 and 2). Their interference with the enzymes of GA biosynthesis reduces stem elongation resulting in more compact plants. The plants with reduced height utilized their saved energy for an increase in branching, reproduction, flowering, and induction of stress tolerance. The applications of paclobutrazol (PBZ) reduced plant height, increased stem strength, and decreased lodging in maize. The foliar and leaf treatments of uniconazole (UNI) reduced GA and increased ABA, resulting in reduced lodging in maize.[3,4]

Triazole-based plant growth-regulating agents play vital roles in regulating the growth, physiological, and biochemical processes leading to improved crop yields (Figure 3). Their applications modulate antioxidant potential, physiological, and biochemical processes that help to alleviate the negative effect of several abiotic stresses including drought, salinity, chilling, and heat stress.[5,6]

Advances in Triazole Chemistry
https://doi.org/10.1016/B978-0-12-817113-4.00008-1

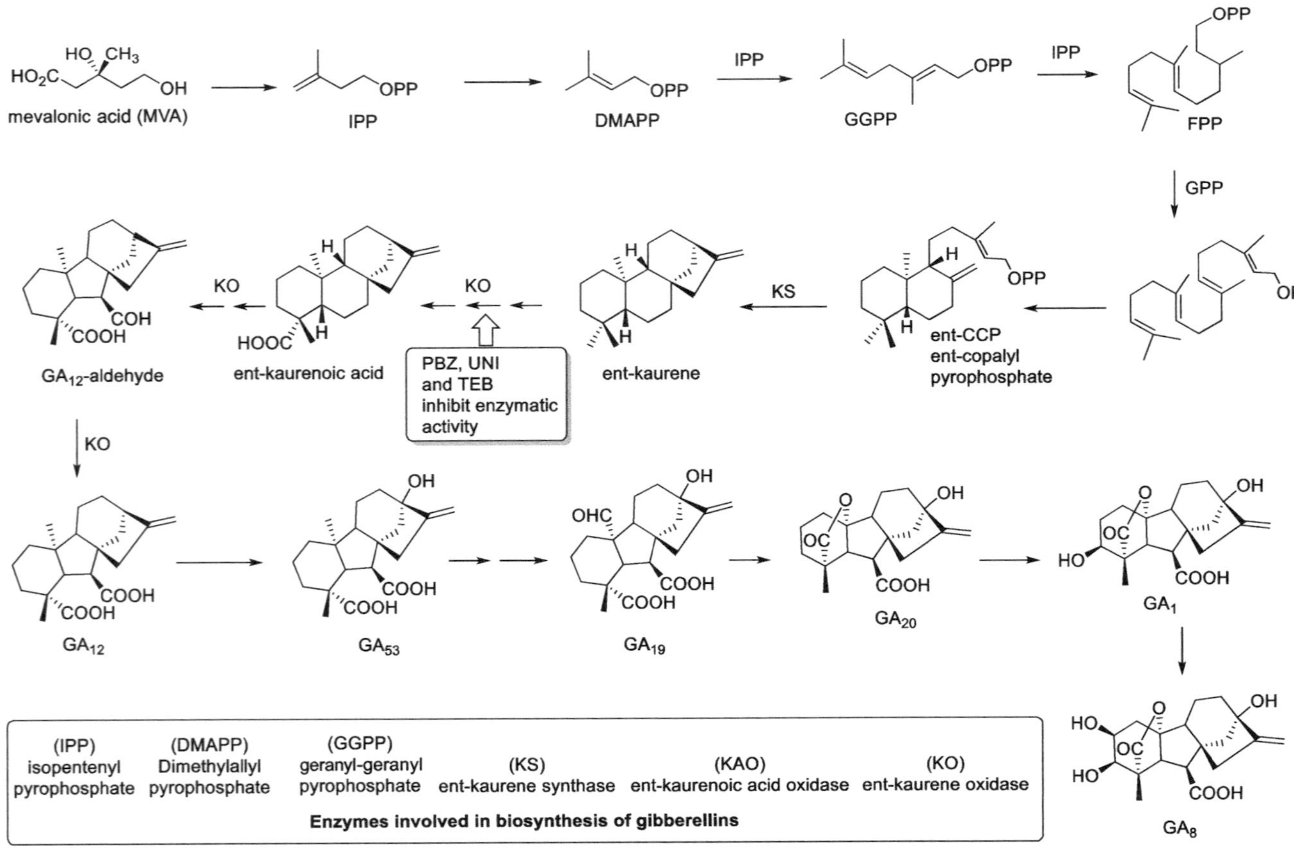

Figure 1 Known mechanism for gibberellin biosynthesis and its inhibition by triazole-based PGRs.

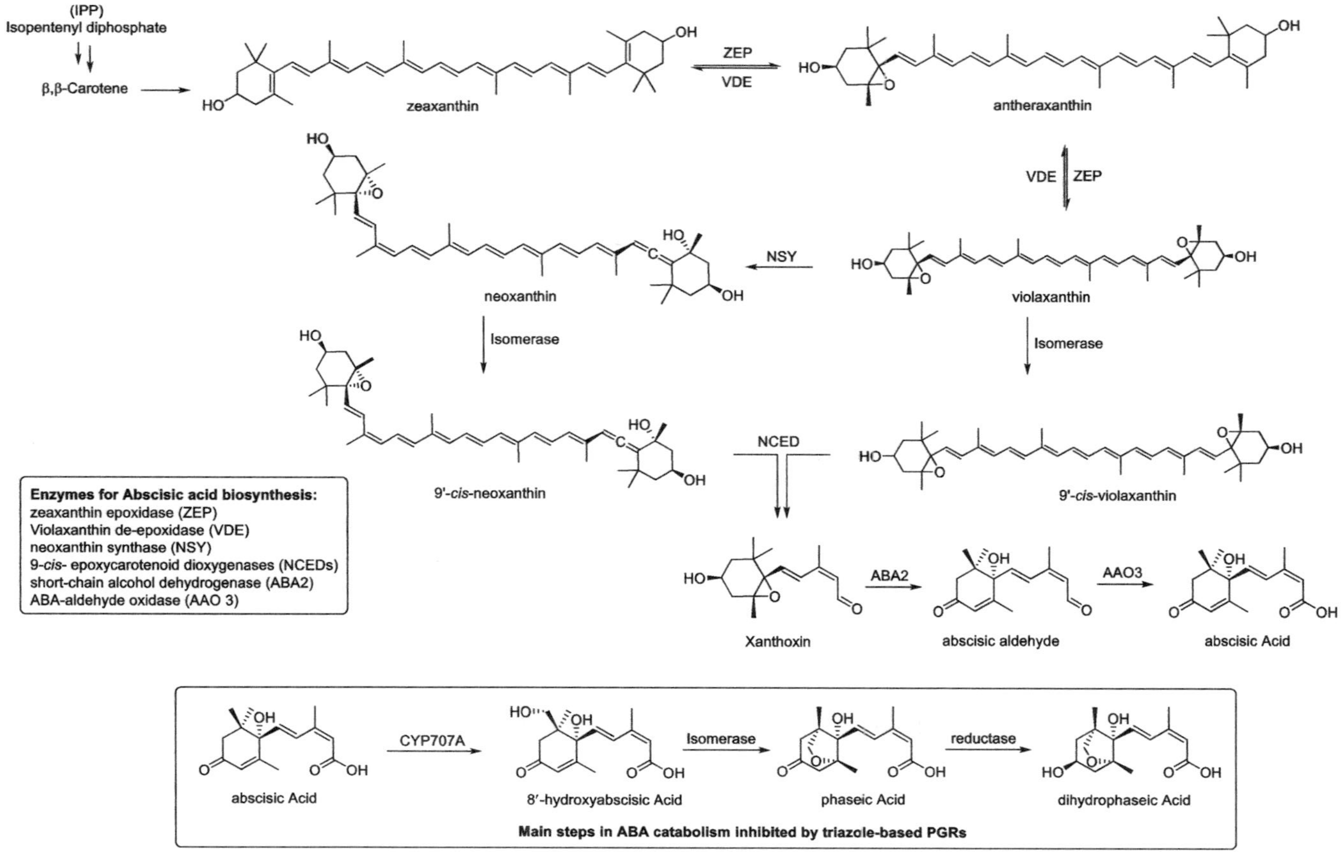

Figure 2 Biosynthesis and catabolism of abscisic acid.

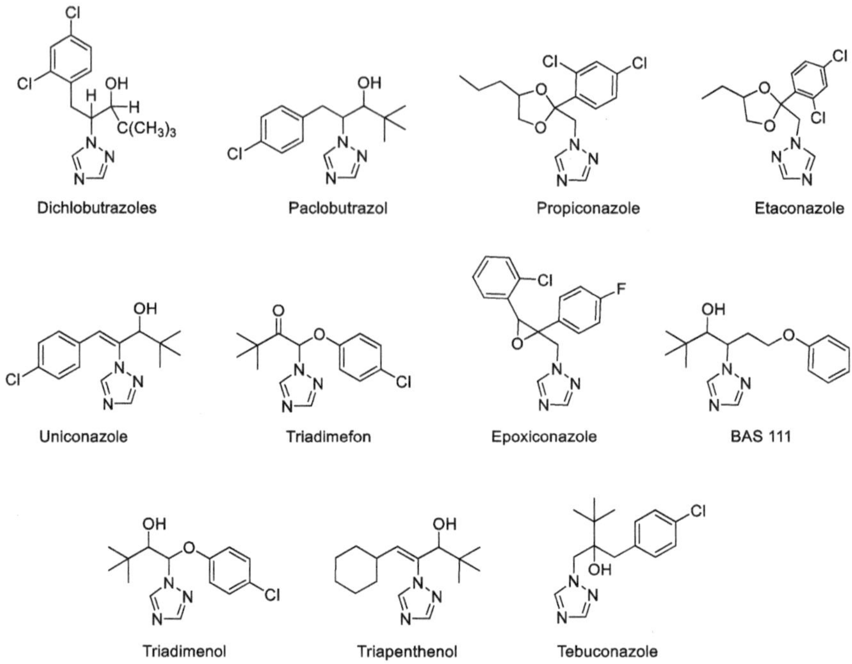

Figure 3 1,2,4-Triazole-based known herbicides, fungicides, and PGRs.

Paclobutrazol

PBZ alters the level of phytohormones like cytokinins, gibberellins, and ABA as a mechanistic approach to regulate growth in plants. In fact, it interferes with the enzymes involved in the isoprenoid pathway resulting in the inhibition of gibberellin synthesis, an increase in ABA and cytokinin, and a decrease of ethylene production. It restricts the oxidation of ent-kaurene to ent-kaurenoic acid by inhibiting the activities of cytochrome P450 oxidase and ent-kaurene oxidase.[1] The PBZ-mediated unbalancing of phytohormones reduce internode elongation, shoot length, and leaf area thus decreasing plant height. The anti-gibberellin nature of PBZ affects cell division and enlargement processes resulting in the suppression of plant growth. It could also interfere with the normal catabolic process of ABA, which results in its extra accumulation in leaves. The ABA causes the closing of stomata and improves water relations by reducing water loss in PBZ-treated plants. It also affects the level of zeatin, zeatin riboside, IAA, GA_3, etc. Its applications also enhance the contents of polyamines.

The PBZ applications up-regulate the activities of antioxidant enzymes like peroxidase (POX), ascorbate peroxidase (APX), catalase (CAT), and superoxide dismutase (SOD) resulting in the counterbalancing of deleterious oxidative stress. It enhances the antioxidant potential of the plants thus minimizing the chances of lipid peroxidation and membrane damage.[7, 8] The ascorbic acid, reduced glutathione, and α-tocopherol are important non-enzymatic antioxidants available in cytosol and chloroplast in plants and are involved in scavenging the excessive production of hazardous free radicals. The PBZ applications are known to increase the contents of these non-enzymatic antioxidants enabling treated crop plants to control the active oxygen species.[9] The contents of proline in cytoplasm regulate the turgor and help to maintain the composition of macromolecules and cellular organelles including the membrane.[10] The PBZ applications cause an increment in proline contents that specifically help to tolerate water deficit stress conditions. The PBZ also induce an increment in photosynthetic pigments and carotenoids in plants. The enhancement in chlorophyll biosynthesis is attributed to the PBZ-mediated increase in cytokinin in PBZ-treated plants. Its applications have also been able to accelerate the rate of photosynthesis in crop plants.[11]

As described earlier, the PZB applications inhibit the biosynthesis of GA and increase cytokinin, ABA, and proline contents in plants under normal and a range of abiotic stresses. It enhances the antioxidant potential by up-regulating antioxidant enzymes and by increasing non-enzymatic antioxidants. Its applications enhance photosynthetic pigments and protect photosynthetic machinery by putting positive effects on xanthophylls' pigment cycle. All previously discussed PBZ-induced physiological, morphological, and biochemical changes enhance the yield of the various crops under normal and stress conditions. They help to modulate tolerance against salinity, drought, heat, and chilling stress in crop plants.[12, 13] The applications of PBZ induce positive effects on the quality and yield of fruit crops. It could induce early and off-season flowerings with a common carryover effect.[14]

Uniconazole

Primarily, UNI blocks a P450-enzyme kaurene oxidase involved in the isoprenoid pathway, which subsequently restricts the P450 monooxygenases-mediated production of ent-kaurenoic acid leading to the inhibition of gibberellins biosynthesis.[15, 16] The UNI alters the endogenous level of plant hormones like cytokinins, gibberellins, and ABA to regulate growth in plants. The UNI-mediated unbalancing of phytohormones reduces internode elongation, shoot length, and leaf area thus decreasing plant height.

The gibberellin inhibitory activity of UNI affects cell division and enlargement processes resulting in the suppression of plant growth. It could also interfere with the normal catabolic process of ABA, which results in its extra accumulation in leaves. The ABA causes the closing of stomata and improves water relations by reducing water loss in UNI-treated plants.[17]

The UNI applications cause overexpression of antioxidant enzymes to enhance the antioxidant potential of the plants enabling them to control the reactive oxygen species for the integrity of biological membranes.[18] The UNI could also enhance the contents of non-enzymatic antioxidants thus boosting antioxidant potential against oxidative stress in plants.[19] The UNI applications cause an increment in proline contents, which specifically help to tolerate water deficit stress conditions. It could also induce increment in photosynthetic pigments and enhance photosynthetic efficiency in a variety of crop plants under normal as well as stress conditions.[20, 21]

The UNI applications inhibit the biosynthesis of GA and increase cytokinin, ABA, and proline contents in plants under normal and a range of abiotic stresses. It enhances the antioxidant potential by up-regulating antioxidant enzymes and by increasing non-enzymatic antioxidants.[22, 23] Its applications enhance photosynthetic pigments and protect photosynthetic machinery by putting positive effects under normal and stress conditions.[24] All these UNI-mediated physiological, morphological, and biochemical changes help mechanistically to modulate tolerance against salinity, drought, heat, and chilling stress in crop plants.[6, 25, 26] The applications of UNI induce early and off-season flowerings and also produce positive effects on the quality and yield of fruits.[27]

Tebuconazole

Just like other triazole-based PGRs, the tebuconazole (TEB) also inhibits gibberellin biosynthesis thus acting as plant growth-regulating agents by altering the levels of endogenous phytophormones.[28] Its exogenous applications up-regulate the antioxidant potential by modulating the enzymatic and non-enzymatic antioxidants and increasing the photosynthetic pigments even under stress conditions.[29, 30] It enhances the proline and ABA contents as a mechanistic approach to cope with conditions of drought.[31] The application of TEB regulates the endogenous production of reactive oxygen species. Its applications also help to mitigate the negative effects of a variety of abiotic stresses including salt, drought, heat, and chilling stress.[29, 32] However, its applications under low-temperature conditions produce some phytotoxic effects on maize seedlings.[33, 34] The TEB applications modulate physiological and biochemical parameters in plants to counter undesirable abiotic stresses.[35–37]

New trends in the applications of triazole-based plant growth-regulating agents

Recently, there has been great interest in the applications of nanotechnology in the field of agriculture especially for the nanoscale delivery of plant growth-regulating agents, herbicides, fertilizers, and pesticides. The nanodelivery systems facilitate the controlled release of active ingredients, which reduces the chances of ecotoxicology by avoiding the rampant use of chemicals.[38, 39] Nanocarriers have acted as safer and effective delivery systems for the controlled and sustained release of agrochemicals for a longer time period with increased safety profiles. The nanocarriers could be designed with an inherent ability to develop an affinity with soil structure or plant roots for long-sustaining and controlled release of agrochemicals including plant growth-regulating agents. The polymeric nanocarriers manage to sustain release of active agents through their entrapment or encapsulation.[40, 41]

In a very recent study, Maluin *et al.* used chitosan, a natural biopolymer, as a nanocarrier for encapsulating a well-known fungicide and plant growth-regulating agent called hexaconazole (HEX).[42] They used sodium tripolyphosphate in different concentrations to prepare chitosan-hexaconazole nanoparticles (NPs) by controlling loading content and particle size distribution employing the ion gelation method. The size of prepared chitosan-hexaconazole NPs could be controlled by increasing tripolyphophate (TPP) concentration. The NPs prepared from 5 mg/mL of TPP exhibited a prolonged release time for 86 hours with sustained release of HEX (99.9%) and highest loading efficiency. The prepared nanodelivery system was suggested as a promising anti-fungal system with low toxicity, high efficiency, and sustain-releasing profile. The chitosan was used as a nanocarrier due to its inherent ability to encapsulate both hydrophobic as well as hydrophilic active agents for their sustained and controlled delivery to plants. The nanodelivery systems for fungicides have also been developed using silica, cyclodextrin, and other polymers.[43, 44] The polymeric NPs have been regarded as promising candidates for the development of sustainable and ecofriendly operations for modern agriculture. They maintain a controlled release and improved and targeted uptake minimizing the chances of ecotoxicity. Such polymeric nanocarriers also represent extra benefits of biocompatibility and biodegradability showing their ecofriendly nature.[45]

Campos *et al.* developed polymeric nanocapsules and solid lipid NPs for controlled and targeted delivery of TEB and carbendazim for fungal disease management in plants.[46] They studied the cytotoxicities and stability of these nanoformulations. The developed nanosystems exhibited controlled release of fungicides due to their high affinity with nanocarriers. The new formulations reduced the cytotoxicity of bioactive compounds in cytotoxicity assays and were suggested as promising candidates for fungal disease management in plants for sustainable productivity of crops.

The encapsulation of agrochemicals with biodegradable polymeric materials avoids the use of organic solvents for their applications in field crops, minimizing pollution-related issues. Very recently, Barrera Mendez *et al.* used copolymer poly(lactic-*co*-glycolic) acid and polylactic acid as nanocarriers for controlled release of propiconazole with enhanced anti-fungal efficiency without the use of any emulsifier.[47] The nanoencapsulation of propiconazole prolonged the time of availability of the active anti-fungal agent. The encapsulating agents were selected due to the non-toxicity of their biodegraded products.

Various natural and synthetic materials have been used as carriers of pesticides in recent microencapsulation strategy for pesticides. The microencapsulation approach enhances their efficiency with the controlled release over an extended time period. However, single-walled polymeric carriers produce a high initial burst due to their thin membrane structure. Therefore, double-walled carrier systems reduce the initial burst through the better control of bioactive agents. Over the last few years, micro-organisms have successfully been employed as biodegradable and biocompatible carrier systems for the microencapsulation of several bioactive molecules.[48, 49] The cyanobacteria represent an ideal membrane structure with an outer membrane, peptidoglycan layer, and a plasma membrane for controlled release microencapsulation applications. Zhang *et al.* encapsulated TEB using cyanobacteria as an ecofriendly encapsulating material followed by coating with urea-formaldehyde resins.[50] The prepared monodispersed microencapsulating system very effectively controlled the initial burst of TEB and prolonged the time of its efficiency. Thus, cyanobacteria proved to be an effective medium for encapsulation of bioactive agents.

Over the last few years, the metal-organic frameworks (MOFs) have found interesting applications in chemical sensors, enantioselective separations, and heterogeneous catalysis. The MOFs exhibit unique inherent properties due to their porous crystalline nature with an open framework of organic linkers and metal ions. The highly extended surface area with high porosity has made them highly efficient and versatile drug carriers. Furthermore, they have been attractive in applications in biohybrid sensors in plants, encapsulation of plant hormones, and controlled release of fertilizers.[51–53] The TEB is mainly formulated as a suspension, wettable powder, or emulsifiable concentrate due to its poor water solubility. The available formulations of TEB not only reduce its effective period but utilization as well, which results in their repeated applications with limited success. Thus, a highly efficient formulation of TEB has been desirous according to new standards of precision agriculture.

In a recent approach, Tang *et al.* used Zr (IV) metal ions and meso-tetra(4-carboxyphenyl)porphine as an organic linker for the preparation of porphyrinic MOFs as the carrier of TEB.[54] The control releasing encapsulation of TEB was finally ready after the layer-by-layer assembly of chitosan and pectin. The prepared microcapsules exhibited stimuli-responsive controlled release of fungicide, and the MOFs showed high loading efficiency. The prepared MOFs-based encapsulating system did not show any toxic effects when applied to Chinese cabbage. Thus, the prepared stimuli-responsive delivery system was suggested as an efficient sustainable formulation of TEB with adjustable control-releasing characteristics.

Over the last few years, much attention has been diverted for the preparation of a new generation of PGRs, fungicides, and pesticides with controlled-releasing behavior and biodegradable carriers. As described earlier, such formulations are desirous to control the rampant use of agrochemicals to reduce environmental toxicity issues. To this end, several natural and synthetic materials have been exploited considering their compatibility with bioactive agents, their non-toxic biodegraded products, and safety profiles for living and non-living components of the ecosystem. The materials used for this purpose include natural materials like clays and clay-like silicates, natural and synthetic polymers and polymeric blends, and low molecular weight components, etc.[45, 55] Over the last few years, the degradable microbial polymers like polyhydroxyalkanoates (PHAs) have shown potential application in electronics, medicine, and pharmacology. The most common PHA, the homopolymer poly(3-hydroxybutyrate) P(3HB), has also been utilized for encapsulating TEB for the preparation of biodegradable pellets and films.[56]

Very recently, Volova *et al.* used P(3HB) as biodegradable encapsulating material and natural fillers like peat, wood flour, and clay for the preparation

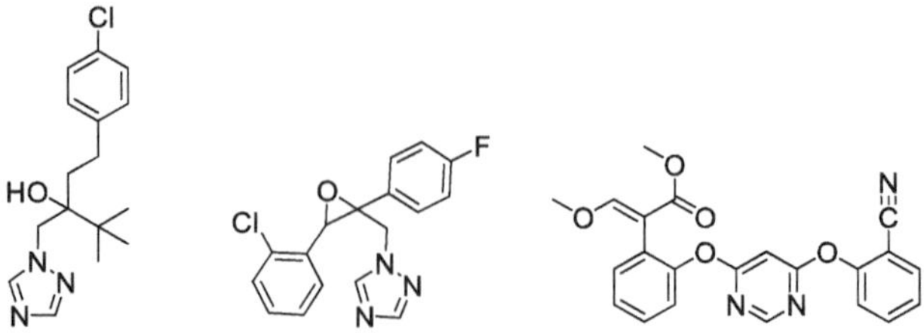

Figure 4 Bioactive materials encapsulated in biodegradable P(3HB).

of slow-releasing formulations of TEB, epoxiconazole, and azoxystrobin (Figure 4).[57] During the study, the biodegradation of delivery systems was found to be influenced by the chemistry of fillers as there was no chemical bonding between the filler and matrices. However, the biodegradation behavior was dependent on the shape of the encapsulating system, i.e., pellets or granular. The as-prepared formulations ensured the sustained delivery of bioactive agents over a longer time period.

New targets of triazoles in plants

The root development, branching initiating and other developmental processes in plants are controlled by terpenoid-derived plant hormones called strigolactones (SLs). These phytohormones manage to control several agronomic traits in plants including leaf senescence, shoot branching, and induction of stress tolerance. They are terpenoid-derived signaling molecules. They are also acting as hyphae-branching factors and rhizosphere-signaling molecules for arbuscular mycorrhizal fungi and root parasitic weeds, respectively. In rice and Arabidopsis, the biosynthesis of SLs is mediated by several enzymes. During the biosynthetic process, 9-cis-β-carotene is produced from trans-β-carotene catalyzed by carotenoid isomerase. Similarly, carotene cleavage dioxygenase 7 and 8 mediate the synthesis of caprolactone, which is an important intermediate for SL biosynthesis. It has been envisioned that the yield of crops could be increased by manipulating the SL biosynthesis. Kawada *et al.* came up with new triazole-based compounds as SL biosynthesis inhibitors. Out of the prepared SL inhibitors, T1S108 turned out to be a highly efficient synthetic compound for the inhibition of SL biosynthesis (Figure 5).[58]

Very recently, the same group utilized a structure-activity relationship and found a new triazolic compound (KK5), a more potent inhibitor than T1S108 for controlling the level of 4-deoxyorobanchol in roots (Figure 6).[59] The applications of KK5 increased branching in Arabidopsis

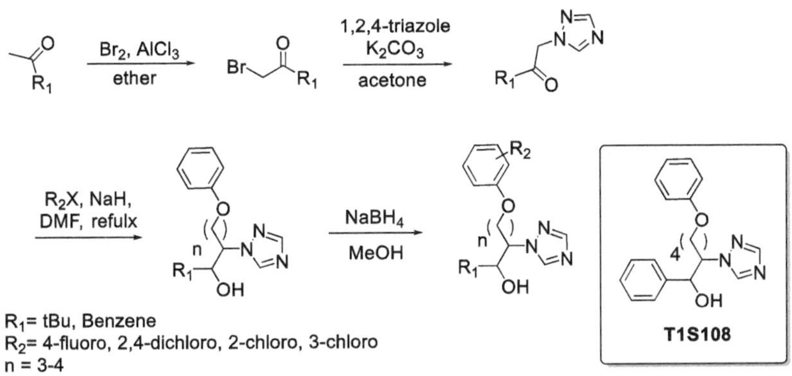

Figure 5 Synthesis of triazole-based strigolactone inhibitors.

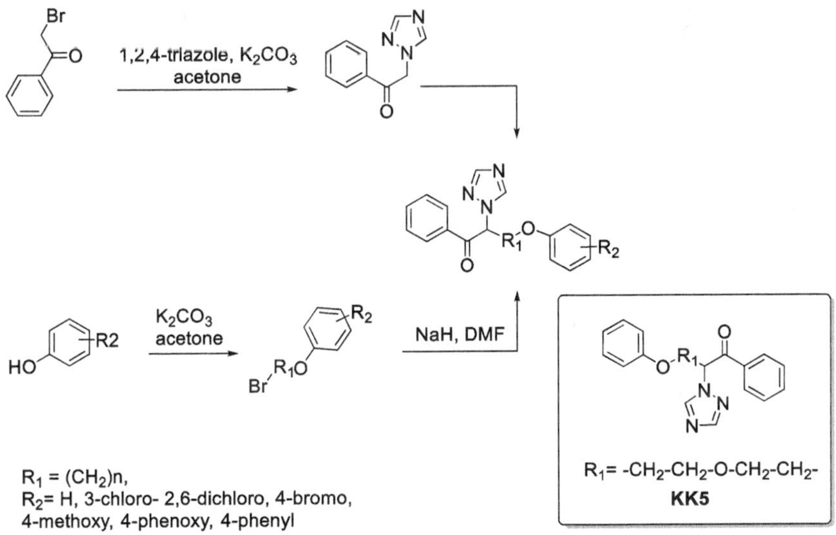

Figure 6 Recently developed strigolactone inhibitors.

and also acted as a highly specific SL biosynthesis in rice. The prepared triazoles did not interfere with the brassinosteroid and gibberellin synthesis in rice and Arabidopsis, respectively. Thus, these novel triazoles were suggested as promising candidates with the potential to specifically inhibit SL biosynthesis. They represent an alternative approach for controlling of plant growth without inhibiting the gibberellin synthesis. Furthermore, the SL production levels induce salt stress tolerance.[60]

In another very recent approach, Nakamura *et al.* prepared 1,2,3-triazole-based urea compounds as a potent SL antagonist. The prepared triazolic compounds were able to develop a covalent bond with SL

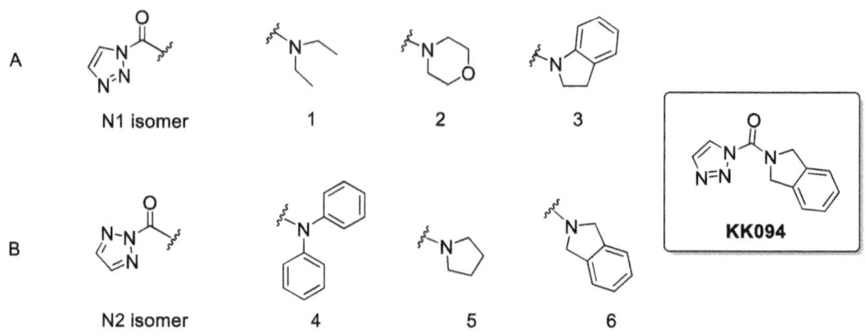

Figure 7 Series of 1,2,3-triazole-based urea compounds as potent SL antagonists.

D14 receptor (Figure 7).[61] Among the tested compounds, the KK094 triazole proved to be a potent inhibitor of SL receptor in yeast two-hybrid and rice-tillering assays. The D14 receptor hydrolyzed the KK094, and the products developed covalent bonds with Ser residue of the receptor. The hydrolysis process and the corresponding products were identified by X-ray crystallography and LC-MS/MS. Thus, blocking of hydrolytic activity of D14 has become a targeted research area for the development of SL inhibitors. The tested triazolic-urea compounds were suggested as potential candidates for promising applications in agriculture.

As discussed earlier, the triazolic compounds have a wide range of applications including herbicides, pesticides, and PGRs. However, their synthetic nature has also been a subject of controversy due to their toxicity toward living organisms, micro-organisms, and the environment. Over the years, there has been an increasing demand to develop environmentally safer agrochemicals from natural sources.[62] The natural product-based agrochemicals could help to attain the main goals of sustainable agriculture. Many of the known plant-based compounds have shown their potential for plant growth-modulation properties.[63–65]

Very recently, Nejma *et al.* used natural maslinic acid and oleanolic acids for the preparation of 1,2,3-triazolic-based semi-synthetic herbicides (Figure 8).[66] They reacted a series of phthalimide azides with propargylated-MA under click conditions. The click reactions provided corresponding 1,4-disubstituted 1,2,3-triazoles. The prepared semi-synthetic triazoles inhibited seed germination and were suggested as promising leads for the development of potent fungicides. The triazole bridge was regarded as the main source of herbicidal activity, which ranged from 91 to 100% in preliminary assays. Thus, the prepared hybrid triazoles were found to be promising candidates for weed management and high crop production.

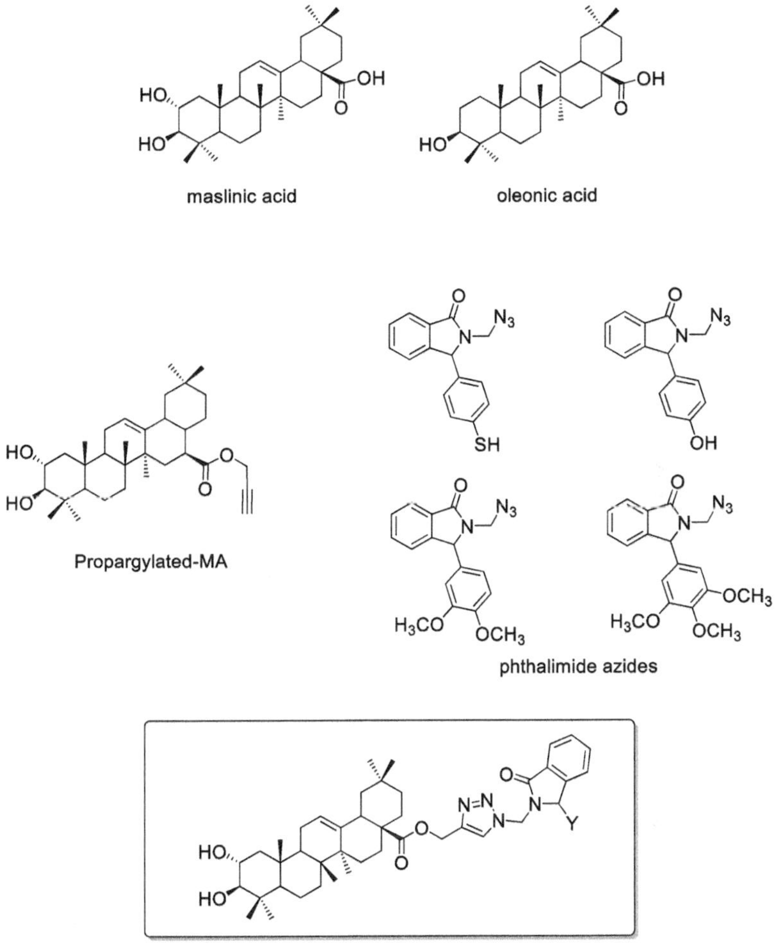

Figure 8 1,2,3-Triazolic-based semi-synthetic herbicides.

Conclusion

Triazoles modulate molecular, physiological, and biochemical processes in plants for growth promontory and induction of tolerance against a range of biotic and abiotic stresses. They generally control the endogenous level of phytohormones to exert their influence on plant growth. Over the last few years, there is an increasing trend to develop their controlled-release formulations to enhance their efficacy with minimum toxic effects. The chapter has highlighted some recent developments in the preparation and applications of triazole-based plant growth-regulating agents for sustainable agriculture.

References

1. Rademacher W. Chemical regulators of gibberellin status and their application in plant production. In: *Annual plant reviews online*; 2018. p. 359–403.
2. Jiang K, Asami T. Chemical regulators of plant hormones and their applications in basic research and agriculture. *Biosci Biotechnol Biochem* 2018;**82**(8):1265–300.
3. Rademacher W. Plant growth regulators: backgrounds and uses in plant production. *J Plant Growth Regul* 2015;**34**(4):845–72.
4. Fletcher RA, Gilley A, Sankhla N, Davis TD. Triazoles as plant growth regulators and stress protectants. *Hortic Rev* 2000;**24**:55–138.
5. Yuan Z, Wang B, Jiang Y, Xie B, Zhang H, Dong S, et al. Effects of uniconazole on physiological and biochemical properties of roots of different sweetpotato cultivars at seedling stage. *Agric Sci Technol* 2015;**16**(4):629.
6. Keshavarz H, Khodabin G. The role of uniconazole in improving physiological and biochemical attributes of bean (Phaseolus vulgaris L.) subjected to drought stress. *J Crop Sci Biotechnol* 2019;**22**(2):161–8.
7. Yadav DK, Hemantaranjan A. Mitigating effects of paclobutrazol on flooding stress damage by shifting biochemical and antioxidant defense mechanisms in mungbean (Vigna radiata L.) at pre-flowering stage. *Legum Res* 2017;**40**(3):453–61.
8. Li X, Yang W, Wan J, Liang J. Effects of paclobutrazol on bud, plant height and antioxidant enzyme activities of Cymbidium hybridum. *Earth Environ Sci* 2020;**440**(2), 022038.
9. Hajihashemi S. Physiological, biochemical, antioxidant and growth characterizations of gibberellin and paclobutrazol-treated sweet leaf (Stevia rebaudiana B.) herb. *J Plant Biochem Biotechnol* 2018;**27**(2):237–40.
10. Tesfahun W. A review on: response of crops to paclobutrazol application. *Cogent Food Agric* 2018;**4**(1), 1525169.
11. Kamran M, Ahmad S, Ahmad I, Hussain I, Meng X, Zhang X, et al. Paclobutrazol application favors yield improvement of maize under semiarid regions by delaying leaf senescence and regulating photosynthetic capacity and antioxidant system during grain-filling stage. *Agronomy* 2020;**10**(2):187.
12. Chandra S, Roychoudhury A. Penconazole, paclobutrazol, and triacontanol in overcoming environmental stress in plants. In: Roychoudhury A, Tripathi DK, editors. *Protective chemical agents in the amelioration of plant abiotic stress: biochemical and molecular perspectives.* Wiley Blackwell; 2020. p. 510–34.
13. Fan Z, Li S, Sun H. Paclobutrazol modulates physiological and hormonal changes in Amorpha fruticosa under drought stress. *Russ J Plant Physiol* 2020;**67**(1):122–30.
14. Ashraf N, Ashraf M. Response of growth inhibitor paclobutrazol in fruit crops. In: *Prunus.* IntechOpen; 2020.
15. Duan W, Zhang H, Xie B, Wang B, Hou F, Li A, et al. Foliar application of uniconazole improves yield through enhancement of photosynthate partitioning and translocation to tuberous roots in sweetpotato. *Arch Agron Soil Sci* 2019;**66**:316–29.
16. Alami Ů, Karimi M. Effect of cycocel and uniconazole on some morphological and biochemical properties of zinnia. *Int J Hortic Sci Technol* 2020;**7**(1):81–91.
17. Yan W, Yanhong Y, Wenyu Y, Taiwen Y, Weiguo L, Xiaochun W. Responses of root growth and nitrogen transfer metabolism to uniconazole, a growth retardant, during the seedling stage of soybean under relay strip intercropping system. *Commun Soil Sci Plant Anal* 2013;**44**(22):3267–80.
18. Ahmad I, Kamran M, Yang X, Meng X, Ali S, Ahmad S, et al. Effects of applying uniconazole alone or combined with manganese on the photosynthetic efficiency, antioxidant defense system, and yield in wheat in semiarid regions. *Agric Water Manag* 2019;**216**:400–14.

19. Gomathinayagam M, Jaleel C, Azooz M, Panneerselvam R. Triazole induced alterations in the peroxidation of membrane lipids and antioxidant status of Manihot esculenta Crantz. *Glob J Mol Sci* 2008;**3**(2):80–5.
20. Bakhita MA, Hussein MM. Uniconazole effect on endogenous hormones, proteins and proline contents of barley plants (Hordeum vulgare) under salinity stress (NaCl). *Nusantara Biosci* 2014;**6**(1).
21. Ahmad I, Kamran M, Su W, Haiqi W, Ali S, Bilegjargal B, et al. Application of uniconazole improves photosynthetic efficiency of maize by enhancing the antioxidant defense mechanism and delaying leaf senescence in semiarid regions. *J Plant Growth Regul* 2019;**38**(3):855–69.
22. Ahmad I, Kamran M, Ali S, Cai T, Bilegjargal B, Liu T, et al. Seed filling in maize and hormones crosstalk regulated by exogenous application of uniconazole in semiarid regions. *Environ Sci Pollut Res* 2018;**25**(33):33225–39.
23. Zhang M, Duan L, Tian X, He Z, Li J, Wang B, et al. Uniconazole-induced tolerance of soybean to water deficit stress in relation to changes in photosynthesis, hormones and antioxidant system. *J Plant Physiol* 2007;**164**(6):709–17.
24. Hussein M, Bakheta M, Zaki S. Influence of uniconazole on growth characters, photosynthetic pigments, total carbohydrates and total soluble sugars of Hordium vulgare L. plants grown under salinity stress. *Int J Sci Res* 2014;**3**(12):2208–14.
25. Li N, Wang J, Shi Y. Regulation of salt tolerance by uniconazole (S3307) on Petunia hybrida seedlings. *J Shenyang Agric Univ* 2011;**42**(6):668–71.
26. Zhao J, Feng N, Wang X, Cai G, Cao M, Zheng D, et al. Uniconazole confers chilling stress tolerance in soybean (Glycine max L.) by modulating photosynthesis, photoinhibition, and activating oxygen metabolism system. *Photosynthetica* 2019;**57**(2):446–57.
27. Lima GMdS, Pereira MCT, Oliveira MB, Nietsche S, Mizobutsi GP, Públio Filho WM, et al. Floral induction management in 'Palmer' mango using uniconazole. *Cienc Rural* 2016;**46**(8):1350–6.
28. Child R, Evans D, Allen J, Arnold G. Growth responses in oilseed rape (Brassica napus L.) to combined applications of the triazole chemicals triapenthenol and tebuconazole and interactions with gibberellin. *Plant Growth Regul* 1993;**13**(2):203–12.
29. Mohsin SM, Hasanuzzaman M, Bhuyan M, Parvin K, Fujita M. Exogenous tebuconazole and trifloxystrobin regulates reactive oxygen species metabolism toward mitigating salt-induced damages in cucumber seedling. *Plants* 2019;**8**(10):428.
30. Arivalagan M, Somasundaram R. Alteration of photosynthetic pigments and antioxidant systems in tomato under drought with tebuconazole and hexaconazole applications. *J SciAgric* 2017;146–57.
31. Rajasekar M, Rabert GA, Manivannan P. Triazole induced changes on biochemical and antioxidant metabolism of Zea mays L.(Maize) under drought stress. *J Plant Stress Physiol* 2015;35–42.
32. Robert GA, Rajasekar M, Manivannan P. Triazole-induced drought stress amelioration on growth, yield, and pigments composition of Helianthus annuus L.(sunflower). *Int Multidiscip Res J* 2015;**5**:6–15.
33. Shishatskaya E, Menzyanova N, Zhila N, Prudnikova S, Volova T, Thomas S. Toxic effects of the fungicide tebuconazole on the root system of fusarium-infected wheat plants. *Plant Physiol Biochem* 2018;**132**:400–7.
34. Zhang C, Wang Q, Zhang B, Zhang F, Liu P, Zhou S, et al. Hormonal and enzymatic responses of maize seedlings to chilling stress as affected by triazoles seed treatments. *Plant Physiol Biochem* 2020;**148**:220–7.
35. Rogach V, Voytenko L, Shcherbatiuk M, Kosakivska I, Rogach T. Morphogenesis, pigment content, phytohormones and productivity of eggplants under the action of gibberellin and tebuconazole. *Regul Mech Biosyst* 2020;**11**(1):116–22.

36. Mohsin SM, Hasanuzzaman M, Nahar K, Hossain MS, Bhuyan M, Parvin K, et al. Tebuconazole and trifloxystrobin regulate the physiology, antioxidant defense and methylglyoxal detoxification systems in conferring salt stress tolerance in Triticum aestivum L. *Physiol Mol Biol Plants* 2020;**26**:1139–54.
37. Rajasekar M, Rabert GA, Manivannan P. The effect of triazole induced photosynthetic pigments and biochemical constituents of Zea mays L. (Maize) under drought stress. *Appl Nanosci* 2016;**6**(5):727–35.
38. Cao L, Zhang H, Cao C, Zhang J, Li F, Huang Q. Quaternized chitosan-capped mesoporous silica nanoparticles as nanocarriers for controlled pesticide release. *Nanomaterials* 2016;**6**(7):126.
39. Talreja N, Chauhan D, Rodríguez CA, Mera AC, Ashfaq M. Nanocarriers: an emerging tool for micronutrient delivery in plants. In: *Plant micronutrients.* Springer; 2020. p. 373–87.
40. Wani TA, Masoodi F, Baba WN, Ahmad M, Rahmanian N, Jafari SM. Nanoencapsulation of agrochemicals, fertilizers, and pesticides for improved plant production. In: *Advances in phytonanotechnology.* Elsevier; 2019. p. 279–98.
41. Chariou PL, Ortega-Rivera OA, Steinmetz NF. Nanocarriers for the delivery of medical, veterinary, and agricultural active ingredients. *ACS Nano* 2020;**14**(3):2678–701.
42. Maluin FN, Hussein MZ, Yusof NA, Fakurazi S, Idris AS, Zainol Hilmi NH, et al. Preparation of chitosan–hexaconazole nanoparticles as fungicide nanodelivery system for combating Ganoderma disease in oil palm. *Molecules* 2019;**24**(13):2498.
43. Campos EVR, de Oliveira JL, Fraceto LF, Singh B. Polysaccharides as safer release systems for agrochemicals. *Agron Sustain Dev* 2015;**35**(1):47–66.
44. Abdelrahman TM, Qin X, Li D, Senosy IA, Mmby M, Wan H, et al. Pectinase-responsive carriers based on mesoporous silica nanoparticles for improving the translocation and fungicidal activity of prochloraz in rice plants. *Chem Eng J* 2020; 126440.
45. Shakiba S, Astete CE, Paudel S, Sabliov CM, Rodrigues DF, Louie SM. Emerging investigator series: polymeric nanocarriers for agricultural applications: synthesis, characterization, and environmental and biological interactions. *Environ Sci Nano* 2020;**7**(1):37–67.
46. Campos EVR, De Oliveira JL, Da Silva CMG, Pascoli M, Pasquoto T, Lima R, et al. Polymeric and solid lipid nanoparticles for sustained release of carbendazim and tebuconazole in agricultural applications. *Sci Rep* 2015;**5**(1):1–14.
47. Méndez FB, Sánchez DAM, Rangel DS, Landa IB, Haas JBR, Villanueva JLM, et al. Propiconazole nanoencapsulation in biodegradable polymers to obtain pesticide controlled delivery systems. *J Mex Chem Soc* 2019;**63**(1).
48. Paramera EI, Konteles SJ, Karathanos VT. Microencapsulation of curcumin in cells of Saccharomyces cerevisiae. *Food Chem* 2011;**125**(3):892–902.
49. Shi G, Rao L, Yu H, Xiang H, Yang H, Ji R. Stabilization and encapsulation of photosensitive resveratrol within yeast cell. *Int J Pharm* 2008;**349**(1–2):83–93.
50. Zhang B, Zhang T, Wang Q, Ren T. Microorganism-based monodisperse microcapsules: encapsulation of the fungicide tebuconazole and its controlled release properties. *RSC Adv* 2015;**5**(32):25164–70.
51. Lazaro IA, Forgan RS. Application of zirconium MOFs in drug delivery and biomedicine. *Coord Chem Rev* 2019;**380**:230–59.
52. Chansi RB, Hadwani K, Basu T. Role of metal–organic framework (MOF) for pesticide sensing. *Nanosci Sustain Agric* 2019;75.
53. Wu K, Du C, Ma F, Shen Y, Liang D, Zhou J. Degradation of metal-organic framework materials as controlled-release fertilizers in crop fields. *Polymers* 2019;**11**(6):947.
54. Tang J, Ding G, Niu J, Zhang W, Tang G, Liang Y, et al. Preparation and characterization of tebuconazole metal-organic framework-based microcapsules with dual-microbicidal activity. *Chem Eng J* 2019;**359**:225–32.

55. Sanzari I, Leone A, Ambrosone A. Nanotechnology in plant science: to make a long story short. *Front Bioeng Biotechnol* 2019;**7**:120.
56. Volova T, Zhila N, Vinogradova O, Shumilova A, Prudnikova S, Shishatskaya E. Characterization of biodegradable poly-3-hydroxybutyrate films and pellets loaded with the fungicide tebuconazole. *Environ Sci Pollut Res* 2016;**23**(6):5243–54.
57. Volova T, Prudnikova S, Boyandin A, Zhila N, Kiselev E, Shumilova A, et al. Constructing slow-release fungicide formulations based on poly (3-hydroxybutyrate) and natural materials as a degradable matrix. *J Agric Food Chem* 2019;**67**(33):9220–31.
58. Ito S, Umehara M, Hanada A, Kitahata N, Hayase H, Yamaguchi S, et al. Effects of triazole derivatives on strigolactone levels and growth retardation in rice. *PLoS One* 2011;**6**(7):e21723.
59. Kawada K, Takahashi I, Arai M, Sasaki Y, Asami T, Yajima S, et al. Synthesis and biological evaluation of novel triazole derivatives as strigolactone biosynthesis inhibitors. *J Agric Food Chem* 2019;**67**(22):6143–9.
60. Kong C-C, Ren C-G, Li R-Z, Xie Z-H, Wang J-P. Hydrogen peroxide and strigolactones signaling are involved in alleviation of salt stress induced by arbuscular mycorrhizal fungus in Sesbania cannabina seedlings. *J Plant Growth Regul* 2017;**36**(3):734–42.
61. Nakamura H, Hirabayashi K, Miyakawa T, Kikuzato K, Hu W, Xu Y, et al. Triazole ureas covalently bind to strigolactone receptor and antagonize strigolactone responses. *Mol Plant* 2019;**12**(1):44–58.
62. Zhi X-y, Jiang L-y, Li T, Song L-l, Wu L-j, Cao H, et al. Natural product-based semisynthesis and biological evaluation of thiol/amino-Michael adducts of xanthatin derived from Xanthium strumarium as potential pesticidal agents. *Bioorg Chem* 2020;, 103696.
63. Loiseleur O. Natural products in the discovery of agrochemicals. *CHIMIA Int J Chem* 2017;**71**(12):810–22.
64. Vurro M, Miguel-Rojas C, Pérez-de-Luque A. Safe nanotechnologies for increasing the effectiveness of environmentally friendly natural agrochemicals. *Pest Manag Sci* 2019;**75**(9):2403–12.
65. Sparks TC, Hahn DR, Garizi NV. Natural products, their derivatives, mimics and synthetic equivalents: role in agrochemical discovery. *Pest Manag Sci* 2017;**73**(4):700–15.
66. Nejma AB, Znati M, Daich A, Othman M, Lawson AM, Jannet HB. Design and semisynthesis of new herbicide as 1, 2, 3-triazole derivatives of the natural maslinic acid. *Steroids* 2018;**138**:102–7.

CHAPTER 8

Triazoles in Peptidomimetics: A Recent Update

Tahir Farooq*
Department of Applied Chemistry, Government College University, Faisalabad, Pakistan
*Corresponding author. E-mail: tahirfarooqfsd@gmail.com

Introduction

The click triazole-based peptidomimetics have emerged as a leading approach for the preparation of a variety of structurally stable peptidomimetics with enhanced biopotential through the selective access to 1,4- and 1,5-disubstituted 1,2,3-triazoles via Cu- and Ru-catalyzed Huisgen cycloadditions. The benign reaction conditions tolerate a variety of substituents, peptides, and proteins. The structural and electronic similarity between peptide bond and 1,2,3-triazole has made this moiety a privileged functionality for applications in peptidomimetics.[1] In fact, strong dipole, planner nature, hydrogen bond donating, and accepting properties has made 1,4- and 1,5-disubstituted 1,2,3-triazoles analogues to the peptide bond. Furthermore, the metabolic and proteolytic stability of triazoles has represented them as a better alternative for amide bond in peptidomimetics.[2] The precursor azide and alkyne functionality could easily be incorporated into amino acids to achieve corresponding acyclic and cyclic peptidomimetics (Figure 1).[3, 4] However, site-specific incorporation of these functionalities into proteins and peptides often becomes challenging. Over the last decade, a number of approaches have been developed for an easy, controlled, and site-specific incorporation of azido and alkyne moieties.

The peptides and proteins are involved in a number of vital structural and functional processes in living organisms, thus they have become key targets for drug designing and pharmaceutical development.[5] However, their direct usage in research activities had often been encumbered due to their sensitivity toward metabolic enzymes. The unprecedented advancements in triazole-based peptidomimetics have served well to settle such structural and conformational issues by providing the mimics with high stability and enhanced *in vitro* and *in vivo* bioactivities.[6]

Advances in Triazole Chemistry
https://doi.org/10.1016/B978-0-12-817113-4.00007-X

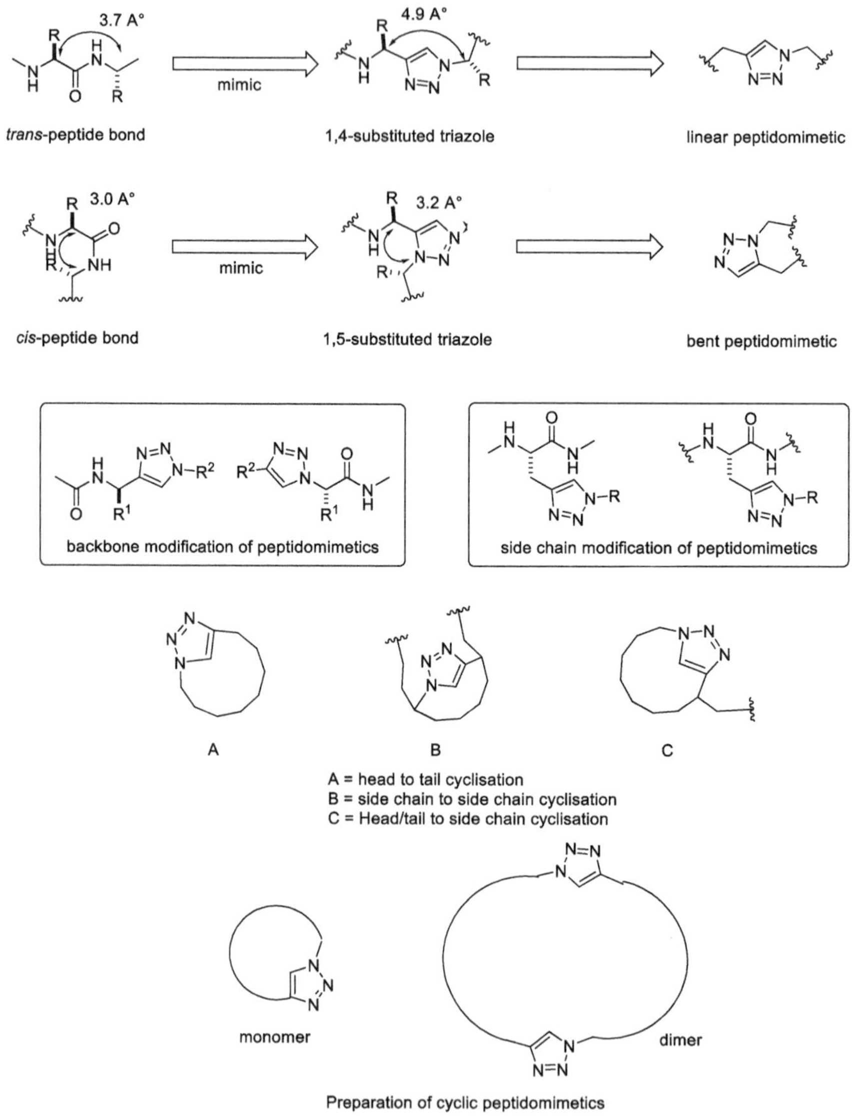

Figure 1 Known applications of triazoles as amide bond isosteres in peptidomimetics.

In this chapter, the latest developments in the field of 1,2,3-triazole-peptidomimetics have been presented to highlight the scope of this ever-advancing approach in protein science. The recent examples presented in subsequent sections show the use of triazole-peptidomimetics in the development of fluorescent probe, peptide-based inhibitors for neurodegenerative diseases, novel foldamers, and bioimaging agents.

1,2,3-Triazoles in peptidomimetics

Fluorogenic probes are special compounds employed for detectable interactions with biological targets, cell tracking, and bioimaging applications. Such molecules exhibit structural-based specific photophysical properties that change when they interact with *in vivo* or *in vitro* targets. This sort of change is exhibited as an emission that is subsequently used for target identification.[7,8] For their preparation, a labeling unit is linked with a signaling moiety using a suitable spacer. Over the years, the commercially available fluorescent probes have shown limited applications due to some serious issues like low protease stability, low cell permeability, and insufficient brightness. These problems arise due to highly labile functionalities of commonly used signaling units like cyanine, benzopyrans, and rhodamine. A replacement strategy has become very successful for the introduction of more stable isosteres in place of easily cleavable functionalities. The recent advancements in click chemistry have revolutionized the isostere-based replacement strategy because the protease-stable 1,2,3-triazole could be used to replace keto, ester, or amide functionality. In a recent report, Mohan *et al.* prepared coumarin-based fluorescent probes employing isostere substitution method. They used click reaction to link peptide fragments with coumarin scaffolds prepared through multicomponent reactions (MCRs). In fact, they combined click with MCR and prepared type 1 and type 2 chromene peptidomimetics (Figure 2).[9] The chromenes are privileged heterocyclic molecules displaying a wide spectrum of therapeutic potential, especially anti-HIV, anti-viral, and anti-cancer properties.[10] During the preparation, the azide-functionalized peptide was synthesized using Ugi- or Mannich-type MCRs, and a three-component reaction was used to prepare alkyne-functionalized chromene. The as-prepared peptidomimetics exhibited photophysical properties equivalent to commercially available fluorophores. Some of the promising chromene peptidomimetics were suggested as attractive candidates for flow cytometry and bioimaging applications. Soumya *et al.* also prepared a series of fluorescent coumarin-triazole-pyridine peptidomimetics employing MCR-click strategy (Figure 3).[11] They synthesized six azido-coumarins and two new alkyne-pyridines using MCR strategy. The prepared peptidomimetics with notable photophysical properties were suggested as suitable candidates for the development of inhibitors for new targets and novel bioimaging agents.

Peptides are capable of expressing a wide spectrum of biological potential; however, their poor pharmacokinetic profiles pose serious challenges to their successful applications, especially as oral drugs. They often experience

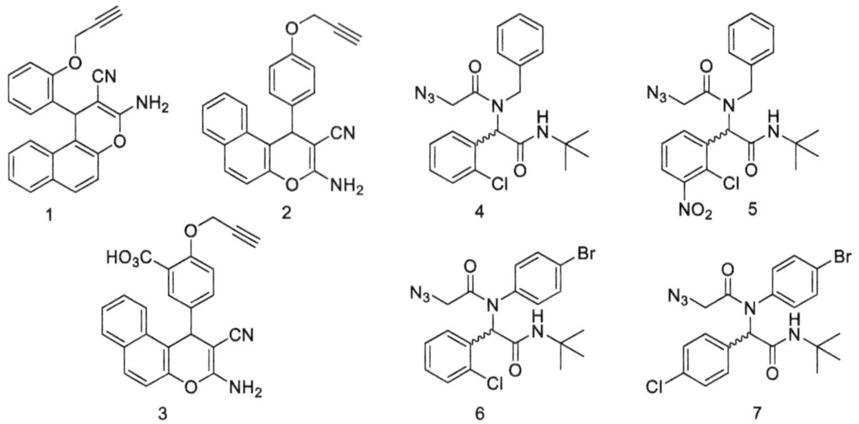

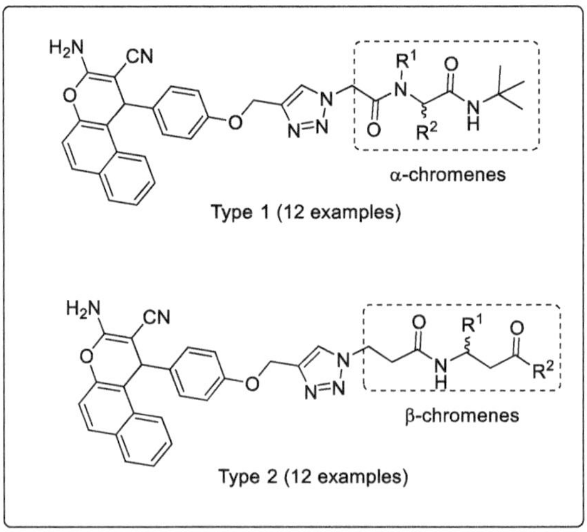

Figure 2 Type 1 and type 2 chromene peptidomimetics with fluorescence.

the issues of insufficient systemic exposure due to low permeability across biological membranes and high susceptibility to proteolytic cleavage. Over the years, such limitations have been encountered by the incorporation of non-structural amino acids or other functionalities onto peptide scaffolds. Such derivatizations improve their physiochemical properties and binding affinities thus inducing drug-like properties.[12] Coffey *et al.* considered click methodology as an effective approach for the preparation of peptidomimetics owing to its simplicity and high substrate tolerance. They aimed to incorporate aliphatic, aryl, and heteroaryl substituents on *N*-termini of privileged peptides. They planned to encounter the permeability issues by replacing *N*-terminal amide with 1,2,3-triazole. Accordingly, the

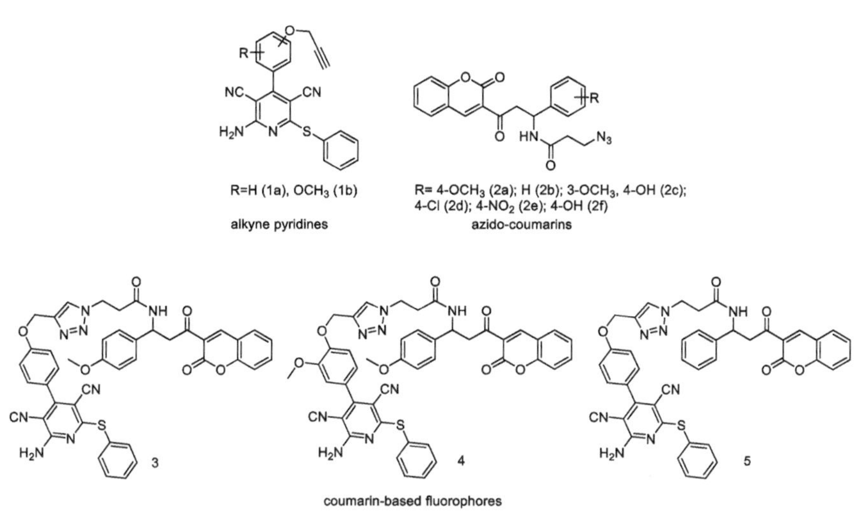

Figure 3 Fluorescent coumarin-triazole-pyridine peptidomimetics.

diversely substituted azides were prepared employing *in situ* method to diversify the synthesis of peptidomimetics. The preparation of *N*1-substituted triazole-peptidomimetics was managed through a one-pot two-step solution phase click chemistry approach showing tolerance for peptide functionalities (Figure 4).[13] The protocol was suggested as general with applications to modify the C-termini for diversification of peptidomimetic synthesis.

The human neuronal synapses contain a scaffolding protein called postsynaptic density protein-95 (PSD-95) with three postsynaptic density protein-95/disks-large/zonula occludens-1 (PDZ) domains. These domains mediate protein-protein interactions in cell signaling thus are abundant in human proteome for highly important cellular functions.[14] They interact with nitric oxide synthase, the neuronal signaling protein, and a receptor called *N*-methyl-D-aspartate (NMDA).[15] Over the years, PSD-95 has become a promising drug target for the development of peptide-based inhibitors against neurodegenerative diseases and ischemic stroke. It has been established that the pentapeptides and longer peptides exhibit similar affinities while targeting PSD-95. The stability and affinity of tetrapeptides has

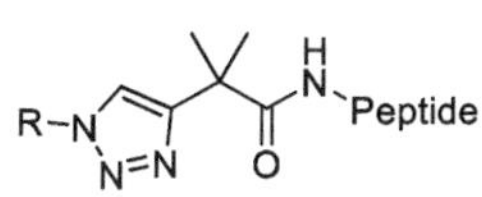

Figure 4 *N*1-substituted triazole-peptidomimetics.

been increased by replacing amides with thioamides and by *N*-alkylation, respectively. The affinities of hexapeptides had also been increased by cyclization of P1 side chain. Bach *et al.* improved the stability, selectivity, and affinity of short PSD-95 peptide ligands by incorporating a rigid scaffold into a short PSD-95. The SAR-identified tripeptide Thr-Ala-Val (TAV), having medium affinity for binding with PDZ1 and PDZ2, was taken as a starting point to achieve small molecule-like compounds. Previously, protease-stable potent compounds have been prepared using triazole as a bioisostere of amide due to similarity of size and electronic properties. Thus, they prepared a series of 1,2,3-triazole-containing analogues of TAV from corresponding alkyne and azide analogues of amino acids and dipeptides of TAV. This strategy involved the incorporation of triazole in PSD-95 PDZ domain peptide inhibitors. They were able to prepare a promising novel peptidomimetic inhibitor with triazolic functionality (Figure 5).[16]

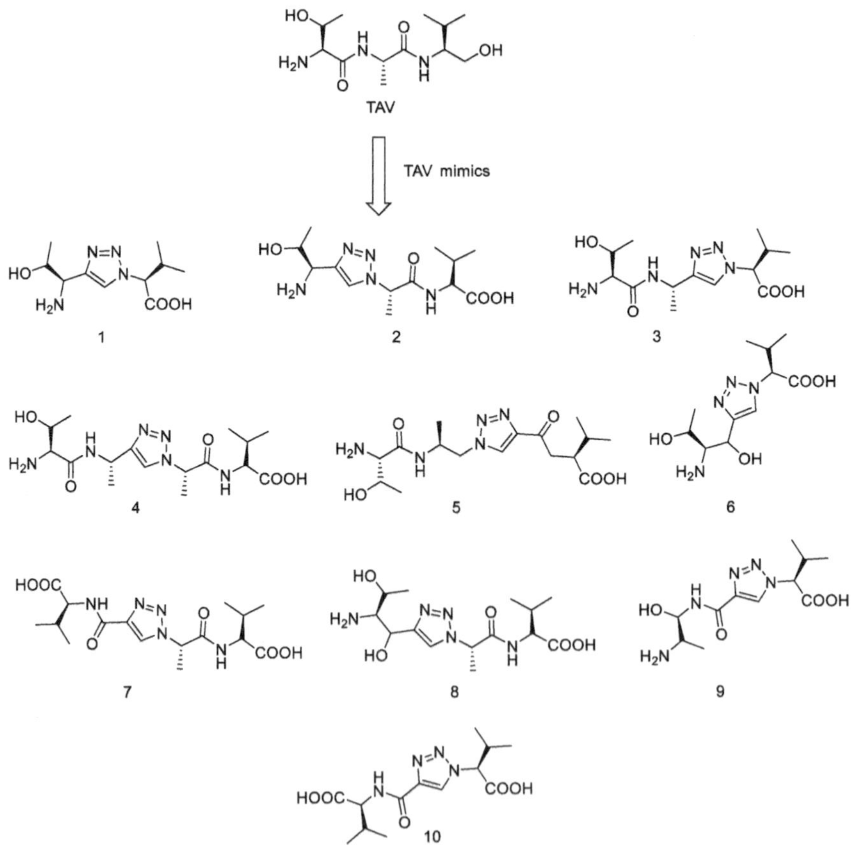

Figure 5 1,2,3-Triazole-containing analogues of TAV.

The well-defined structural conformations of peptides and proteins are important for biological functions; therefore, the incorporation of constrained units enhance their pharmacological efficacy. Over the last few years, the electronic and structural properties of triazole has greatly been exploited as bioisosteres of amide and peptide bond to induce secondary conformational attributes like α-helix and β-turns. Furthermore, the confirmation, lipophilicity, and acidity of molecules have also been modulated by replacing H-atoms with F for the improvement of their pharmacological profiles. These bioisostere-related attributes have been well attracted by the peptidomimetic-based drug development.[17, 18] In regard to these consideration, Mamone *et al.* decided to incorporate synthetic *N*-difluoromethyltriazolylamino acid into tetrapseudopeptide for studying the influence of CF_2N on fluorinated pseudopeptides. The tunable tetrapeptides were easily synthesized from azidodifluoroacetamides of amino acids and propargylamides using click chemistry approach (Figure 6).[19] The as-prepared non-fluorinated pseudotetrapeptides achieved β-turn conformation while fluorinated analogues exhibited extended conformation. It was concluded that the introduction of *N*-difluoromethyltriazole induced β-strand mimics for medicinal chemistry applications.

The disulfide bonds are vital for controlling the structural conformations, stability, and biological functions of peptides and proteins. The instability of disulfide bonds in reducing environment affects the *in vivo* activity and metabolic stability.[20] To encounter such issues, the disulfide bonds have

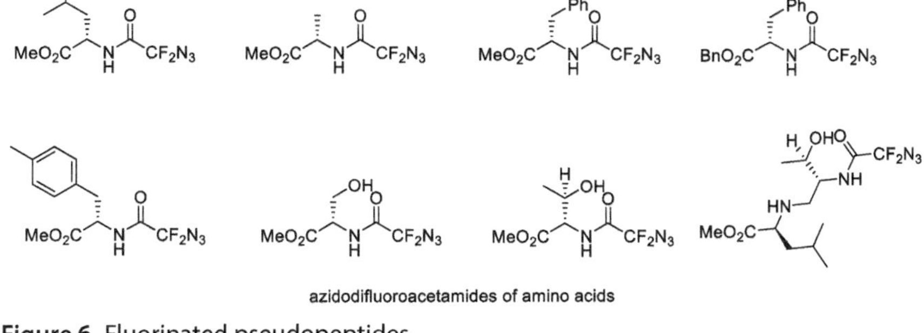

Figure 6 Fluorinated pseudopeptides.

been replaced with olefins, thioethers, lactam bridges, and diselenides; however, such replacements as mimics reduced the bioactivity of peptidomimetics. Furthermore, the introduction of such mimetic moieties involved complex chemical reactions thus are less attractive.[21] In a relatively recent report, Tala *et al.* employed click chemistry approach for the replacement of disulfide bridge with triazole and synthesized a series of three peptides as agouti-related protein (AGRP) chimeric peptide agonists at mouse melanocortin receptors (Figure 7).[22] The melanocortin system regulates a number of physiological functions like pigmentation, sexual functions, inflammation, feeding behavior, weight homeostasis, and energy. The endogenous agonist and antagonists regulate five melanocortin receptors that belong to superfamily of G protein-coupled receptors. The mouse melanocortin receptors were used to evaluate the pharmacological activity of as-prepared 1,2,3-triazole-bridged peptidomimetics. The triazole-carrying mimics showed almost the same activity on melanocortin receptors. Both the 1,5- and 1,4-disubstituted 1,2,3-triazoles exhibited selective interactions with receptors leading to the designing of melanocortin ligands with specific selectivity toward receptor subtypes. The as-prepared peptidomimetics were

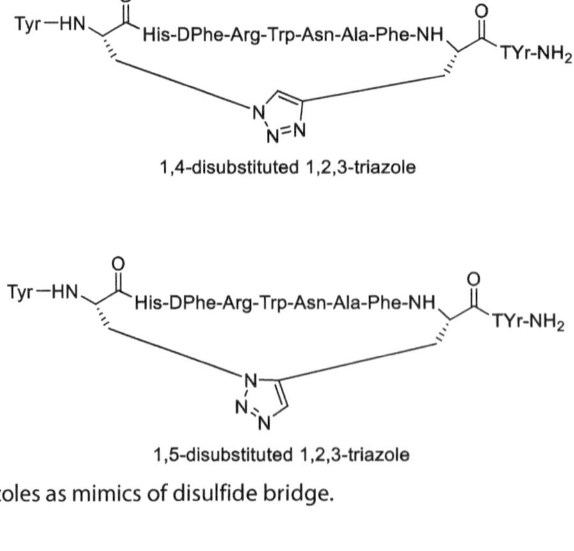

Figure 7 Triazoles as mimics of disulfide bridge.

suggested to be more stable due to the replacement of disulfide bridge with unnatural triazole having inherent redox stability. This synthesis of model peptidomimetics opened up a new dimension for the designing and preparation of novel chemical neuronal probes.

Over the several years, the conotoxin peptides have shown promising clinical options for the treatment of type 2 diabetes, hypertension, and neuropathic pain as they selectively target ion channels and receptors. However, their clinical applications are often limited due to redox instability and poor resistance to protease.[23,24] According to a very recent report, Knuhtsen *et al.* replaced disulfide bond with 1,5-disubstituted 1,2,3-triazole as its surrogate in conotoxin peptidomimetics employing the Ru-catalyzed azide-alkyne cycloaddition reactions (Figure 8).[25] The prepared mimic A exhibited an enhanced activity with relatively higher stability in blood plasma establishing the worth of triazole as an excellent replacement of degradable disulfide

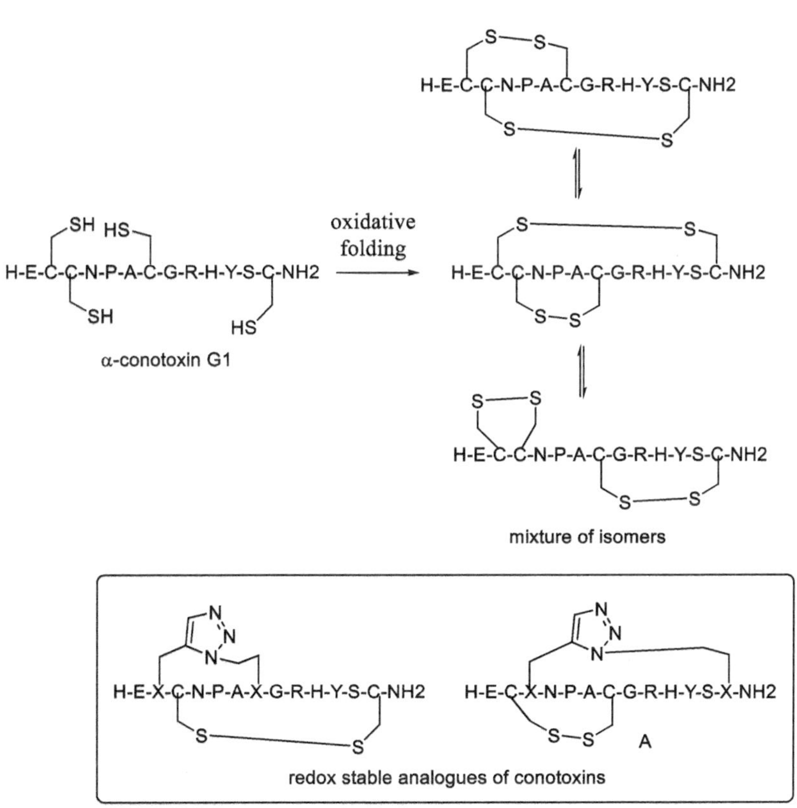

Figure 8 1,5-Disubstituted 1,2,3-triazole as isostere of disulfide bridge.

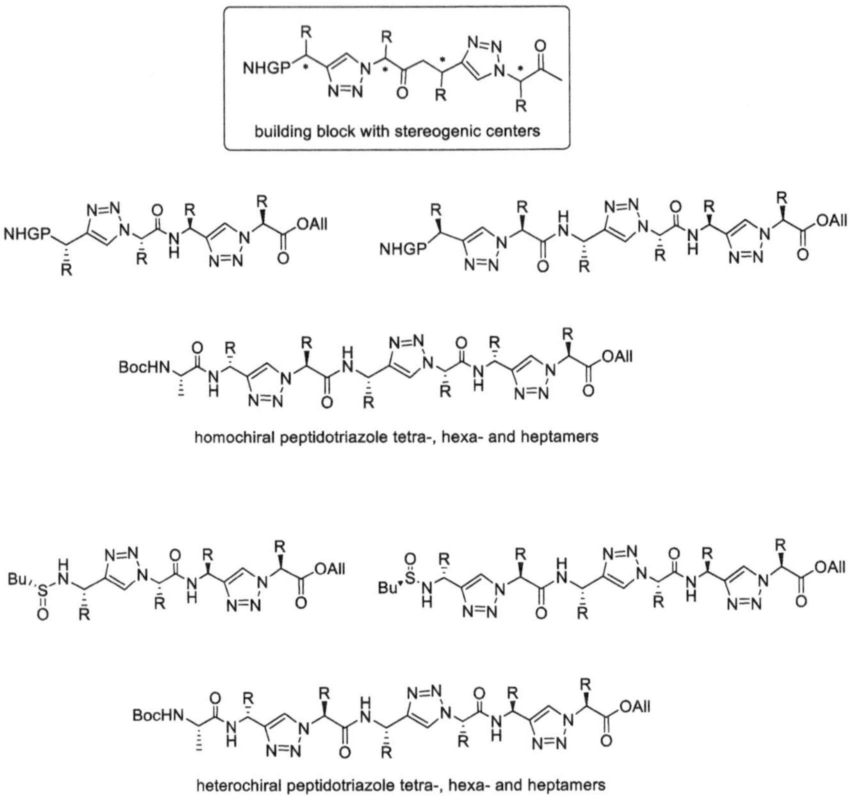

Figure 9 Homo- and heterochiral peptidotriazolamers.

bridge. The study was performed in human CN 21 cells thus providing a background stage for the designing of therapeutic options for humans.

The alternate presence of amide bonds and 1,4-disubstituted 1*H*-1,2,3-triazole provides features of peptides and triazolamers in hybrid structures called peptidotriazolamers. Last year, Schroder *et al.* used 1,4-disubstituted 1*H*-1,2,3-triazoles alternating to amide bonds for the preparation of a new class of peptidomimetics and studied their properties in solution (Figure 9).[26] The CuAAC approach was employed to develop building blocks from azido-esters and enantiomerically pure propargylamine with stereogenic centers. A series of homo- and heterochiral peptidotriazolamers were prepared and evaluated for their solvation and conformational properties. The heterochiral ones formed an S shape while homochiral compounds formed regular helical structures. However, the glycine-substituted oligomers displayed an unclear secondary structure.

The protein engineering and peptidomimetics improve the physiochemical features of peptides by incorporating synthetic, non-natural amino acids into peptides. Such insertions afford extra resistance against enzymatic and redox degradation thus inducing novelty in mimics. Over the last few decades, the foldamer development with non-natural amino acids has emerged as a leading peptidomimetic approach with a wide spectrum of applications.[27, 28] Over the last two decades, the field of foldamers have really been revolutionized due to the involvement of triazoles as bioisosteres of amide functionality in amino acids. The easy access to 1,5- and 1,4-disustituted triazole via Ru-catalyzed and Cu-catalyzed click reactions have made them attractive entities for the designing and development of peptidic foldamers with improved physiochemical properties. However, amino acid synthesis with chiral mom- and disubstituted triazole has not been well-exploited until recently. Stålsmeden *et al.* employed Ru-catalyzed click approach for the preparation of eight possible chiral derivatives of a triazole monomer (Figure 10).[29] All as-prepared monomers were able to form

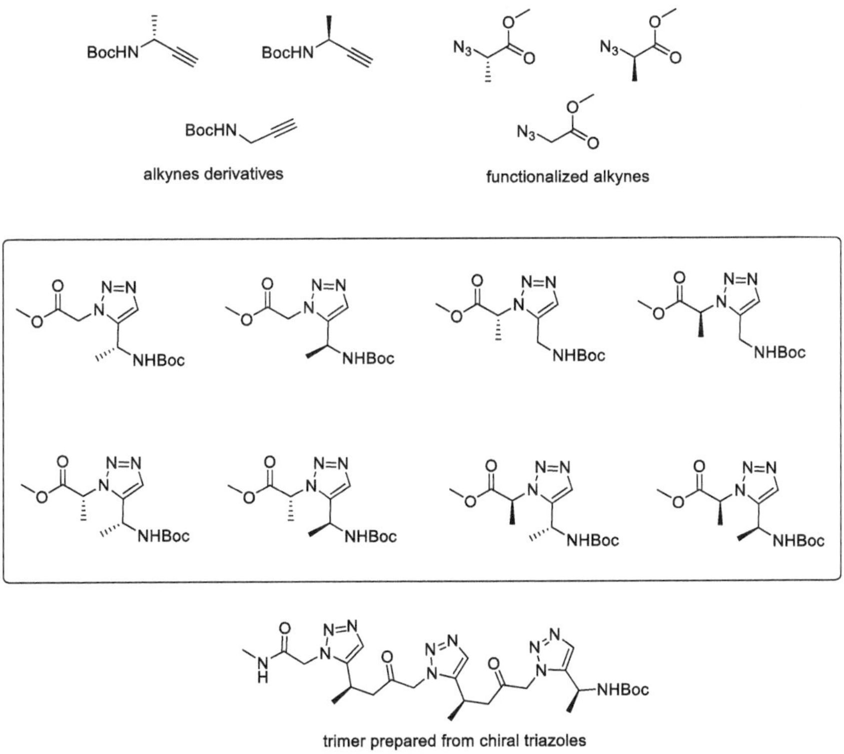

Figure 10 Chiral derivative of a triazole monomer.

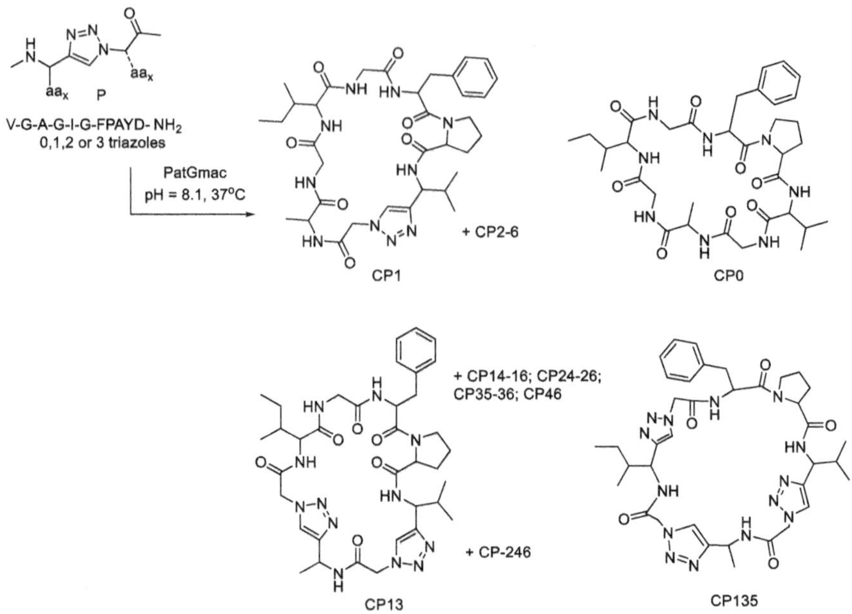

Figure 11 Cyclic peptides with triazoles.

several low energy conformers in a systematic quantum chemical study. This study opened up new options for the applications of chiral 1,5-disubstituted 1,2,3-triazoles for the development of novel non-natural foldamers with interesting properties valuable for biotechnology and pharmaceutics.

Linear peptides show significant improvement in their stability and biological activities as a result of macrocyclization. To this end, the enzyme-mediated macrocyclization is highly preferred over chemical methodologies. The known enzyme-based methodologies have been found active on peptides and proteins.[30,31] However, Oueis *et al.* presented an enzyme-mediated methodology for macrocyclizing peptides incorporated with triazoles (Figure 11).[32] As described earlier, the physiochemical attributes and bioactive potential of macrocyclic peptides is tuned with the introduction of non-peptidic scaffolds like triazoles. The enzymes of the cyanobactin family (PatGmac) were found effective for the cyclization of triazole-containing peptides. Furthermore, the study provided a platform for the preparation of macrocycle peptidomimetics with chemical diversity.

Conclusion

For a normal file, the peptides and proteins are vital for a number of functional processes, thus they have become key targets for drug designing and development. However, their sensitivity toward metabolic enzymes restricts their direct applications in research activities. The recent advancements in triazole-based peptidomimetics provide mimics with high stability and enhanced bioactivities. The recent examples of triazole-peptidomimetics for the development of fluorescent probe, peptide-based inhibitors for neurodegenerative diseases, novel foldamers, and bioimaging agents highlight it as an emerging approach in the field of protein science.

References

1. Kracker O, Góra J, Krzciuk-Gula J, Marion A, Neumann B, Stammler HG, et al. 1, 5-disubstituted 1, 2, 3-triazole-containing peptidotriazolamers: design principles for a class of versatile peptidomimetics. *Chem A Eur J* 2018;**24**(4):953–61.
2. Diness F, Schoffelen S, Meldal M. Advances in merging triazoles with peptides and proteins. In: *Peptidomimetics I.* Springer; 2015. p. 267–304.
3. Pedersen DS, Abell A. 1, 2, 3-Triazoles in peptidomimetic chemistry. *Eur J Org Chem* 2011;**2011**(13):2399–411.
4. Angell YL, Burgess K. Peptidomimetics via copper-catalyzed azide–alkyne cycloadditions. *Chem Soc Rev* 2007;**36**(10):1674–89.
5. Mabonga L, Kappo AP. Peptidomimetics: a synthetic tool for inhibiting protein–protein interactions in cancer. *Int J Pept Res Ther* 2020;**26**(1):225–41.
6. Piras M, Testa A, Fleming IN, Dall'Angelo S, Andriu A, Menta S, et al. High affinity "click" RGD peptidomimetics as radiolabeled probes for imaging αvβ3 integrin. *ChemMedChem* 2017;**12**:1142–51.
7. Zhang J, Chai X, He X-P, Kim H-J, Yoon J, Tian H. Fluorogenic probes for disease-relevant enzymes. *Chem Soc Rev* 2019;**48**(2):683–722.
8. Chyan W, Raines RT. Enzyme-activated fluorogenic probes for live-cell and in vivo imaging. *ACS Chem Biol* 2018;**13**(7):1810–23.
9. Mohan TJ, Bahulayan D. Design, synthesis and fluorescence property evaluation of blue emitting triazole-linked chromene peptidomimetics. *Mol Divers* 2017;**21**(3):585–96.
10. Costa M, Dias TA, Brito A, Proenca F. Biological importance of structurally diversified chromenes. *Eur J Med Chem* 2016;**123**:487–507.
11. Soumya T, Ajmal CM, Bahulayan D. Synthesis of bioactive and fluorescent pyridine-triazole-coumarin peptidomimetics through sequential click-multicomponent reactions. *Bioorg Med Chem Lett* 2017;**27**(3):450–5.
12. Boutureira O, Bernardes GJL. Advances in chemical protein modification. *Chem Rev* 2015;**115**(5):2174–95.
13. Coffey SB, Aspnes G, Londregan AT. Expedient synthesis of N1-substituted triazole peptidomimetics. *ACS Comb Sci* 2015;**17**(12):706–9.
14. Chi CN, Bach A, Strømgaard K, Gianni S, Jemth P. Ligand binding by PDZ domains. *Biofactors* 2012;**38**(5):338–48.
15. Bach A, Eildal JN, Stuhr-Hansen N, Deeskamp R, Gottschalk M, Pedersen SW, et al. Cell-permeable and plasma-stable peptidomimetic inhibitors of the postsynaptic density-95/N-methyl-D-aspartate receptor interaction. *J Med Chem* 2011;**54**(5):1333–46.
16. Bach A, Pedersen TB, Strømgaard K. Design and synthesis of triazole-based peptidomimetics of a PSD-95 PDZ domain inhibitor. *Med Chem Commun* 2016;**7**(3):531–6.
17. Mohammed I, Kummetha IR, Singh G, Sharova N, Lichinchi G, Dang J, et al. 1, 2, 3-triazoles as amide bioisosteres: discovery of a new class of potent HIV-1 Vif antagonists. *J Med Chem* 2016;**59**(16):7677–82.
18. Bonandi E, Christodoulou MS, Fumagalli G, Perdicchia D, Rastelli G, Passarella D. The 1, 2, 3-triazole ring as a bioisostere in medicinal chemistry. *Drug Discov Today* 2017;**22**(10):1572–81.

19. Mamone M, Gonçalves R, Blanchard F, Bernadat G, Ongeri S, Milcent T, et al. N-Difluoromethyl-triazole as a constrained scaffold in peptidomimetics. *Chem Commun* 2017;**53**(36):5024–7.
20. Bulaj G. Formation of disulfide bonds in proteins and peptides. *Biotechnol Adv* 2005;**23**(1):87–92.
21. Gori A, Gagni P, Rinaldi S. Disulfide bond mimetics: strategies and challenges. *Chem A Eur J* 2017;**23**(60):14987–95.
22. Tala SR, Singh A, Lensing CJ, Schnell SM, Freeman KT, Rocca JR, et al. 1, 2, 3-Triazole rings as a disulfide bond mimetic in chimeric AGRP-melanocortin peptides: design, synthesis, and functional characterization. *ACS Chem Nerosci* 2017;**9**(5):1001–13.
23. Mir R, Karim S, Amjad Kamal M, Wilson CM, Mirza Z. Conotoxins: structure, therapeutic potential and pharmacological applications. *Curr Pharm Des* 2016;**22**(5):582–9.
24. Wilson DT, Bansal PS, Carter DA, Vetter I, Nicke A, Dutertre S, et al. Characterisation of a novel A-superfamily conotoxin. *Biomedicine* 2020;**8**(5):128.
25. Knuhtsen A, Whitmore C, McWhinnie FS, McDougall L, Whiting R, Smith BO, et al. α-Conotoxin GI triazole-peptidomimetics: potent and stable blockers of a human acetylcholine receptor. *Chem Sci* 2019;**10**(6):1671–6.
26. Schröder DC, Kracker O, Fröhr T, Góra J, Jewginski M, Nieß A, et al. 1, 4-Disubstituted 1H-1, 2, 3-triazole containing peptidotriazolamers: a new class of peptidomimetics with interesting foldamer properties. *Front Chem* 2019;**7**:155.
27. Jongkees SA, Hipolito CJ, Rogers JM, Suga H. Model foldamers: applications and structures of stable macrocyclic peptides identified using in vitro selection. *New J Chem* 2015;**39**(5):3197–207.
28. Lee M, Shim J, Kang P, Choi M-G, Choi SH. Stabilization of 11/9-helical α/β-peptide foldamers in protic solvents. *Chem Commun* 2016;**52**(35):5950–2.
29. Stålsmeden AS, Paterson AJ, Szigyártó IC, Thunberg L, Johansson JR, Beke-Somfai T, et al. Chiral 1, 5-disubstituted 1, 2, 3-triazoles–versatile tools for foldamers and peptidomimetic applications. *Org Biomol Chem* 2020;**18**(10):1957–67.
30. Schmidt M, Toplak A, Quaedflieg PJ, van Maarseveen JH, Nuijens T. Enzyme-catalyzed peptide cyclization. *Drug Discov Today Technol* 2017;**26**:11–6.
31. Houssen WE. Peptide cyclization catalyzed by cyanobactin macrocyclases. In: *Enzyme-mediated ligation methods.* Springer; 2019. p. 193–210.
32. Oueis E, Jaspars M, Westwood NJ, Naismith JH. Enzymatic macrocyclization of 1, 2, 3-triazole peptide mimetics. *Angew Chem* 2016;**128**(19):5936–9.

CHAPTER 9

Triazoles in Coordination Complexes

Tahir Farooq*
Department of Applied Chemistry, Government College University, Faisalabad, Pakistan
*Corresponding author. E-mail: tahirfarooqfsd@gmail.com

Introduction

The diverse prevalence of triazolic moiety across a number of scientific domains inspired coordination chemists to use them as promising ligands for the development of triazole-based metal complexes. Both 1,2,4-triazole and 1,2,3-triazole show high biding affinity toward metals under the influence of nature of substituents and protonation of triazolic ring. These *N*-heterocycles with different metal-binding modes have further been exploited for the preparation of metal-organic frameworks (MOFs) and coordination polymers.[1,2]

The presence of three nitrogen-donor atoms capacitates the triazole rings to act as polydentate ligands, and the heterocycles may act as bridging ligands. The deprotonated 1,2,4-triazolate could coordinate two or three metal ions while protonated 1,2,4-triazole functions as a dinucleating ligand. Thus, five different coordination modes could be adopted by a 1,2,4-triazole ring for bridging metal ions (Figure 1).[1] These coordination possibilities have really been exploited for the designing and preparation of a large variety of organic/inorganic hybrid architectures. Similarly, five different coordination modes could be opted by the 1,2,3-triazoles for bridging with metal ions. The deprotonated ring could function as a trinucleating or dinucleating ligand while its protonated form acts as a dinucleating ligand. The CuAAC protocol show tolerance to a variety of functional groups thus allowing the incorporation of other donor moieties creating possibilities for the preparation of a variety of chelating ligands. Furthermore, the high polarizable C_5–H could be cleaved to generate anionic triazolides. The 1,3,4-trisubstituted 1,2,3-triazolium salts are prepared by the selective alkylation at N(3), which further activates the C_5–H bond furnishing easy access to abnormal mesoionic *N*-heterocyclic carbene (NHC) ligands.[3,4]

Advances in Triazole Chemistry
https://doi.org/10.1016/B978-0-12-817113-4.00001-9

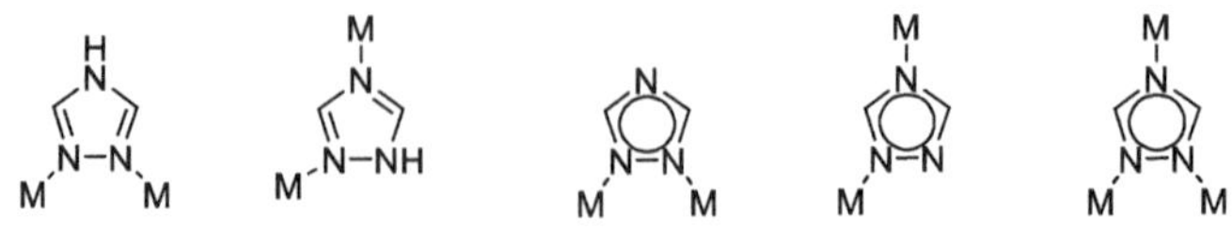

Bridging coordination modes of 1,2,4- triazole with metal (M)

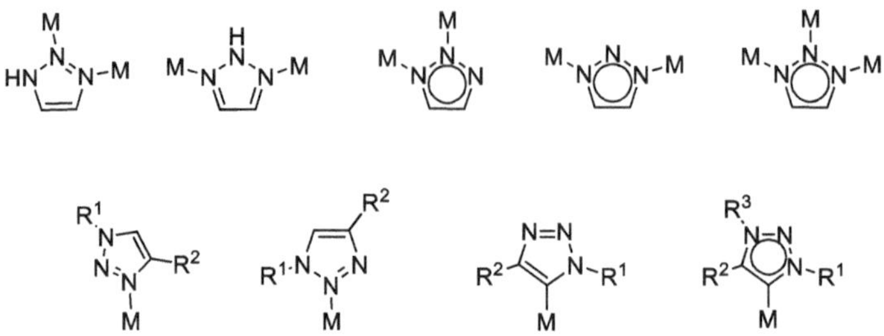

Bridging coordination modes of 1,2,3- triazole with metal (M)

Figure 1 Known coordination modes of triazoles with metal.

The late transition metals like Pt, Pd, Ir, Ru, and Re have frequently been used for the preparation of various types of complexes using triazole as ligand for a wide spectrum of applications including electroluminescent devices, dye-sensitized solar cells, nanosensors, anticancer therapeutics, and catalysis, etc. The efficient ligand designing rendered the tuning of electronic and photophysical properties thus widening their scope of applications. Until recently, the pyridyl moiety remained one of the promising donor motifs available for incorporation in chelate ligands. Over the last decade, the triazole-based ligands have become an attractive choice for the development of metal complexes due to their aforesaid electronic and structural features (Figure 2).

The subsequent sections of this chapter do highlight the recent development of triazole-based metal complexes and their subsequent applications. To this end, a few specific examples have been mentioned to signify their role in catalysis, photoelectronics, and biological activity.

Triazole-based metal complexes and their applications in catalysis

1,2,4-Triazoles-based metal complexes as catalysts

Over the last few years, the Ni-based catalysis has attracted attention due its several accessible oxidation states, high nucleophilicity, and inexpensive availability. The homogenous catalysis has widely explored the NHC

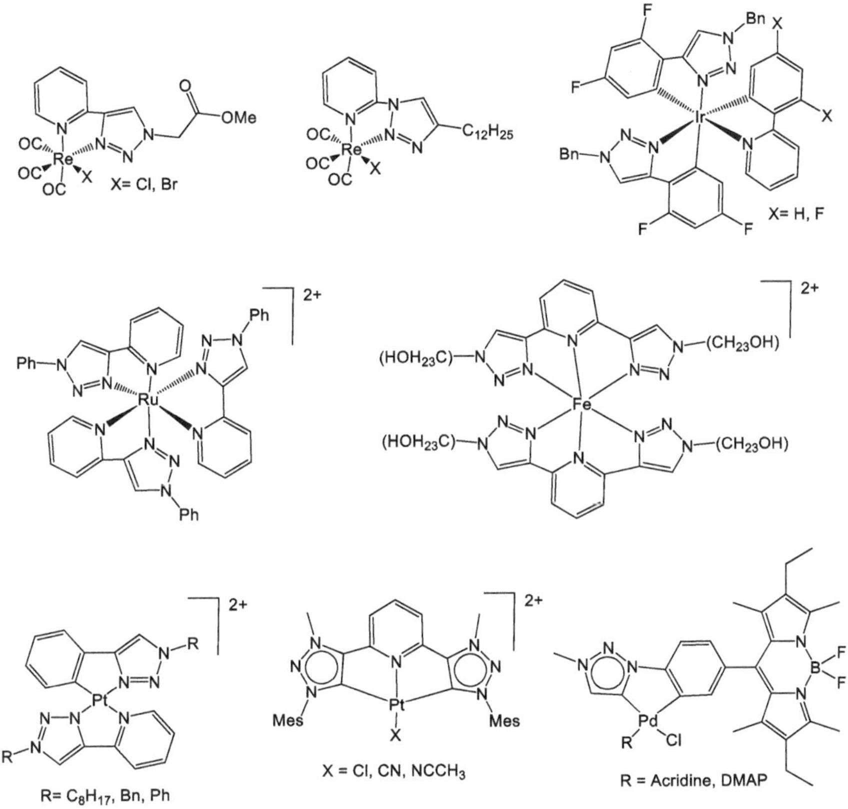

Figure 2 Different known triazole-based metal complexes.

ligands as an emerging choice over the last few years. Keeping these facts in mind, they were put together to prepare nickel NHC complexes. These complexes exhibited superior catalytic efficiency in hydroamination reactions compared with their palladium analogs.[5] The high value intermediates for Suzuki-Miyaura cross-coupling reactions and for other routine synthesis could be achieved by an important borylation reaction. Considering the importance of the reaction, Kumar *et al.* used amidofunctionalized NHC ligands of 1,2,4-triazoles for the preparation of a series of nickel complexes. In the presence of a base, $NiCl_2 \cdot 6H_2O$ and triazolium salts underwent direct reaction to provide corresponding complexes (Figure 3).[6] According to the DFT, the NHC-Ni σ-bonding interaction was found to be highly polar where the Ni atom showed minimum contribution (4%) and the NHC ligand showed maximum contribution (about 69%). A moderate catalytic activity was exhibited by the as-prepared complexes in a borylation reaction (Figure 4).

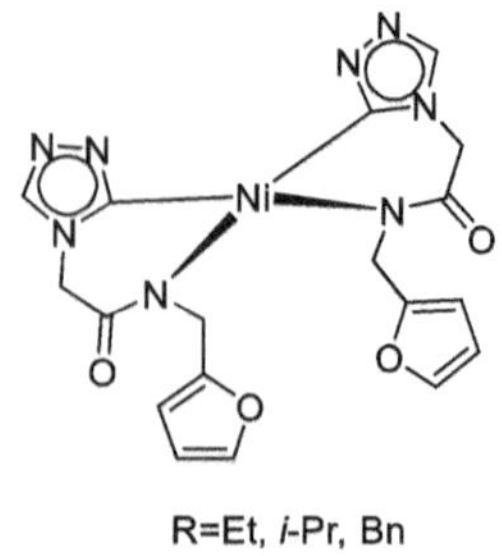

Figure 3 Ni-complexes from triazolium salts.

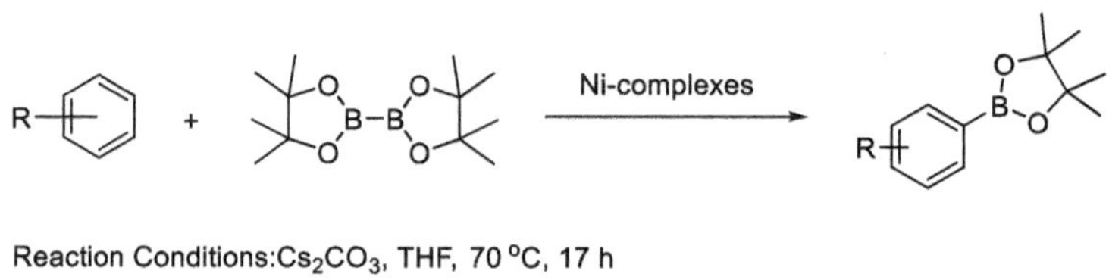

Figure 4 Use of Ni-complexes in borylation reaction.

Over the years, the NHC ligands have been found superior to the other types of ligands like cyclopentadienyl and phosphine ligands owing to the highly stable metal-NHC bond and widely variable steric and electronic parameters. Accordingly, the moisture-resistance complexes of Pd-halides with pyridine and NHC ligands as Pd-PEPPSI became quite common in use, especially when these complexes of benzimidazole and imidazole turned out to be more attractive.[7] Thus, the applications of 1,2,4-triazole-3-ylidene ligands for the preparation of corresponding Pd-PEPPSI complexes became an attractive domain yet remained unexplored until very recently. In 2018, Chernenko *et al.* used NHCs of 1,2,4-triazoles for the preparation of Pd-PEPPSI for the first time with good to excellent yield (Figure 5).[8] In pyridine, the 1,4-di-alkyl-1,2,4-triazolium salts were made to react with $PdCl_2$ using KI/KBr as a source of halide. The use of phase-transfer catalyst provided these complexes in a high yield. There was no difference in

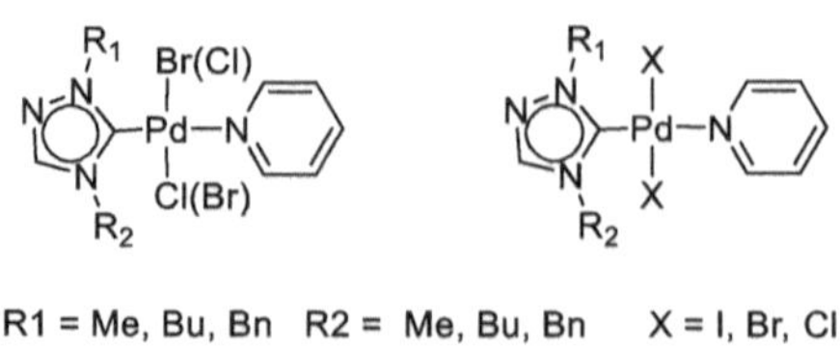

Figure 5 1,2,4-Triazole-based Pd-PEPPSI complexes.

catalytic efficiency in as-prepared complexes and commercially available Pd-PEPPSI in Suzuki-Miyaura cross-coupling reactions.

The Suzuki-Miyaura reaction has emerged as a leading methodology with a wide spectrum of applications in the field of functional materials, pesticides, and drug synthesis.[9] This methodology usually utilizes Pd or its complexes with various ligands as catalysts. As described earlier, Ni-based complexes have been well explored as efficient and cheaper alternatives over the last few years. In this regard, the Ni-NHC complexes display a high level of catalytic efficiency. However, their synthesis usually requires multistep reactions making their preparation laborious, low-yielding, and thus uneconomical. In last few years, the Ni-bis-NHC complexes have found considerable attention as their synthesis involved a single-step reaction between azolium salts and Ni-salts.[10] To this end, very recently, Soliev *et al.* used 1,2,4-triazoles, benzimidazole, and imidazoles for the preparation of NCH complexes of Pd and Ni (Figure 6).[11] Furthermore, the as-prepared Ni/Pd-NCH complexes were utilized in cross-coupling reaction of arylboronic acids with aryl halides for a comparative evaluation of their catalytic efficiency. Consequently, the $(NHC)_2NiX_2$ and Ni-bis-NHC complexes exhibited catalytic efficiency equally well in Suzuki-Miyaura reactions compared with the similar Pd-based complexes.

1,2,3-Triazoles-based metal complexes as catalysts

Over the last few years, the triazole ligand has become an attractive, effective, and powerful tool in the field of catalysis for rapid introduction of molecular complexity. In fact, it has caused a paradigm shift in catalysis influencing the lab-based synthesis as well as reaction methodologies at industrial scale productions. In this context, the triazole-ligand-based complexes

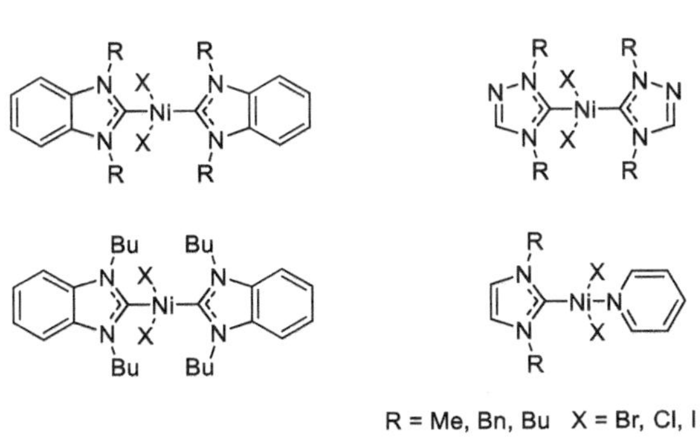

Figure 6 Preparation of NCH complexes of Pd and Ni.

as catalysts have also been anticipated as promising candidates for green chemistry, atom economy, and atom-efficient reactions.

Over the last few years, hydrogen-borrowing reaction have emerged as an environmentally benign and economical methodology for the preparation of densely substituted molecules.[12] These reactions show high atom economy with water as the only by-product. In connection to the hydrogen-borrowing reactions, Wu *et al.* presented a facile synthesis of highly stable pyridyltriazole-ligand ruthenium (II) complexes (Figure 7).[13] The as-prepared complexes were employed as catalysts in hydrogen-borrowing reactions for selective alkylation and one-pot C–H hydroxylation of 2-oxindole with various primary alcohols. This methodology presented cheaper access to novel catalysts for hydrogen-borrowing reactions affording excellent yields of quaternary α-hydroxy carbonyl compound, 3-functionalized-3-hydroxy-2-oxindoles, and 3-functionalized-2-oxindoles.

Over the last few decades, the multicomponent reactions have received special attention in synthetic organic chemistry due to its ecofriendly nature owing to its potential to construct multiple bonds in a one-pot reaction.[14]

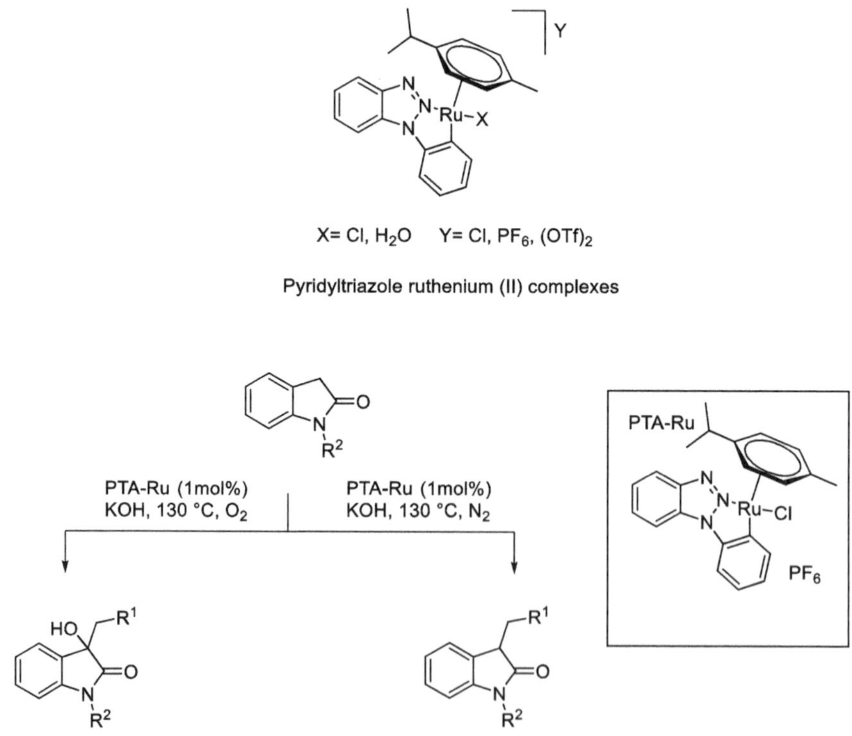

Figure 7 Preparation and applications of pyridyltriazole Ru(II) complexes.

At the same time, the click chemistry approach has also emerged as a powerful synthetic tool for the construction of a variety of molecules with triazolic moiety. Over the years, attempts have been made to combine a multicomponent reaction approach with click chemistry methodology for the synthesis of a variety of compounds including metal complexes.

Ramírez-Palma *et al.* employed click reaction using azide, alkyne, and Cu-salt for the preparation of bis-triazole Cu-complex under a multicomponent approach (Figure 8).[15] The reaction involved *in situ* formation of a triazole ligand. Subsequently, the as-prepared complexes were employed as catalysts in click reactions for the preparation of triazoles in good yields. The catalysts were reusable at least four times.

The homogenous catalysis has also attracted 1,2,3-triazoles as versatile ligands for coordination complexes. Over the years, the 1,4-disubstituted 1,2,3-triazoles bearing NHC ligands have also been exploited for the preparation of metal complexes.[16] Gu *et al.* reacted Cu-powder with imidazolium salts for the preparation of bis- and trinuclear Cu-complexes of 1,2,3-triazole-tethered NHC ligands.[17] The variety of Cu-NHC complexes depends upon the *N*-substituents and imidazolium backbone. The as-prepared triazole-based polynuclear complexes were employed as catalysts for the preparation of 1,2,3-triazoles in click reactions, and complex 4 was found as the most efficient one.

Similarly, Pretorius *et al.* prepared a variety of new 1,2,3-triazolylidene Au(I) chloride complexes, and they were evaluated for their potential as a gold-based catalyst precursor (Figure 9).[18] The metal complexes were produced using MeOTf or potassium salts as Cl^{-1} scavengers. However, the

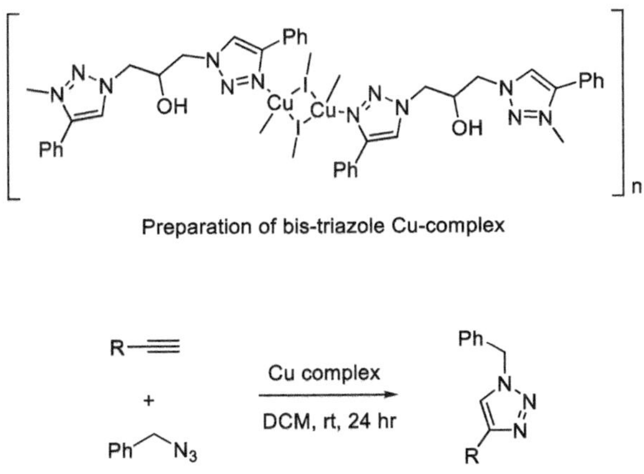

Figure 8 Preparation and applications of bis-triazole Cu-complex.

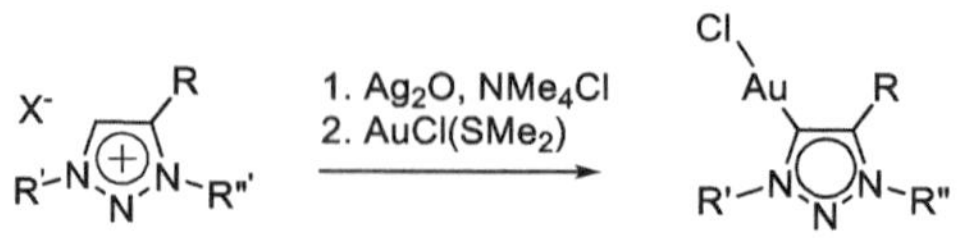

Figure 9 Preparation of 1,2,3-triazolylidene Au(I) chloride complexes.

Ag-based complexes were found less efficient than Au-based complexes for the synthesis of oxazoline. The catalytic efficiency was found to be slightly influenced by the nature of ligand substituents.

Triazole-based metal complexes and their applications in photoelectronics

1,2,4-Triazoles-based metal complexes as phosphorescent

Over the last decade, the organic light-emitting diode (OLED) has emerged as an alternative technology with ever-widening scope in phosphorescent applications. However, some of the later developments like OLED solid-state lighting (SSL) are operationally less efficient for achieving high power.[19] The primary emitters, namely blue, red, and green, could emit light across the entire visible spectrum with high-rendering index. The long-lasting green and red phosphorescent emitters are commercially available; however, production of highly efficient blue emitters have remained a challenge over the last few years. The Ir(III) complexes have shown promising performance as phosphorescent blue emitters, but most of these exhibit low quantum efficiencies. The 5-aryl-4*H*-1,2,4-triazole ligand containing homoleptic Ir-complexes also displayed the tuning of color with the steric hindrance of the ligands resulting in chirality.[20]

Very recently, Feldman *et al.* prepared a sky blue emitter using 5-aryl-4*H*-1,2,4-triazole ligands for the preparation of homoleptic *fac*-Ir(III) L_3 (Figure 10).[21] The atropisomerism was exhibited by the triazolic ligands with unsymmetric substituents. Furthermore, they produced four diastereoisomers upon cyclometalation with Ir. The as-prepared emitters exhibited emission peaks in the sky blue range with high quantum efficiencies thus suggestive as promising candidates for SSL applications. Similarly, Wang *et al.* used phtz-type C^N ligands for the preparation of a series of cationic iridium complexes (Figure 11).[22] The non-radiative deactivation of the emitting triplet state is managed by the metal-centered states in blue-emitting complexes. The yellow-green, blue, and blue-green electroluminescence was displayed by the SSL electrochemical cells with 1–3 complexes.

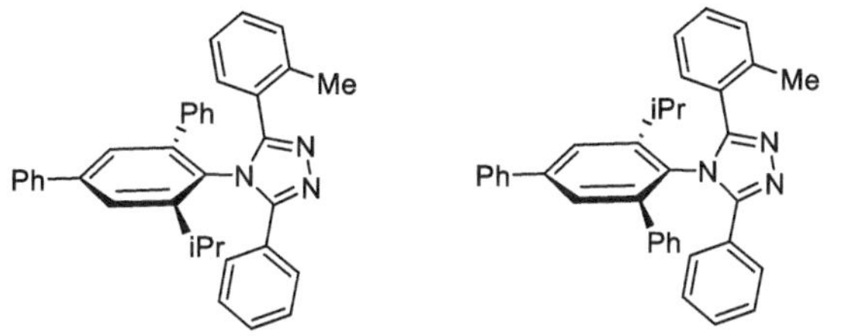

atropisomerism of triazolic ligands

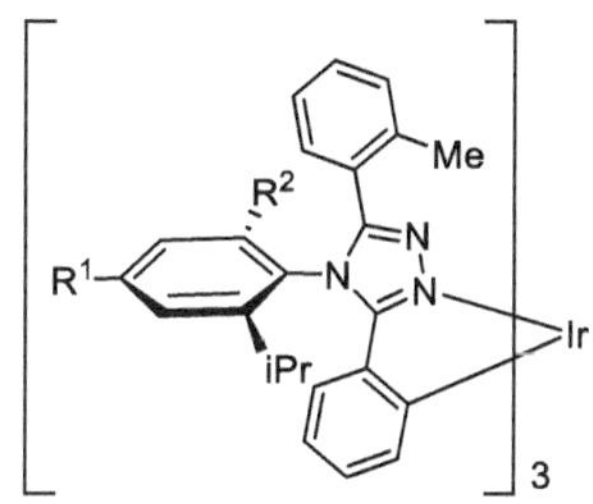

R1 = H, R2 = *i*Pr (1)
R1, R2 = Ph (2)

Figure 10 Blue-emitting Ir(III) complexes.

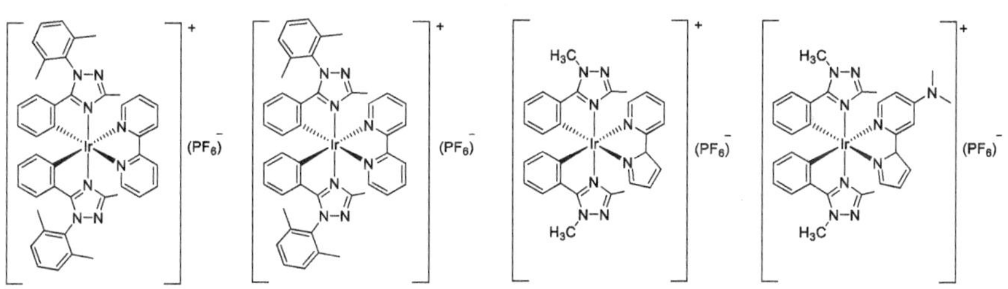

1 2 3 4

Figure 11 Preparation of cationic iridium complexes.

Tang *et al.* synthesized $[(TPTA)_2\ Ir(dPPOA)]PF_6$ as a thermally stable novel orange cationic Ir(III) complex (Figure 12).[23] Its relative emission intensity at 100°C was about 90% of that at room temperature, and its decomposition temperature was 375°C. The blue GaN (450 nm) chips could be used for an efficient excitation of the prepared complex. It was suggested as a promising phosphor candidate for warm white light emitting diods (WLEDs), and it can improve their red light component.

In a very recent report, Song *et al.* observed marked differences in quantum efficiencies in four structurally similar Ir-complexes with isomerized phenyltriazole (ptz) ligands (Figure 13).[24] Their photophysical properties,

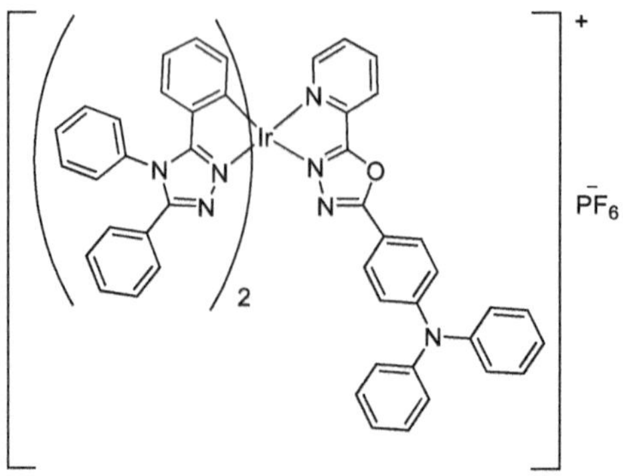

Figure 12 Thermally stable novel orange cationic Ir(III) complex.

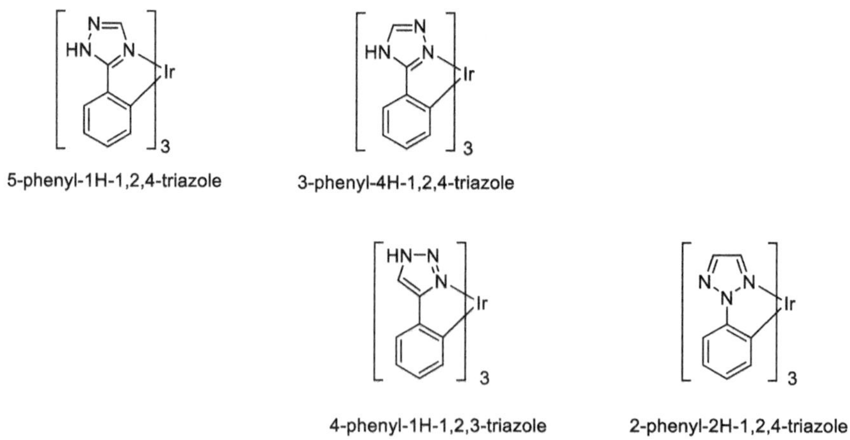

Figure 13 Ir-complexes with isomerized phenyltriazole (ptz) ligands.

including absorption and emission properties and electronic and geometrical structures, were studied using density functional theory. It provided a better understanding about the structure-property relationship for complexes with different ptz isomeric ligands. The difference in quantum efficiencies was found related to non-radiative vibrational relaxation and thermal deactivation pathways via the metal-centered pathways.

1,2,3-Triazole-based metal complexes as phosphorescent

Over the last two decades, the light-emitting electrochemical cells (LEECs) have gained huge attraction in display and white-light applications due to some unique features. Generally, they are composed of a sandwich of semi-transparent anode and an air-stable cathode containing a thin film of a luminescent ionic transition metal complex. They could be prepared in a cost-effective manner by using solutions in a relatively simpler technology. Furthermore, it requires lower operating bias than OLEDs. They are operationally independent from the electrodes suggesting applications of less reactive metals. Despite all these positives, the choice of LEECs have also presented some limitations like slow turn-on voltage, issues of stability, and lifetime of devices that restrict their spectrum of applications.[25]

The cationic Ir(III) complexes of the form [Ir(CN)2(NN)]PF6 have become attractive luminophoric materials over the last few years. In such cationic complex, the N^N represents neutral diamine ancillary ligands like 2,2-bipyridine (bpy) and C^N, and anionic cyclometallating ligands like 2-phenylpyridine (ppy). The photophysical features like tuning of colors by the substituents on ligand, high quantum yields, and phosphorescence with shorter excited-state lifetimes have put them in the limelight as a promising choice for LEEDs.

In a notable study, a blue-shift in emission was promoted when 1,2,3-triazole was used instead of a pyridine ring in C^N ligand.[26] Afterward, in another study, a high-energy emission was observed when pyridyl-1,2,3-triazole (pytl) was replaced bpy ligand.[27] Subsequently, Jesus *et al.* considered these structure-based developments for the production of next-generation blue emitters for SSL applications (Figure 14).[28] They

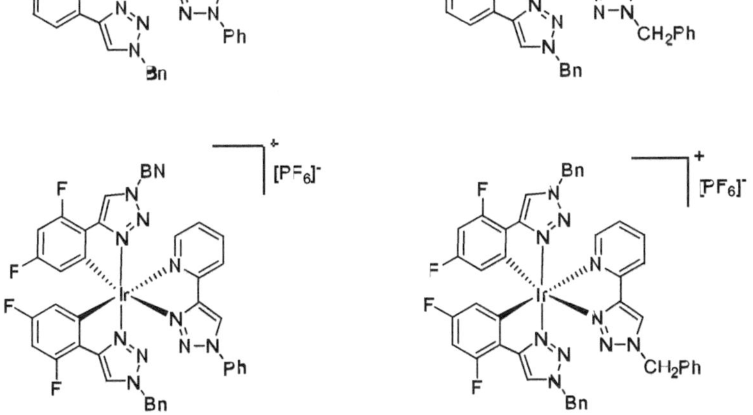

Figure 14 Next-generation blue emitters.

used pytl ancillary ligands for the preparation of four fluorinated cationic iridium complexes. The as-prepared complexes were true blue-emitters with λ_{em} 452–487 nm at 298 K. The emission from these complexes was displayed by thin films in PMMA, powder form, and solution in acetonitrile. They underwent quasi-irreversible reduction and reversible oxidation, and the multiexponential level of decay kinetic was observed in the excited state. The photoluminescent quantum efficiencies were low both in solid and solution states with mixed nature of emission due to ligand-to-ligand or metal-to-ligand charge transfer state. The prepared complexes were used to fabricate LEEC, and 2a exhibited true blue emission with λ 487 nm.

As explained earlier, the steric modifications and electronic features control the photophysical behavior of the complexes. The Ir(III)- and Pt(II)-based complexes with aryl-substituted NHC systems and phenylpyridines have become ideal choices for the development of high-quality dopants. The abnormal NHCs form carbene in the backbone of heterocycles leading to mesoionic structures.[29] Following these lines, the 1,2,3-triazolylidenes, being mesoionic in nature, have also been well explored in homogenous catalysis; however, their use as ligands in luminescence is still in its infancy. In a recent study, Soellner *et al.* prepared diaryl-1,2,3-triazole-based MIC platinum(II)-complexes bearing C^C* cyclometalated mesoionic carbene (Figure 15).[30] The as-prepared complexes exhibited rigid geometry owing to cyclometalation and excellent photophysical properties due to strongly donating heterocycles. The arylation protocols were adopted for the preparation of MIC ligand precursors from 1-phenyl-1,2,3-triazoles. The cyclic voltammetry and DFT calculations were found in agreement with real-time emission properties performed at 77 K and room temperature. In PMMA films, all complexes showed quantum efficiencies from 70 to 84% with emission maxima with 502–534 nm.

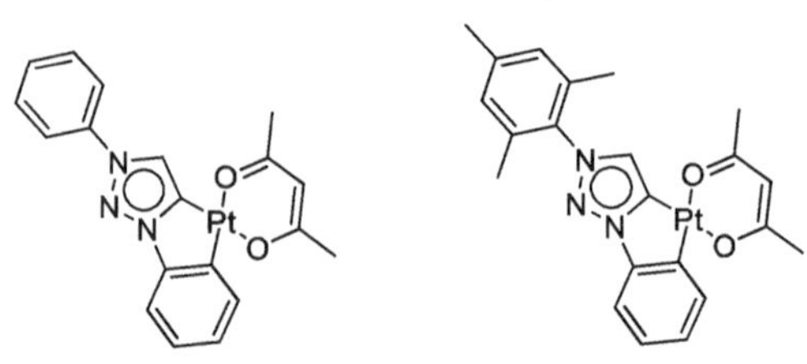

R= Me, Mes, Dur

Figure 15 Diaryl-1,2,3-triazole-based MIC platinum(II)-complexes.

Triazole-based metal complexes and their bioactive potential

1,2,4-Triazole-based metal complexes and their bioactivity

Over the last two decades, drug development has gone through a major transformation due to unprecedented advancements in synthetic medicinal chemistry. The metal-complexation approach has emerged as a leading strategy for the preparation of highly potent drug candidates. The complexes of transition metals with azole derivatives has been used to produce novel drugs with high biocompatibility and safety profiles.[31] In many cases, it has been observed that such complex-based drugs exhibit higher bioactivities compared with free azole ligands.[32] In azoles, the triazoles have emerged as highly potent molecules with a wide spectrum of biopotential ranging from antimicrobial to anticancer activities. In recent past years, the biopotential of triazoles have been well explored by complexing it with different transition metals.

In one such recent study, Murcia *et al.* initially prepared a novel triazole ligand (1) by reacting 1*H*-1,2,4-triazole with 1,3-bis(bromomethyl)toluene in the presence of a phase transfer catalyst. Afterward, they used this ligand for the preparation of six metal complexes using Co(II) and Cr(III) and subsequently characterized them by gravimetric analyses, electrical conductivity, and Raman spectroscopy (Figure 16).[33] The antimicrobial studies demonstrated that the air-stable complexes were more active against fungi and bacteria than the corresponding ligand. Furthermore, they exhibited better activity against fungi than bacteria. Furthermore, Cr-complexes displayed better results against fungus strains. This study provided a sound platform for designing of novel antimicrobial agents with low cytotoxic profiles.

Over the last few years, the cancer therapeutics have seen great advancement due to the involvement of inorganic compounds in drug development. In this context, development of metal-based anticancer drugs has been a focus for improved oral administration and better safety profiles. Accordingly, a number of transition and nontransition metals are under investigation in the domain of anticancer therapy. In this regard, ruthenium has become an attractive choice owing to its selectivity toward cancerous cells, variable oxidation states, and low toxicity. Its complexes have become promising candidates for next-generation anticancer drugs because of high *in vitro* and *in vivo* activities.[34, 35] The anticancer activities have also been exhibited by the Ru half-sandwich complex. It has been observed that the type of arene and ancillary ligands control the anticancer activity of

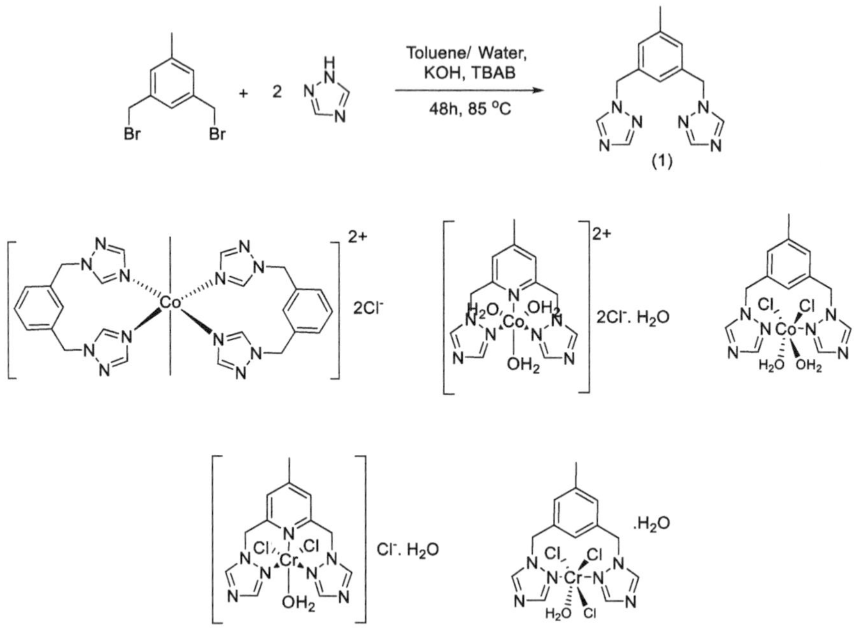

Figure 16 1,2,4-Triazole-based air-stable Co(II) and Cr(III) complexes.

half-sandwich Ru(II) arene compounds. Thus, systematic variation of the substituents could be used to optimize their pharmacological profiles.

Gichumbi *et al.* employed dimeric Ru-arene precursor and a series of dipyridyl-triazole ligands for the preparation of a novel series of cationic mononuclear complexes (Figure 17).[36] Furthermore, they were tested for cytotoxicities against noncancer cell line (KEK 293) and cancer cell line (Caco-2). The complex 6 exhibited antiproliferative property better than standard while 1 and 7 were inactive and 2–5 displayed moderate anticancer activities. Furthermore, the ligand and compound 6 were used for antimicrobial studies.

In cancer therapeutics, the development of platinum-based drugs started when a known anticancer drug, cisplatin, exhibited cytotoxicity against a range of human-derived cancer cells. With the passage of time, the Pt-based drugs also displayed severe side effects, which ultimately forced researchers to focus other metals, especially Au and Ag.[37,38] These metals manage to initiate cellular apoptosis by targeting thioredoxin reductase, which executes a number of metabolic processes. The Ag could target multiple pathways causing inhibition of glycolysis, induction of oxidative stress, and inhibition of topoisomerase I/II. Thus, the Ag-based anticancer drugs were expected

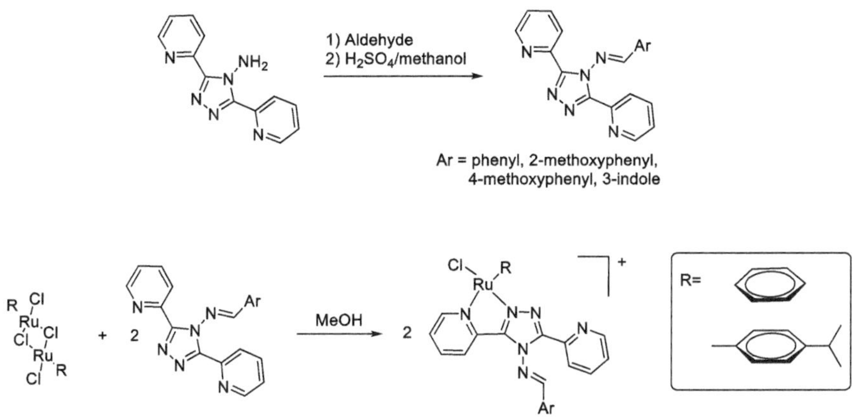

Figure 17 Dipyridyl-triazole ligands for the preparation of cationic mononuclear complexes.

to be highly efficient against tumor cells owing to their inherent ability to target a number of vital cellular pathways.[39]

As described earlier, the NHCs have become promising ligands due to the ease of preparation, high stability, and strong stabilizing effects. Furthermore, the presence of pi-donating substituents facilitates the formation of a stronger NHC-metal bond. Accordingly, the bio-organometallic compounds developed from NHC-based metal complexes have found a range of applications across a number of scientific disciplines. Over the last decade, the NHC-Au and Ag complexes have also found considerable applications in biomedical applications including cancer therapeutics. Such complexes have mainly been prepared using benzimidazole or imidazole as ligands; however, no or little attention has been paid to 1,2,4-triazole-based NHC ligands.

Very recently, Achar *et al.* used coumarin-functionalized 1,2,4-triazol-5-ylidene ligands for the preparation of three sterically tuned types of Au(I) and Ag(I) complexes (Figure 18).[40] The *in situ* deprotonation method was adopted for the preparation of Ag-complexes employing AgOAC and Ag_2O as metal. The as-prepared Ag-complexes were made to undergo transmetallation for the preparation of Au-complexes. A linear coordination geometry with inversion of metal atoms was adopted by the prepared Ag-complexes. The CH of NHC ligands and the metal atom showed anagostic-like interactions. Additionally, the hexafluorophosphate anions and cations developed hydrogen-bonding interactions leading to unique frameworks with different supramolecular architectures. The anticancer activity of the samples

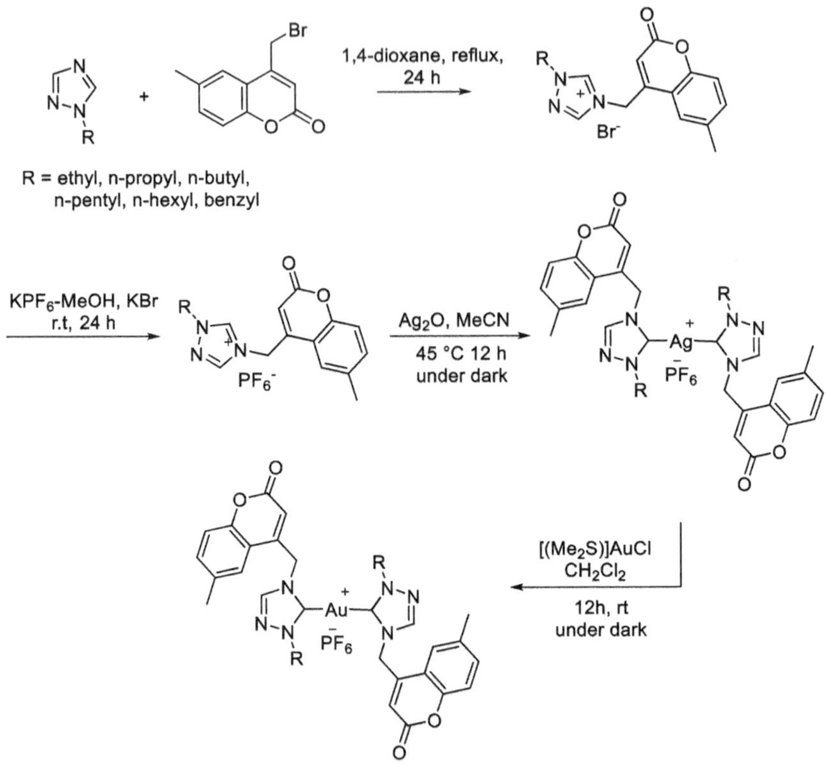

Figure 18 Coumarin-functionalized Au(I) and Ag(I) complexes.

was better against HT-29 while moderate against MCF 7. The structure-dependent activity highlighted the vital role of Ag-centers in biopotential of as-prepared complexes.

1,2,3-Triazole-based metal complexes and their bioactivity

As described earlier, the Ru-based complexes have shown a broad spectrum of biological activities. Similarly, the 1,2,3-triazole also possess a number of biological properties including anticancer potential. In recent years, there has been great interest to exploit the synergistic effect of coordinating biopotent ligands with metals.[41] Accordingly, the anticancer therapeutics have attracted Ru-based metallodrugs as promising alternatives to existing low-profile drugs over the last decade. In hypoxic tumor tissues, the Ru(III) is reduced to its activated Ru(II) form. Over the years, the Ru(II)-arene complexes have emerged as promising candidates with tunable physiobiological properties according to the nature of the ligand substituents.[42, 43]

Riedl *et al.* prepared Os(II) and Ru(II) arene metallodrugs by cyclometalation of 1-substituted 4-phenyl 1,2,3-triazoles as tunable ligands (Figure 19).[44] They studied the antiproliferative differences due to the change of metal centers, substituents on position 1 of the triazole and nature of solubilizer for biotests. They further reported their mode of action, interactions with biomolecules, and *in vivo* anticancer efficacy using human cancer cell lines (CH1/PA-1, SW480, A549). The lipophilicity of the ligands influenced their cytotoxicity in *in vitro* studies. The ease of synthetic protocol, tunable substituents, and high *in vitro* activity profiles signified their worth as valuable candidates for future drug development.

The photophysical and photochemical properties of rhenium (1) tricarbonyl complexes have made them promising candidates with a wide spectrum of applications in luminescence sensors, OLEDs, CO_2 reduction catalysis, and photosensitizers in solar cells.[45, 46] Over the last few years, they have also attracted medicinal chemists for the development of novel therapeutic agents.

Aimene *et al.* employed click chemistry approach for the preparation of pharmacophore (4-substituted benzenesulfonamide) containing two bidentates 2-pyridyl-1,2,3-triazole ligands and subsequently used them for the preparation of Re-complexes (Figure 20).[47] The as-prepared ligands and their corresponding complexes were evaluated for their potential against

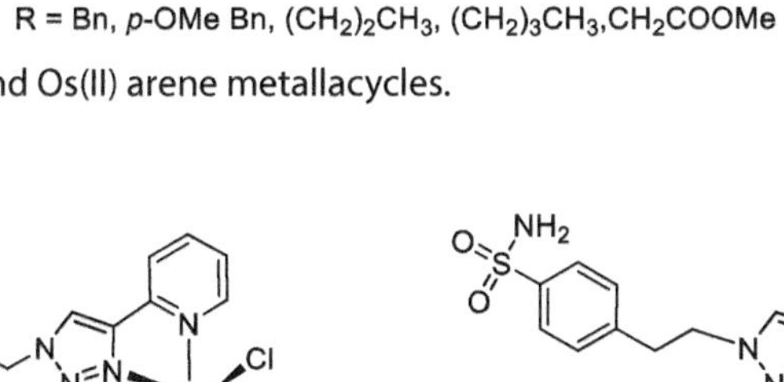

Figure 19 Ru(II) and Os(II) arene metallacycles.

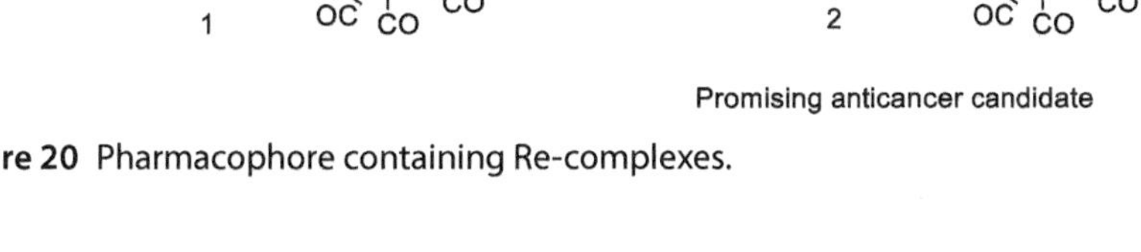

Figure 20 Pharmacophore containing Re-complexes.

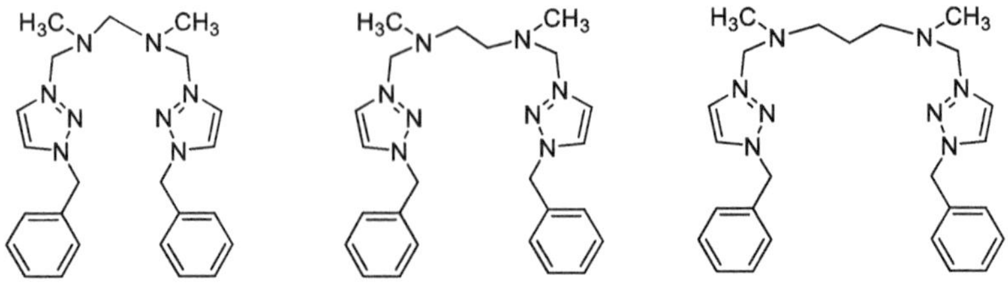

Figure 21 1,4-Disubstituted 1,2,3-triazole as tetradentate ligands.

carbonic anhydrase isoform IX. Compound 2 turned out to be promising anticancer candidate due to its selective inhibitory activity against hCA IX.

In a recent study, De Paula *et al.* used 1,4-disubstituted 1,2,3-triazole as tetradentate ligands for the preparation of Cu(II)-complexes (Figure 21).[48] The prepared complexes were evaluated for their interactions with HSA and thermodynamic stability in comparison to free ligands. Much weaker interactions were exhibited by the free ligands. The ROS produced by the complexes induced oxidative damage of proteins. The study exhibited that the triazole ligands were competitive toward Cu(II) in biological media.

Conclusion

The presence of three donor N atoms enables the triazole rings to act as polydentate ligands, and the heterocycles may act as bridging ligands. The triazole ring could adopt at least five coordination modes for bridging metal ions. These coordination possibilities have been well exploited for the designing and preparation of a large variety of organic/inorganic hybrid architectures. Some of the recent notable examples of triazole-based metal complexes have been presented to highlight their valuable biological, photoelectronic, and catalytic properties.

References

1. Aromí G, Barrios LA, Roubeau O, Gamez P. Triazoles and tetrazoles: prime ligands to generate remarkable coordination materials. *Coord Chem Rev* 2011;**255**(5–6):485–546.
2. Scattergood PA, Sinopoli A, Elliott PI. Photophysics and photochemistry of 1, 2, 3-triazole-based complexes. *Coord Chem Rev* 2017;**350**:136–54.
3. Nolte C, Mayer P, Straub BF. Isolation of a copper (I) triazolide: a "click" intermediate. *Angew Chem Int Ed* 2007;**46**(12):2101–3.
4. Donnelly KF, Petronilho A, Albrecht M. Application of 1, 2, 3-triazolylidenes as versatile NHC-type ligands: synthesis, properties, and application in catalysis and beyond. *Chem Commun* 2013;**49**(12):1145–59.
5. Dash C, Shaikh MM, Butcher RJ, Ghosh P. A comparison between nickel and palladium precatalysts of 1, 2, 4-triazole based N-heterocyclic carbenes in hydroamination of activated olefins. *Dalton Trans* 2010;**39**(10):2515–24.

6. Kumar A, Bheeter LP, Gangwar MK, Sortais J-B, Darcel C, Ghosh P. Nickel complexes of 1, 2, 4-triazole derived amido-functionalized N-heterocyclic carbene ligands: synthesis, theoretical studies and catalytic application. *J Organomet Chem* 2015;**786**:63–70.
7. Akkoç S, Gök Y, İlhan İÖ, Kayser V. N-Methylphthalimide-substituted benzimidazolium salts and PEPPSI Pd–NHC complexes: synthesis, characterization and catalytic activity in carbon–carbon bond-forming reactions. *Beilstein J Org Chem* 2016;**12**(1):81–8.
8. Chernenko AY, Astakhov A, Pasyukov D, Dorovatovskii P, Zubavichus YV, Khrustalev V, et al. Pd-PEPPSI complexes based on 1, 2, 4-triazol-3-ylidene ligands as efficient catalysts in the Suzuki—Miyaura reaction. *Russ Chem Bull* 2018;**67**(1):79–84.
9. Devendar P, Qu R-Y, Kang W-M, He B, Yang G-F. Palladium-catalyzed cross-coupling reactions: a powerful tool for the synthesis of agrochemicals. *J Agric Food Chem* 2018;**66**(34):8914–34.
10. Astakhov AV, Khazipov OV, Degtyareva ES, Khrustalev VN, Chernyshev VM, Ananikov VP. Facile hydrolysis of nickel (II) complexes with N-heterocyclic carbene ligands. *Organometallics* 2015;**34**(24):5759–66.
11. Soliev S, Astakhov A, Pasyukov D, Chernyshev V. Nickel (ii) N-heterocyclic carbene complexes as efficient catalysts for the Suzuki—Miyaura reaction. *Russ Chem Bull* 2020;**69**:683–90.
12. Mastalir M, Tomsu G, Pittenauer E, Allmaier G, Kirchner K. Co (II) PCP pincer complexes as catalysts for the alkylation of aromatic amines with primary alcohols. *Org Lett* 2016;**18**(14):3462–5.
13. Wu Q, Pan L, Du G, Zhang C, Wang D. Preparation of pyridyltriazole ruthenium complexes as effective catalysts for the selective alkylation and one-pot C–H hydroxylation of 2-oxindole with alcohols and mechanism exploration. *Org Chem Front* 2018;**5**(18):2668–75.
14. Cioc RC, Ruijter E, Orru RV. Multicomponent reactions: advanced tools for sustainable organic synthesis. *Green Chem* 2014;**16**(6):2958–75.
15. Ramírez-Palma MT, Segura-Arzate J, López-Téllez G, Cuevas-Yañez E. Ligand synthesis catalyst and complex metal ion: multicomponent synthesis of 1, 3-bis (4-phenyl-[1,2,3] triazol-1-yl)-propan-2-ol copper (I) complex and application in copper-catalyzed alkyne-azide cycloaddition. *J Chem* 2016;**2016**.
16. Sluijter SN, Elsevier CJ. Synthesis and reactivity of heteroditopic dicarbene rhodium (I) and iridium (I) complexes bearing chelating 1, 2, 3-triazolylidene–imidazolylidene ligands. *Organometallics* 2014;**33**(22):6389–97.
17. Gu S, Du J, Huang J, Xia H, Yang L, Xu W, et al. Bi-and trinuclear copper (I) complexes of 1, 2, 3-triazole-tethered NHC ligands: synthesis, structure, and catalytic properties. *Beilstein J Org Chem* 2016;**12**(1):863–73.
18. Pretorius R, Fructos MR, Müller-Bunz H, Gossage RA, Pérez PJ, Albrecht M. Synthesis and catalytic applications of 1, 2, 3-triazolylidene gold (I) complexes in silver-free oxazoline syntheses and C–H bond activation. *Dalton Trans* 2016;**45**(37):14591–602.
19. Bender VC, Marchesan TB, Alonso JM. Solid-state lighting: a concise review of the state of the art on LED and OLED modeling. *IEEE Ind Electron Mag* 2015;**9**(2):6–16.
20. Dobbs KD, Feldman J, Marshall WJ, McLain SJ, McLaren CD, Meth JS, et al. Phosphorescent iridium (III) complexes of cyclometalated 5-aryl-1 H-1, 2, 4-triazole ligands: structural, computational, spectroscopic, and device studies. *J Phys Chem C* 2014;**118**(48):27763–71.
21. Feldman J, Vo GD, McLaren CD, Gehret TC, Park K-H, Meth JS, et al. Highly quantum efficient phosphorescent sky blue emitters based on diastereomeric iridium (III) complexes of atropisomeric 5-aryl-4 H-1, 2, 4-triazole ligands. *Organometallics* 2015;**34**(15):3665–9.
22. Wang X, Wang S, Pan F, He L, Duan L. Cationic iridium complexes with 5-phenyl-1 H-1, 2, 4-triazole type cyclometalating ligands: toward blue-shifted emission. *Inorg Chem* 2019;**58**(18):12132–45.

23. Tang H, Meng G, Chen Z, Wang K, Zhou Q, Wang Z. Warm white light-emitting diodes based on a novel orange cationic iridium (iii) complex. *Materials* 2017;**10**(6):657.
24. Song C, Chen Y, Li J, Zhao F, Zhang H. Unraveling the marked differences of the phosphorescence efficiencies of blue-emitting iridium complexes with isomerized phenyltriazole ligands. *Inorg Chem Front* 2019;**6**(10):2776–87.
25. Hu T, He L, Duan L, Qiu Y. Solid-state light-emitting electrochemical cells based on ionic iridium (iii) complexes. *J Mater Chem* 2012;**22**(10):4206–15.
26. Donato L, Abel P, Zysman-Colman E. Cationic iridium (iii) complexes bearing a bis (triazole) ancillary ligand. *Dalton Trans* 2013;**42**(23):8402–12.
27. Mydlak M, Bizzarri C, Hartmann D, Sarfert W, Schmid G, De Cola L. Positively charged iridium (III) triazole derivatives as blue emitters for light-emitting electrochemical cells. *Adv Funct Mater* 2010;**20**(11):1812–20.
28. Fernández-Hernández JM, Ladouceur S, Shen Y, Iordache A, Wang X, Donato L, et al. Blue light emitting electrochemical cells incorporating triazole-based luminophores. *J Mater Chem C* 2013;**1**(44):7440–52.
29. Gruendemann S, Kovacevic A, Albrecht M, Faller JW, Crabtree RH. Abnormal ligand binding and reversible ring hydrogenation in the reaction of imidazolium salts with IrH5 (PPh3) 2. *J Am Chem Soc* 2002;**124**(35):10473–81.
30. Soellner J, Strassner T. Diaryl-1, 2, 3-triazolylidene platinum (II) complexes. *Chem Eur J* 2018;**24**(21):5584–90.
31. Kljun J, Scott AJ, Lanišnik Rižner T, Keiser J, Turel I. Synthesis and biological evaluation of organoruthenium complexes with azole antifungal agents. First crystal structure of a tioconazole metal complex. *Organometallics* 2014;**33**(7):1594–601.
32. Yousef T, El-Reash GA, Al-Jahdali M, El-Rakhawy E-BR. Synthesis, spectral characterization and biological evaluation of Mn (II), Co (II), Ni (II), Cu (II), Zn (II) and Cd (II) complexes with thiosemicarbazone ending by pyrazole and pyridyl rings. *Spectrochim Acta A Mol Biomol Spectrosc* 2014;**129**:163–72.
33. Murcia RA, Leal SM, Roa MV, Nagles E, Muñoz-Castro A, Hurtado JJ. Development of antibacterial and antifungal triazole chromium (III) and cobalt (II) complexes: synthesis and biological activity evaluations. *Molecules* 2018;**23**(8):2013.
34. Lin K, Zhao Z-Z, Bo H-B, Hao X-J, Wang J-Q. Applications of ruthenium complex in tumor diagnosis and therapy. *Front Pharmacol* 2018;**9**:1323.
35. Thota S, Rodrigues DA, Crans DC, Barreiro EJ. Ru (II) compounds: next-generation anticancer metallotherapeutics? *J Med Chem* 2018;**61**(14):5805–21.
36. Gichumbi JM, Friedrich HB, Omondi B, Lazarus GG, Singh M, Chenia HY. Synthesis, characterization, anticancer and antimicrobial study of arene ruthenium (II) complexes with 1, 2, 4-triazole ligands containing an α-diimine moiety. *Z Naturforsh B* 2018;**73**(3–4):167–78.
37. Ndagi U, Mhlongo N, Soliman ME. Metal complexes in cancer therapy—an update from drug design perspective. *Drug Des Devel Ther* 2017;**11**:599.
38. Ringhieri P, Morelli G, Accardo A. Supramolecular delivery systems for non-platinum metal-based anticancer drugs. *Crit Rev Ther Drug Carrier Syst* 2017;**34**(2).
39. Li S-X, Chen Y-F, Lu Y, Xu S-H, Liao B-L. Synthesis, fluorescence, and anticancer activity of silver (I) complex based on 2-hydroxyquinoxaline ligand. *Inorg Nano-Met Chem* 2020;**50**(4):315–20.
40. Achar G, Shahini C, Patil SA, Małecki JG, Budagumpi S. Coumarin-substituted 1, 2, 4-triazole-derived silver (i) and gold (i) complexes: synthesis, characterization and anticancer studies. *New J Chem* 2019;**43**(3):1216–29.
41. Golbaghi G, Castonguay A. Rationally designed ruthenium complexes for breast cancer therapy. *Molecules* 2020;**25**(2):265.

42. Caruso F, Pettinari R, Rossi M, Monti E, Gariboldi MB, Marchetti F, et al. The in vitro antitumor activity of arene-ruthenium (II) curcuminoid complexes improves when decreasing curcumin polarity. *J Inorg Biochem* 2016;**162**:44–51.
43. Deacon-Price C, Romano D, Riedel T, Dyson PJ, Blom B. Synthesis, characterisation and cytotoxicity studies of ruthenium arene complexes bearing trichlorogermyl ligands. *Inorg Chim Acta* 2019;**484**:513–20.
44. Riedl CA, Flocke LS, Hejl M, Roller A, Klose MH, Jakupec MA, et al. Introducing the 4-phenyl-1, 2, 3-triazole moiety as a versatile scaffold for the development of cytotoxic ruthenium (II) and osmium (II) arene cyclometalates. *Inorg Chem* 2017;**56**(1):528–41.
45. KomReddy V, Ensz K, Nguyen H, Rillema DP. Synthesis and characterization of rhenium (I) 4, 4′-dicarboxy-2, 2′-bipyridine tricarbonyl complexes for solar energy conversion. *Inorg Chim Acta* 2020;, 119815.
46. McKinnon M, Ngo KT, Sobottka S, Sarkar B, Ertem MZ, Grills DC, et al. Synergistic metal–ligand redox cooperativity for electrocatalytic CO2 reduction promoted by a ligand-based redox couple in Mn and Re tricarbonyl complexes. *Organometallics* 2018;**38**(6):1317–29.
47. Aimene Y, Eychenne R, Mallet-Ladeira S, Saffon N, Winum J-Y, Nocentini A, et al. Novel Re (I) tricarbonyl coordination compounds based on 2-pyridyl-1, 2, 3-triazole derivatives bearing a 4-amino-substituted benzenesulfonamide arm: synthesis, crystal structure, computational studies and inhibitory activity against carbonic anhydrase I, II, and IX isoforms. *J Enzyme Inhib Med Chem* 2019;**34**(1):773–82.
48. de Paula QA, Joly J-P, Selmeczi K, Fonseca DE, Caramori GF, Farrell NP, et al. Binding affinity studies of 1, 2, 3-triazole copper (II) complexes to human serum albumin. *J Coord Chem* 2018;**71**(11–13):1894–909.

CHAPTER 10

Triazoles in Material Sciences

Tahir Farooq*
Department of Applied Chemistry, Government College University, Faisalabad, Pakistan
*Corresponding author. E-mail: tahirfarooqfsd@gmail.com

Introduction

Through exceptional advancements over the last two decades, synthetic triazole chemistry has resulted in a broadening of its scope of applications including in materials. The triazolic moiety has become a promising motif for material sciences due to their high coordination ability, thermal stability, and high electron affinity. They are highly electron deficient and show electron transport and hole-blocking properties.[1] Both 1,2,4- and 1,2,3-triazoles have shown a range of applications including their ever-increasing demand in light emitting diodes, solar cells and high energy materials. Both types of triazoles have also been used for the preparation of various types of functional oligomers and polymers with wide applications as sorbents, supercapacitors, electrochronic devices, and bipolar materials.[2,3]

Triazoles have also been used for the preparation of metal organic frameworks that find applications as high energy materials, green explosives, detonators, and heat-resistant materials in aerospace technology. Herein, I present some noteworthy applications of both 1,2,4- and 1,2,3-triazoles in aforesaid domains of material sciences. These examples highlight the worth of advances in triazole chemistry for the development of highly demanded functional materials. Especially, the copper-catalyzed azide alkyne cycloaddition (CuAAC) reaction for the facile preparation of 1,2,3-triazole has been widely exploited for the construction of a variety of novel functional materials.[4]

The subsequent sections present some relevant examples of functional materials developed by using 1,2,4- and 1,2,3-triazoles as the functional motif.

Light-emitting diodes (LEDs)

1,2,4-Triazoles in LEDs

The modern display and lighting technology heavily rely on organic light-emitting diodes (OLEDs) that utilize organic-based materials between electrodes. The emitting layer (EML) produces light when injected

https://doi.org/10.1016/B978-0-12-817113-4.00002-0

electrons and holes recombine in it due to applied voltage. For highly efficient operations, the emission zone should be as wide as possible providing maximum chances for all electrons to recombine with all holes to form maximum exciton. In recent years, a bipolar EML is preferred to keep carriers and excitons confined in it by putting it between them. In this regard, a "charge balance" condition is achieved by using organic materials as a bipolar host to extend operational life with low driving voltage. The bipolar hosts could be achieved by physically mixing the electron-transporting and hole-transporting materials. For improved operational lifetime, the dopant materials have also been used for broadening of the recombination zone and effective carrier injection and transportation.[5,6]

Recently, Cheng *et al.* used blue phosphorescent OLEDs to study two bipolar hosts by comparing their recombination and carrier transportation characteristics. For one host, the carbazole containing *N,N'*-dicarbazolyl-3,5-benzene (mCP) and triazole containing 3-(Biphenyl-4-yl)-5-(4-*ter*-butylphenyl)-4-phenyl-4*H*-1,2,4-triazole (TAZ) were physically mixed and coevaporated. The other host with bipolar transport characteristics was prepared chemically with two carbazole and one triazole moiety, namely 9,9′-(2-(4,5-diphenyl-4*H*-1,2,4-triazol-3-yl)-1,3-phenylene)bis(9H-carbazole) (2CbzTAZ) (Figure 1).[7] In parallel, the OLED with TAZ and mCP were also studied as single-host organic light-emitting diodes (LEDs). In this comparative study, the OLED with 2CbzTAZ host exhibited lowest efficiency roll-off and highest efficiency with lowest voltage. This phenomenon was observed when bis-[2-(4,6-difluorophenyl)pyridinato-C2-*N*](picolinato)iridium(III) was used as blue phosphor dope. The OLED with 2Cbz-TAZ exhibited the most efficient recombination rate as the recombination zone was well confined in EML. However, a lower recombination rate was observed for OLEDs with mCP:TAZ, TAZ, and mCP as host materials due to carrier accumulation on different organic molecules. There

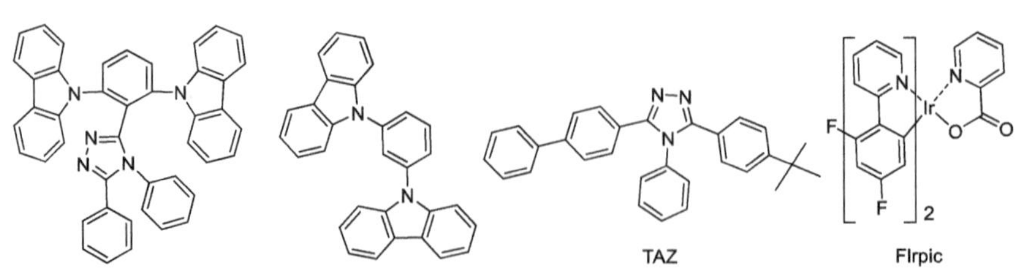

Figure 1 Bipolar materials with 1,2,4-triazoles.

was no carrier accumulation in BPOLED with 2CbzTAZ as host and exhibited highest efficiency compared with three other devices.

The electron deficient character of triazole plays an important role in electron transport, and it enhances the material efficiency by balancing carrier recombination. The introduction of triazole into π-conjugated system can combine electron-transport and light-emitting moiety in a single molecule. This could help the fabrication of bipolar material-based devices.[8] Therefore, Kang *et al.* used [1,2,4]triazolo[4,3-α]pyridine as a core structure for the preparation of bipolar host material. They synthesized tris(4′-([1,2,4]triazolo[4,3-α]pyridin-3-yl)biphenyl-4-yl)amine (TPAPTP) and tris(4′-([1,2,4]triazolo[3,4-α]pyridin-3-yl)phenyl)amine (TPATP) to prepare [1,2,4]triazolo[4,3-α]pyridine-based bipolar red host materials for red phosphorescent OLEDs (Figure 2).[9] The triphenylamine derivatives improve the hole transport and [1,2,4]triazolo[4,3-α]pyridine enhances the electron transport, and the combination produced a bipolar host material. The charge transportation was facilitated by the high triplet energy and appropriate HOMO and LUMO of TPAPTP and TPATP. Furthermore, they were employed to fabricate phosphorescent OLEDs using $Ir(pic)_2acac$ as dopant. The TPAPTP-based PhOLEDs exhibited 5.58 Im/W maximum power efficiency and 9.1% external quantum efficiency (EQE) while the TPATP-based device presented 3.71 Im/W maximum power efficiency and 6.04% EQE. The bipolar molecules were designed by simple phenyl-linkage of electron accepting [1,2,4]triazolo[4,3-α]pyridine and electron-donating triphenylamine moieties.

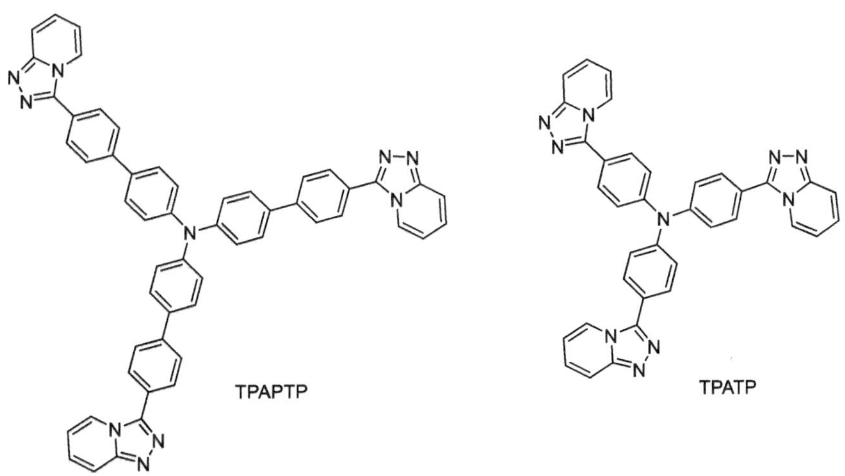

Figure 2 1,2,4-Triazole-carrying bipolar red host materials.

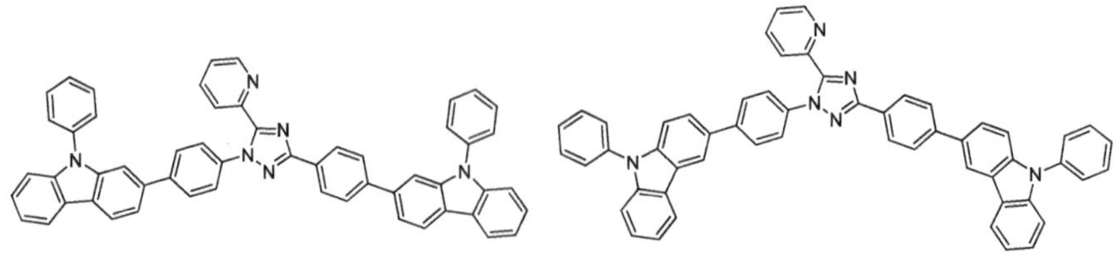

Figure 3 Bipolar host materials with donor-acceptor-donor structures.

In the case of red PhOLEDs, it has been a challenge to achieve high efficiency with small efficiency roll-off due to the issues related to the host materials.[10] It was suggested that the designing of novel bipolar hosts with suitable steric hindrance with torsional structures could decrease the density of triplet excitons and solve the efficiency roll-off problem. In this connection, Wang *et al.* linked electron-withdrawing triazole with C2 and C3 of a carbazole unit to fabricate host materials (2Cz-TAZ-2Cz and 3Cz-TAC-3Cz) with twisted molecular confirmations (Figure 3).[11] Two bipolar host materials were fabricated using donor-acceptor-donor (D-A-D) structures employing phenyl carbazole as electron-donating groups and triazole and pyridine as electron-withdrawing groups. In one molecule, the acceptor is linked with C2 of carbazole while in another molecule, the acceptor is connected at C3 of two donors. Both novel bipolars exhibited excellent thermal stability and carrier-transporting properties. The bis(1-phenylisoquinoline)(acetylacetonate)iridium (III) $Ir(piq)_2acac$ was used as an emitter to fabricate red PhOLEDs using 2Cz-TAZ-2Cz and 3Cz-TAC-3Cz as host molecules. The 2Cz-TAZ-2Cz-based red device exhibited high EQE (16.6%) and maximum current efficiency (12.4 cd/A) with very small efficiency roll-off. This advancement in the conformational-controlled synthesis of bipolar material has been anticipated to show superior applications in PhOLEDS.

1,2,3-Triazoles in LEDs

As discussed earlier, the OLEDs show high device efficiency owing to the balanced charge transfer properties of bipolar host materials. However, a single host molecule with an electron-poor acceptor and an electron-rich donor causes a decrease in the triplet band gap. It causes an undesirable energy transfer from emitter to the host restricting the applications of bipolar hosts as blue emitters with large band gap.[12] However, they are expected to be highly energy efficient for promising applications in OLEDs. Over the years, efforts have been made to control the undesired interactions and energy transfer. To this end, various doable strategies have been proposed to control the donor-acceptor exchange through the conjugated π-systems.[13] It was believed that the overall conjugation in a large π-conjugated system could be decreased by incorporating 1,2,3-triazole as functional linker during the synthesis of such molecules. The recent advances in click chemistry have made it convenient to introduce 1,2,3-triazolic moiety at any position in molecular systems. Kautny *et al.* employed copper-catalyzed alkyne azide cycloaddition (CuAAC) methodology to connect an electron-poor core system with an electron-rich (phenylcarbazole (PCz) or triphenylamine (TPA)) part via 1,2,3-triazole as a functional linker (Figure 4).[14] The electron-accepting pyridine and benzene were selected as core units to study the effect of triazolic link on intramolecular charge transfer and overall conjugation. The triazole was anticipated to establish an electron-accepting subunit with benzene as a core and increase the electron-accepting properties of the core when pyridine is an acceptor. To characterize the linkage mode of triazole on molecular properties, a series of bipolar host materials

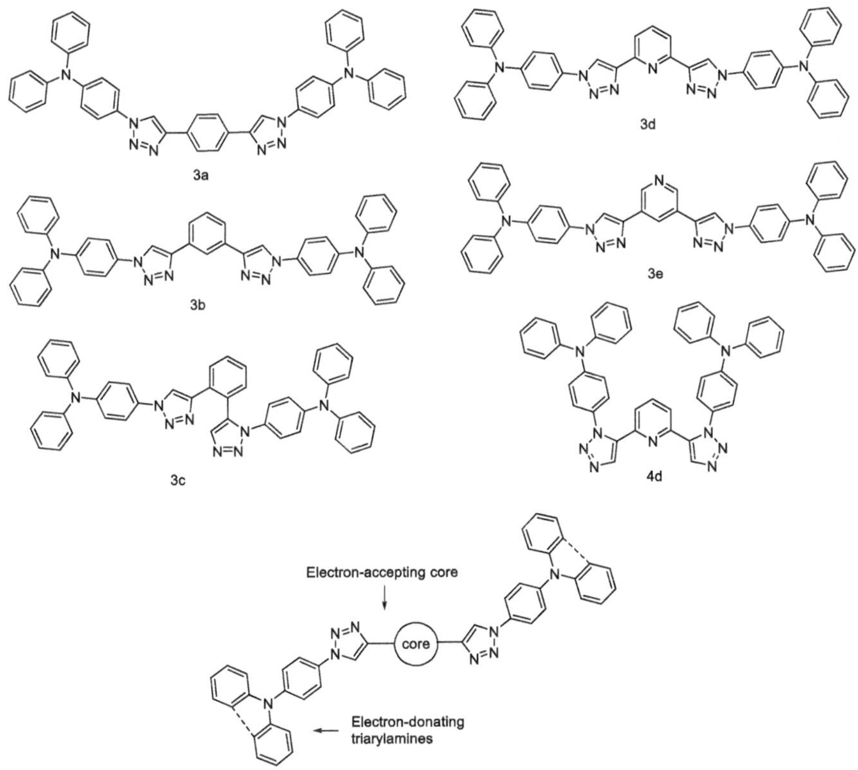

Figure 4 Substituent pattern of triazole and series of host materials.

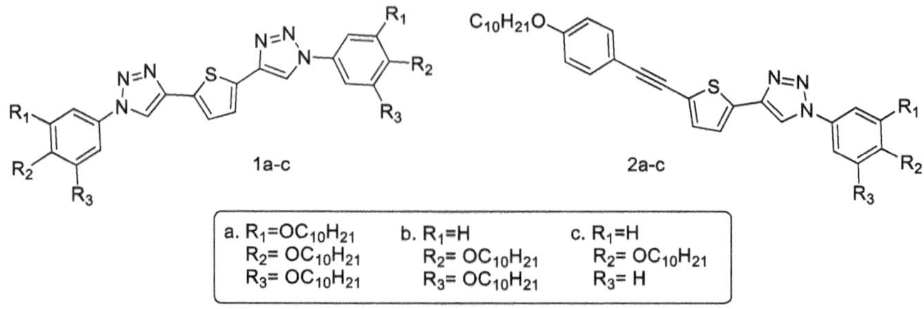

Figure 5 Triazole-containing thiophene-based liquid crystals.

were prepared. The substitution pattern of the triazole was found to control the intramolecular charge transfer. The prepared bipolar materials with different substituent pattern of triazole were used as host materials for the fabrication of PhOLEDs. This study was presented as a novel approach for the preparation of functional π-conjugated bipolar materials for OLEDs.

The interesting physiochemical properties of liquid crystals with 1,2,3-triazolic functionality have attracted attention for applications in high-density memory storage, sensors, solar cells, and OLEDs.[15] They have become attractive due to paramagnetism induced by the metal center and their photoresponsive behavior owing to triazolic linkage with azobenzenes. A dramatic effect is observed on the mesomorphic behavior of the molecule with a change in the position and substituents of the triazole. Girotto *et al.* also prepared a series of triazole-containing thiophene-based molecules (Figure 5).[16] These triazolic compounds exhibited liquid crystal properties with nematic and sematic mesophases. The mesomorphic behavior became unfavorable with increasing alkoxy substituents. The appearance of sematic mesophase of non-symmetrical molecules became more organized with a lowering of melting point when an oxadiazole ring was replaced with alkyne unit as a spacer. This work presented a novel idea for the preparation of 1,2,3-triazole carrying liquid crystals with mesomorphic properties. This report further strengthened the usage of LCs with 1,2,3-triazoles for the fabrication of OLEDs.

Solar cells

1,2,3-Triazoles in solar cells

Over the last few decades, the economical production and high photovoltaic performance of dye-sensitized solar cells (DSSCs) have put them into limelight as the main research area to counter issues related to the energy production. The performance of DSSCs depends on the photosensitizer, which acts as the main structure for injection of electrons and harvesting of light. Thus, a number of attempts have been made to prepare photosensitizers with novel designs and structures to improve the performance of DSSCs.[17] For many years, dyes with ruthenium (Ru) complexes have remained a main choice for the preparation of highly efficient DSSCs; however, the availability issues of Ru have greatly affected its commercial development. Over the last few years, the metal-free organic-dye molecules have become suitable candidates for the preparation of DSSCs due to their high molecular extinction coefficient, facile synthesis, and convenient structural modifications.[18] In this connection, a variety of donor-π-acceptor organic dyes have been produced using different functional groups and their derivatives. The π-linker affects the recombination and transmission of electrons during the photoelectric conversion process by acting as a bridge between donor and acceptor subunits. Well-known synthetic strategies like Stille coupling and Suzuki coupling are employed to connect π-linkers with the donating groups. However, the applications of these synthetic approaches have lost their value due to the involvement of expensive catalysts and harsh reaction conditions.

Over the last two decades, click chemistry has emerged as the leading facile approach for the construction of organic dyes under milder conditions. Furthermore, the triazole has established its worth as an electron-deficient moiety with high potential to increase open-circuit voltage in DSSCs. In this connection, Duan *et al.* employed electron-deficient click triazoles to connect thienyl group and phenoxy group where π-linker as the product used in DSSCs (Figure 6).[19] The prepared dyes showed high molar

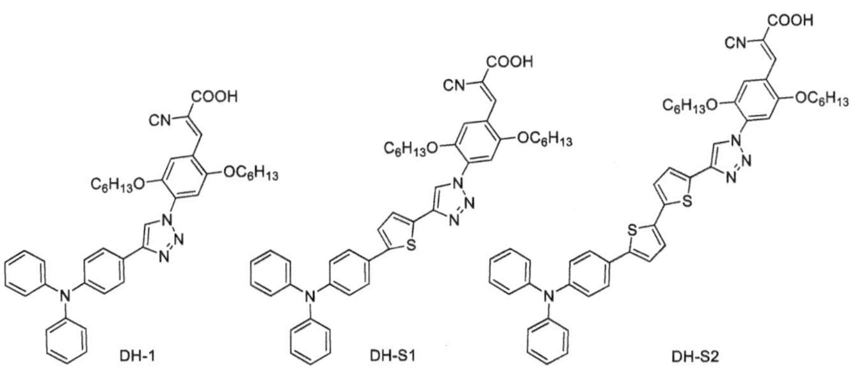

Figure 6 Click triazole-containing dyes for dye-sensitized solar cells.

extinction coefficient and superior intramolecular charge transfer compared with reference dyes. The open-circuit voltage was also improved owing to the presence of triazole, which favors charge separation and power conversion efficiency of dye molecules.

The unique features like mechanical flexibility, light weight, and low production cost have made polymer solar cells attractive choices for energy production. In these cells, an *n*-type semiconductor acceptor and a *p*-type conjugated polymer donor constitute a nanophase-separated bicontinuous network.[20] The synthetic flexibility of non-fullerene acceptors (NFAs) has inspired their applications compared with their fullerene counterparts in the last few years. They have been anticipated to find potential applications in future solar-based energy appliances. The NFAs present high photovoltaic performance, readily tunable energy levels, and low band gaps (LBGs) with good absorptions.[21] From the last few years, attention has been diverted for the preparation of high-performance donor WBG copolymers having strong absorption in short wavelength region. In this connection, very recently, Feng *et al.* prepared PBDTSF-TZNT and PBDTS-TZNT as two novel TZNT-containing WBG polymers (Figure 7).[22] They were prepared with two different side chains. The prepared NF polymers exhibited good molecular packing and high crystallinity due to the presence of rigid backbone of TZNT and BDT. The HOMO levels of both copolymers were lowered after fluorination and sulfuration at side chains of BDT. The device fabricated from PBDTSF-TZNT with IT-4F as acceptor exhibited high V_{oc} (0.93V) and lower E_{loss} (0.59 eV) when ITIC was used as

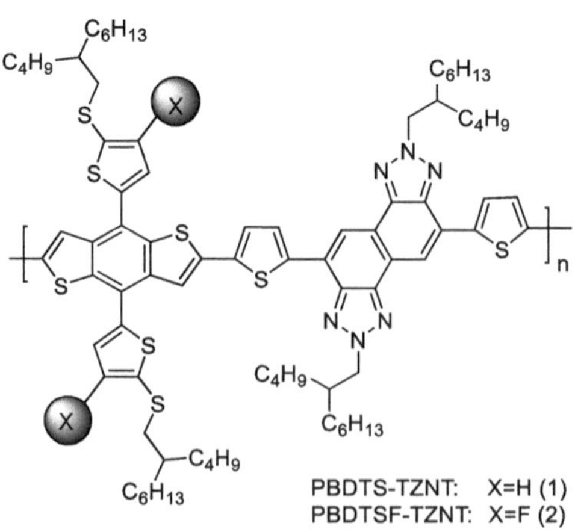

Figure 7 Wide band gap copolymers.

acceptor as it displayed a high V_{oc} (0.98V) and E_{loss} (0.61 eV). The PBDTSF-TZNT:IT-4F devices promoted power conversion efficiency (PCE) from 12.16 to 13.25% due to fine morphology, improved crystallinity, broadened absorption, and low loss of energy. The PCE was boosted to 14.25% with increased light-harvesting potential when HOMO-tandem devices of the same chemistry were used. This presented the best PCE value for non-fullerene polymer solar cells.

1,2,4-Triazoles in solar cells

Over the last decade, attention has been diverted to organometal halide perovskite solar cells due to their ease of availability through a solution-processing technique and they also showed more than 18% PCE. In the last few years, the photovoltaic properties of perovskite solar cells has greatly been influenced by the introduction of novel host-transporting materials (HTMs) instead of conventional band gap hybrid organic-inorganic perovskite structures.[23] A number of HTMs have been developed and utilized for the fabrication of perovskite cells with PCEs more than 15%. However, high production cost of HTMs has restricted their application on a larger scale. The triazine with its electron-deficient nature stabilized the devices by controlling the radical anion formation during light irradiation. The donor-acceptor structures improve hole transport through enhanced intramolecular charge transfer thus stabilizing electrons and holes dissociated from the excitons.[24] Choi *et al.* developed new HTMs using two electron-rich triphenylamine derivatives and triazole as core structure (Figure 8).[25] They incorporated two electron-rich diphenylamino side arms through thiophene bridges or direct linkage to prepare TAZ-[MeOTPATh]$_2$ and TAZ-[MeOPTA]$_2$. These novel HTMs with D-A structures showed effective ICT and high PCE of 10.9 and 14.4%, respectively, representing highly efficient photovoltaic performance in Perovskite-based cells. Hence, the prepared HTMs were found to be highly efficient and economical alternatives for the development of Perovskite solar cells.

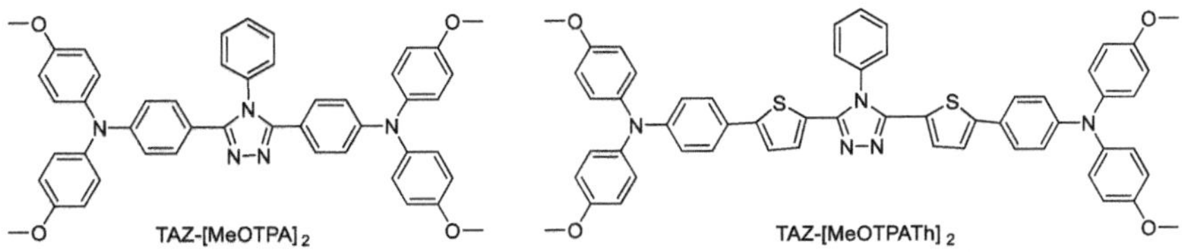

Figure 8 Triazole-containing novel hole-transporting materials.

High energy materials

1,2,4-Triazoles in high energy materials

The heat-resistant energetic materials (EMs) are frequently used in space rocket propellants and underground explosives for the provision of heat resistance even at very high temperatures. However, the commonly used EMs are losing interest due to their complex and non-economical synthetic routes. Over the last few years, it has really been desirous to find novel but economical heat-resistant materials.[26] During the last decade, the coordination bonding between organic linkers and organic metal centers has been well exploited for the preparation of hybrid and crystalline solid state materials known as metal-organic frameworks (MOFs).[27] They have tunable structures, good stability, and high energetic output with their 1D chains, 2D planes, or 3D architectures. Owing to these characteristic features, they are considered as promising candidates for the preparation of next-generation EMs. Accordingly, a variety of their applications in EMs established their value and worth in this regard.[28] Especially, the higher structural reinforcement in 3D-MOFs provide improved energy and heat of detonation with superior thermostability and larger mechanical strength. However, they find limited potential applications as they undergo decomposition at temperatures less than 400°C. Over the last few years, it has been a challenging task to develop a suitable coordination among explosion power, thermostability, and sensitivity by putting these contradictory features together in a single product.

Wang *et al.* prepared novel 3D-MOFs using 4-amino-4*H*-1,2,4-triazole-3,5-diol (ATDO) as a ligand with intentions to achieve a perfect coordination of aforesaid features (Figure 9).[29] They presumed that structural stability could be furnished by the rigidity of a triazolic ring, the explosive properties could be enhanced by the carbonyl oxygen, and the thermal stability could be achieved due to the presence of amino groups with their potential of *H*-bonding. Furthermore, the triazolic moiety with carbonyl oxygen and amino groups was considered as a better choice for the formation of

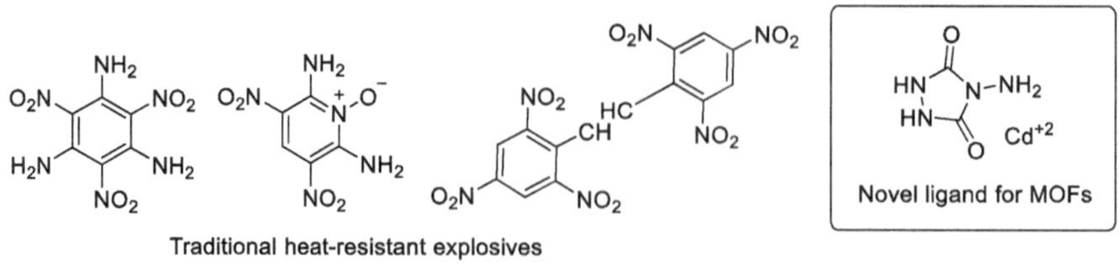

Figure 9 1,2,4-Triazole based-ligands for 3D-metal organic frameworks.

chelating structures rendering structural stability. Thus, they used ATDO as a ligand for the preparation of 3D chelating energetic MOF under milder hydrothermal conditions. The prepared MOF exhibited excellent detonation velocity and pressure, insensitivity to stimulation, good oxygen balance, and high crystal density. The chelating energetic metal-organic framework (CEMOF) was suggested to hold a promising future in heat-resistant explosives due to the powerful explosive performance, thermal stability, and practical synthetic procedure compared with traditional heat-resistant materials used in deep well mining and aerospace technology.

Over the years, density has been observed as one of the key features that could positively influence the performance of EMs. Thus, much interest has been shown for the economical synthesis of high-density energetic MOFs especially using nitrogen-rich ligands.[30] Traditionally, the introduction of nitro or nitroamino groups could easily improve the densities and thermal stabilities of EMs, but these approaches were rarely applied to MOFs. Dong *et al.* prepared two high density MOFs using 5,5′,-dinitro-3,3′-bis-1,2,4-triazole-1-diol (DNBTO) and 5,5′-dinitro-2*H*-2′*H*-3,3′-bi-1,2,4-triazole (DNBT) as rigid polynitro heterocyclic ligands under hydrothermal conditions.[31] The prepared MOFs, namely $[Cu(DNBTO)(ATRZ)_2(H_2O_2)_2]_n$ (1) and $[Cu(DNBT)(ATRZ)_3]_n$ (2), showed higher thermal stability and higher density, especially the introduction of N–O bond in complex 1. Both the MOFs exhibited low sensitivity toward friction and impact while 1 was found more insensitive. They were suggested as potential candidates with promising future applications as high-density energetic materials.

In addition to the previously explained features of MOFs as EMs, they have also been found as solid adsorbents with large capacity, high selectivity, and strong binding affinity. The unique selection of charged linkers and high valence metal centers render them with excellent stability toward acids, bases, and steam.[32] Very recently, Shi *et al.* used triazolate linkers with appending amino groups for the preparation of MOFs with excellent stability in acidic, basic, and aqueous conditions (Figure 10).[33] The best MOF exhibited high CO2 capture under humid condition, low regeneration energy, and notable cycling performance for flue gas mimic. The prepared MOFs showed CO_2/H_2O kinetic adsorption selectivity and CO_2/N_2 thermodynamic adsorption selectivity.

Figure 10 Triazolate linkers metal organic frameworks.

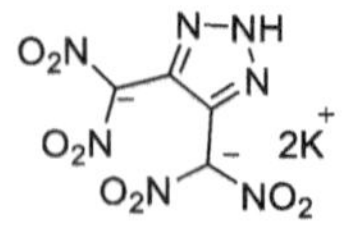

Figure 11 1,2,3-Triazole-based *gem*-dinitromethyl precursors for energetic MOF.

1,2,3-Triazoles in high-energy materials

Even the field of MOFs as EMs has progressed much over the last decade, but it remains challenging to synthesize energetic MOFs at ambient temperature and pressure. In recent years, fluorodinitro-explosophore and *gem*-dinitro groups have emerged as attractive candidates for the preparation of high-density energy materials.[34] Gu *et al.* aimed to prepare 1,2,3-triazole-based *gem*-dinitromethyl precursors for the preparation of *gem*-dinitromethyl EMs (Figure 11).[35] They synthesized gem-dinitromethyl-substituted dipotassium 4,5-bis(dinitromethyl)-1,2,3-triazole as energetic MOF through *in situ* controllable approach. The prepared high-density MOF showed superior detonation velocity comparable to RDX. It was suggested as green EM with promising explosive applications because nitrogen gas was the main detonation product in this case.

Functional oligomers and polymers

1,2,4-Triazoles in oligomers and polymers

The applications of organic semiconductors commonly rely on some physiochemical properties including the valence band (VB), conduction band (CB), and band gap. The band gap gives an idea about the required wavelength that photochemically activates the catalyst. The semiconductors with wide band gap are generally used when photochemical reactions follow reduction or oxidation mechanism. The visible part of sunlight could activate semiconductors with medium band gap for a variety of processes. The band tuning in organic semiconductors is commonly managed by the molecular doping (combining monomers and comonomers of different nature) or by elemental doping using S, B, and F.[36, 37] Savateev *et al.* realized that an organic semiconductor could be used for the preparation of another semiconductor with minimum structural changes from the precursor. Considering this fact, they selected an organic semiconductor 3-amino-1,2,4-triazole tertramer (OATA) with narrow band gap (1.7 eV) (Figure 12).[38] They achieved electronically modified graphitic carbon nitride (OATA-CN) by heating OATA at 550°C. During this transition,

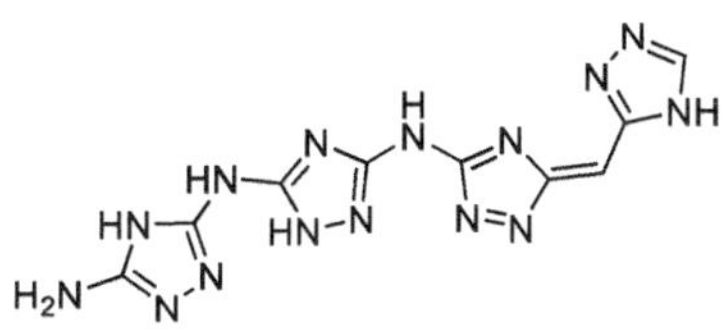

Figure 12 3-Amino-1,2,4-triazole tertramer as precursor of semiconductors.

the resulting carbon nitride (CN) holds the micro- and macrostructures of the precursor with slight morphological changes under heat treatment. In the resulting semiconductor, the optical band gap expanded to 2.2 eV from 1.8 eV while a slight shift was observed in flat band potential (from −0.11 to −0.06 eV). The OATA-CN retained its morphology even when OATA-based nanoparticles, nanobarrels, nanowires, and composite sheets were finally converted into it. The reference bulk carbon nitride prepared from pyrolysis of melamine showed five times less photocatalytic activity in visible light-driven hydrogen evolution than the carbon nitrides obtained from OATA. It was suggested as a promising candidate for the preparation of CN films with controlled morphology and structures for the fabrication of novel heterojunction composites.

Functional polymers containing silicon, acrylate, and azole functionalities show attractive physiochemical properties thus are broadly applied for the sorption of materials, catalysis, optoelectronic, and biomedical implants. Over the years, extensive research efforts have been focused for the extraction of platinum and palladium metals from various sources owing to their extensive utility in catalysis, fuel cells, jewelry, and electrical appliances. The development of polymeric materials for their effective extraction has become an attractive research area. Polymers with *N*-heterocycles have shown a strong tendency for the sorption of these metals owing to their electronic and structural features.[39] The triazole-containing polymers show high sorption capacity due to effective complexing potential for heavy metals. The 1-vinyl-1,2,4-triazole induces many promising properties like thermal stability, chemical stability, non-toxicity, and hydrophilicity in copolymers.[40] Therefore, polymers containing 1-vinyl-1,2,4-triazole functionality find wide applications in a number of scientific domains. Very recently, Prozorova *et al.* used radical polymerization to prepare water-insoluble, cross-linked copolymers using (trimethoxysilyl)methyl-2-methacrylate and 1-vinyl-1,2,4-triazole monomers (Figure 13).[41] The prepared copolymers showed high sorption capacity for Pt(IV) and Pd(II) ions. They exhibited high complexing potential, thermal stability, and resistance to acids. The

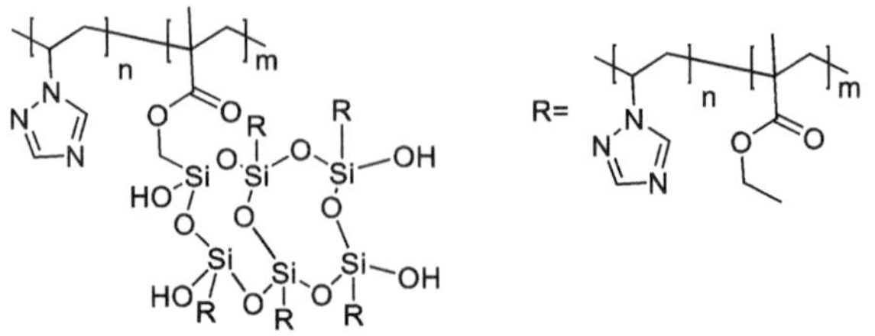

Figure 13 Cross-linked copolymer as sorption agent.

triazole-containing copolymer displayed high sorption capacity even in very dilute samples, making it a promising choice for the fabrication of active sorbents on a large scale.

Over the last few decades, designing and synthesis of functional conducting polymers have received great attention for their potent applications in optical devices, sensors, and detectors of biomolecules. Over the last few years, the designing and synthesizing of polymers with novel functionalities have improved their spectrochemical and electrochemical properties with broadening of electrochromic and technological applications. They find a broad scope of applications as LEDs, batteries, supercapacitors, and solar cells.[42, 43]

Purpald is an attractive material with multi-functionality in its structure. It has not been exploited in the field of functional conductive polymers. Very recently, Gumusay *et al.* synthesized purpald 4-amino-5-((2,5-di(thiophen-2-yl)-1*H*-pyrol-1-yl)amino)-4*H*-1,2,4-triazole-3-thiol (TPTP) with an electroactive monomer (Figure 14).[44] It was converted into functional conductive polymer (pTPTP) by electropolymerization. The amino and thiol groups in the polymeric backbone were functionalized to prepare multifunctional conductive polymer films. These films produced 75% optical contrast when applied in smart windows with good electrochemical stability. The prepared conducting polymer was suggested as a suitable candidate for promising applications in electrochronic devices.

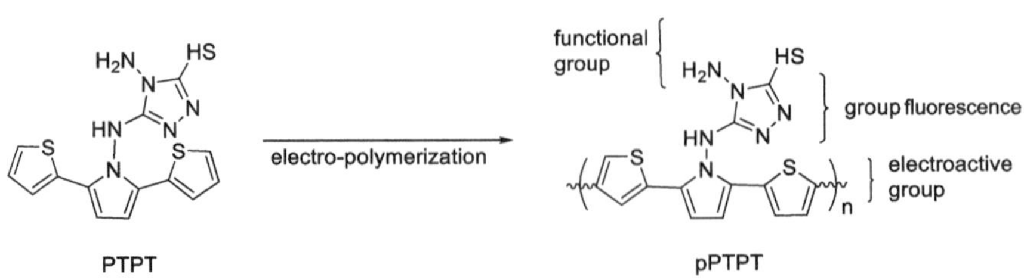

Figure 14 Synthesis of functional conductive polymer.

1,2,3-Triazoles in oligomers and polymers

Over the last decade, great attention has been focused on the development of electrical memory polymers as an alternative to metal-oxide- and silicon-based electrical memory materials. Advantageously, they could be produced as multilayer miniature structures with highly tunable properties controlled by synthetic process. Most of the electrical memory polymers carry non-conjugated backbones with electron donor or donor-acceptor units or they contain fully π-conjugated backbones.[45] The triazole ring with its unique structural and electronic features acts as a charge trap site due to its high hole affinity. The click chemistry-based easy synthesis of triazole along with its electrochemical features prompted researchers to use them for the preparation of high-performance electrical memory polymers. Song *et al.* reacted alkyne-bearing poly(4-azidomethylstyrene) and poly(glycidyl azide) under click conditions for the preparation of triazole containing brush polymers (Figure 15).[46] Their electrical memory performance was studied in the form of thin films. The prepared polymers with triazole, substituted with phenyl or its derivatives exhibited high ON/OFF current ratios, reliable stability, and permanent memory behaviors with low power consumptions. However, as-prepared polymers with alky or alkynylphenyl substituted triazole only show dielectric characteristics. The introduction of substituent on phenyl moiety at triazole could be used to tune *n*- or *p*-type properties of these electrical memory polymers. This methodology

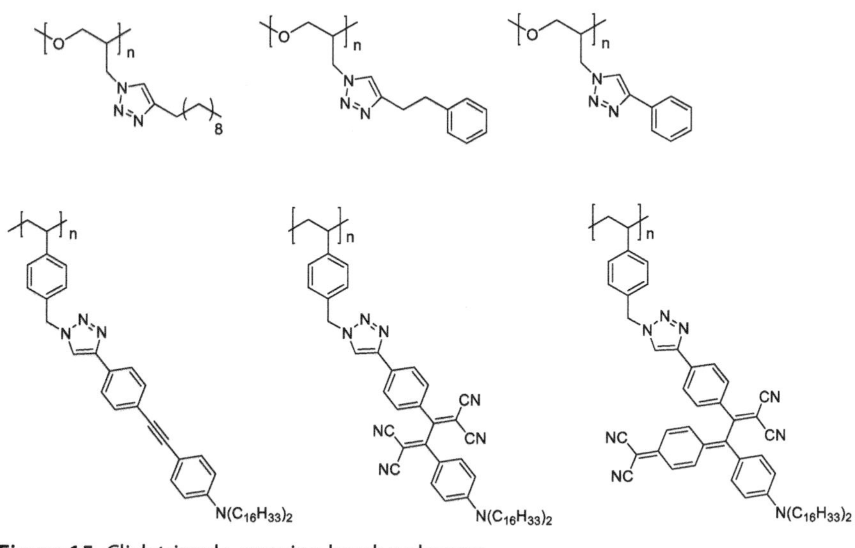

Figure 15 Click triazole-carrying brush polymers.

presented an opportunity to prepare high-performing programmable memory devices on an economical scale.

The click chemistry approach has also been applied for the preparation of fluorescent polymers. In a recent report, Ngororabanga *et al.* employed click chemistry to prepare curcumin-based fluorescent polymer. The polymer was produced using step-growth polymerization of non-fluorescent 3-azidocurcumin-alkyne monomers under click conditions. The prepared polymer exhibited fluorescent properties after the construction of a triazolic ring. An aliphatic comonomer was utilized in a different ratio to tailor the solubility and photophysical properties of the prepared fluorescent polymers.

Ionic liquids

1,2,4-Triazoles in ionic liquids

In the last decade, the ionic liquids have received considerable attention in synthetic chemistry, electrochemical, and DSSCs.[47] They have shown promising applications in other domains of material sciences. Initially, a great degree of flexibility and a wide scope of applicability was exhibited by alkyl- or aryl-substituted imidazolium salts as ionic liquids. Almost a decade ago, Meyer *et al.* came up with a pioneering effort for the preparation of alkyl- or aryl-substituted 1,2,4-triazolium salts as ionic liquids with expansion in their scope of applications (Figure 16).[48] They employed a one-pot synthesis with diformylhydrazine or a coupling reaction for the preparation of phenyl-substituted 4*H* and 1*H*-1,2,4-triazoles. Afterward, they were converted to 1,2,4-triazolium bromides after reaction with alkyl bromides. The inclusion of extra nitrogen in the heterocyclic core (imidazole to triazole) increased the melting point with high temperature stability of the IL. Furthermore, the bis(trifluoromethylsulfonyl)-imide salts were prepared as a result of anion exchange reactions.

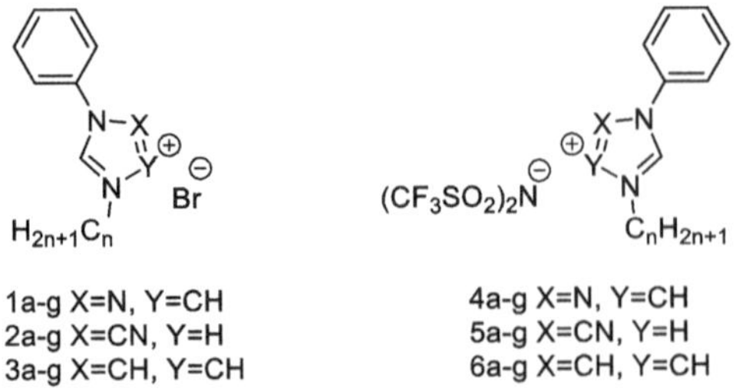

Figure 16 1,2,4-Triazolium salts as ionic liquids.

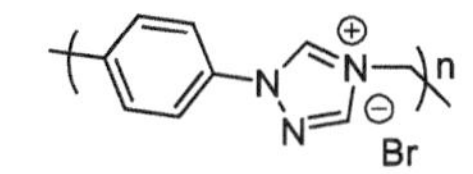

Figure 17 1,2,4-Triazolium-based poly(ionic liquid)s.

In the last few years, the poly(ionic liquid)s, also referred as polymeric ionic liquids (PILs) or polymerized ionic liquids, have received considerable attention in material science and polymer chemistry. They can be produced by the chemical modification of polymer backbone through ligation of reactive ILs or by *N*-alkylation of ion exchange reactions on polymeric materials.[49] The PILs have gained much attention in a shorter period of time owing to the blending of properties and functions of participating polymer and IL. Last year, Yahia *et al.* prepared 1,2,4-triazolium-based poly(ionic liquid)s with linear chain structure through a single-step facile methodology (Figure 17).[50] The commercially available monomer was made to undergo single-component polycondensation reaction at 120°C. Furthermore, different counterions were used to study the solubility and thermal stability profiles of a series of PILs. The anionically charged colloidal cellulose nanocrystals (CNC) were functionalized with PIL to prepare PIL@CNC hybrid showing dispersion in hydrophilic and hydrophobic solvents. This simple method is considered as a welcome advance in PIL synthesis, which will broaden its applications in materials.

1,2,3-Triazole in ionic liquids

The fully organic ionic liquids contain both anions and cations of an organic nature and are usually less common in their applications. A fascinating novel class of ionic liquids has been introduced by using 1,2,3- and 1,2,4-triazoles as cationic and anionic parts. These fully organic ILs offer the advantage of high *H*-bonding with solute due to the presence of many *N*-atoms and also the chemical inertness of triazoles rendering more chemical stability to the system. The task-specific ILs could be prepared using differently substituted triazolium cations and stabilizing them through conjugation with lone pairs of nitrogen. The triazole-based ILs have become an attractive choice as they avoid the deprotonation of C as side reactions, which are observed frequently in the case of imidazolium-based ILs.[51, 52] Savateev *et al.* prepared fully organic ILs employing 1,2,4-triazolium or imidazolium cations along with 1,2,4- or 1,2,3-triaazolide as counter anions (Figure 18).[53] The CN- and NO_2-substituted triazoles were selected as anion precursors as they were able to stabilize −ve charge through mesomeric

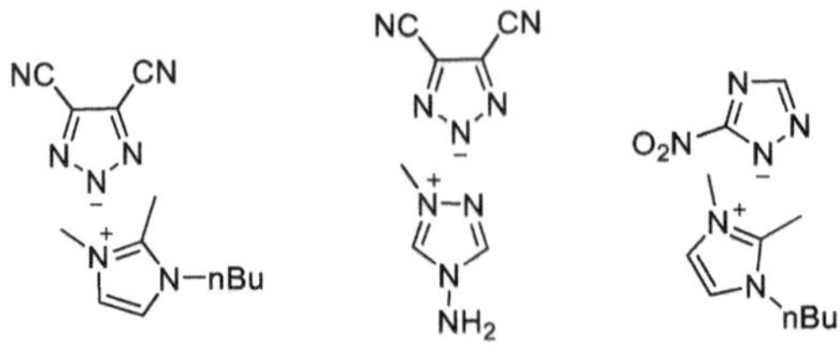

Figure 18 1,2,4-Triazolium-based fully organic ionic liquids.

effect. The as-prepared triazole-based ILs exhibited higher conductivity and thermal and electrochemical stability compared with the known ILs. They were also tested for viscosity, density, and glass transition temperature. The studied properties of the prepared ILs recommended them as a promising choice for energy storage applications.

As described earlier, modern light-emitting technology heavily relies on the promising features of OLEDs. Advantageously, they could be produced over flexible substrates, produce high brightness, and are highly efficient and energetic. Usually, OLEDs are developed as a complex multilayer structure of an emission layer, charge barrier layer, charge transport layer, and charge injection layer. However, fabrication of OLEDs with complex structures produces issues of economical production on a large scale. In recent years, organic electronics have attracted light-emitting electrochemical cells (LEECs) as promising alternatives to OLEDs.[54] In general, LEECs are made using a mixture of polymeric electrolyte and light-emitting material. A variety of emitting materials including complexes of transition metals are commonly used that pose serious pollution threats even at high production costs. In the last few years, the ionic liquids, owing to their promising features, have been employed as alternatives of polymeric blends in LEECs.[55] A variety of ILs are available, and their properties depend on their chemical structures. In this context, recent developments of triazolium-based ILs have broadened their field of applications.

Stroppa *et al.* synthesized 1,2,3-triazolium-based new ILs in good yields and used them as emitting polymer in LEECs (Figure 19).[56] The fabricated device was tested for its performance considering dohmt concentration and active layer thickness. The best performance was exhibited by a device with 4% dohmt concentration and 80 mm of emitting material layer thickness. The prepared ILs managed equal charge distribution in a monolayer LEECs device.

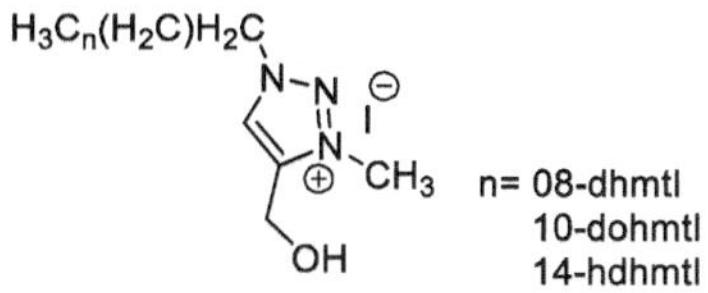

Figure 19 1,2,3-Triazolium-based ionic liquids.

Conclusion

Some very recent and noteworthy examples of functional materials have been presented here that were synthesized using 1,2,4- as well as 1,2,3-triazoles as functional cores. Recent advances in their synthetic methodologies accelerated their usage in solar cell, supercapacitors, bipolar materials, heat-resistant materials, and green explosives, etc. It is envisioned that triazole chemistry has emerged as a leading approach for the fabrication of functional materials inspiring almost all walks of daily life.

References

1. Diaz-Ortiz A, Prieto P, R Carrillo J, Martin R, Torres I. Applications of metal-free 1, 2, 4-triazole derivatives in materials science. *Curr Org Chem* 2015;**19**(7):568–84.
2. Juríček M, Kouwer PH, Rowan AE. Triazole: a unique building block for the construction of functional materials. *Chem Commun* 2011;**47**(31):8740–9.
3. Huo J, Hu H, Zhang M, Hu X, Chen M, Chen D, et al. A mini review of the synthesis of poly-1, 2, 3-triazole-based functional materials. *RSC Adv* 2017;**7**(4):2281–7.
4. Xi W, Scott TF, Kloxin CJ, Bowman CN. Click chemistry in materials science. *Adv Funct Mater* 2014;**24**(18):2572–90.
5. Jou J-H, Kumar S, Agrawal A, Li T-H, Sahoo S. Approaches for fabricating high efficiency organic light emitting diodes. *J Mater Chem C* 2015;**3**(13):2974–3002.
6. Zhao F, Ma D. Approaches to high performance white organic light-emitting diodes for general lighting. *Mater Chem Front* 2017;**1**(10):1933–50.
7. Cheng T-Y, Lee J-H, Chen C-H, Chen P-H, Wang P-S, Lin C-E, et al. Carrier transport and recombination mechanism in blue phosphorescent organic light-emitting diode with hosts consisting of cabazole-and triazole-moiety. *Sci Rep* 2019;**9**(1):1–12.
8. Wu PL, Feng XJ, Tam HL, Wong MS, Cheah KW. Efficient three-photon excited deep blue photoluminescence and lasing of diphenylamino and 1, 2, 4-triazole endcapped oligofluorenes. *J Am Chem Soc* 2009;**131**(3):886–7.
9. Kang HM, Jo SM, Lee D, Shaik B, Han JH, Choi H, et al. Synthesis, physical and electroluminescent properties of [1, 2, 4]-triazolo [4, 3-a]-pyridine based bipolar red host materials and their applications in organic light-emitting diodes. *JOL* 2018;**196**:470–6.
10. Chatterjee T, Hung W-Y, Tang W-F, Chen H-F, Wong K-T. Carbazole-bridged triphenylamine-bipyridine bipolar hosts for high-efficiency low roll-off multi-color PhOLEDs. *Org Electron* 2017;**50**:204–12.
11. Wang F, Zhao Y, Xu H, Zhang J, Miao Y, Guo K, et al. Two novel bipolar hosts based on 1, 2, 4-triazole derivatives for highly efficient red phosphorescent OLEDs showing a small efficiency roll-off. *Org Electron* 2019;**70**:272–8.
12. Tao Y, Yang C, Qin J. Organic host materials for phosphorescent organic light-emitting diodes. *Chem Soc Rev* 2011;**40**(5):2943–70.

13. Kautny P, Lumpi D, Wang Y, Tissot A, Bintinger J, Horkel E, et al. Oxadiazole based bipolar host materials employing planarized triarylamine donors for RGB PHOLEDs with low efficiency roll-off. *J Mater Chem C* 2014;**2**(11):2069–81.
14. Kautny P, Zhao C, Kader T, Stöger B, Horkel E, Chen J, et al. Functional organic click-materials: application in phosphorescent organic light emitting diodes. *RSC Adv* 2017;**7**(20):12150–60.
15. Wang L, He W, Wang M, Wei M, Sun J, Chen X, et al. Effects of symmetrically 2, 5-disubstituted 1, 3, 4-oxadiazoles on the temperature range of liquid crystalline blue phases: a systematic study. *Liq Cryst* 2013;**40**(3):354–67.
16. Girotto E, Bechtold IH, Gallardo H. New liquid crystals derived from thiophene connected to the 1, 2, 3-triazole heterocycle. *Liq Cryst* 2015;**42**(12):1798–807.
17. Zhang S, Yang X, Numata Y, Han L. Highly efficient dye-sensitized solar cells: progress and future challenges. *Energ Environ Sci* 2013;**6**(5):1443–64.
18. Chou CC, Wu KL, Chi Y, Hu WP, Yu SJ, Lee GH, et al. Ruthenium (II) sensitizers with heteroleptic tridentate chelates for dye-sensitized solar cells. *Angew Chem Int Ed* 2011;**50**(9):2054–8.
19. Duan T, Fan K, Li K, Cao W, Zhong C, Chen X, et al. Design of organic dyes for dye-sensitized solar cells: extending π-conjugation backbone via 'Click'reaction to improve photovoltaic performances. *Dyes Pigments* 2015;**117**:108–15.
20. Yan C, Barlow S, Wang Z, Yan H, Jen AK-Y, Marder SR, et al. Non-fullerene acceptors for organic solar cells. *Nat Rev Mater* 2018;**3**(3):1–19.
21. Zhang G, Zhao J, Chow PC, Jiang K, Zhang J, Zhu Z, et al. Nonfullerene acceptor molecules for bulk heterojunction organic solar cells. *Chem Rev* 2018;**118**(7):3447–507.
22. Feng K, Yuan J, Bi Z, Ma W, Xu X, Zhang G, et al. Low-energy-loss polymer solar cells with 14.52% efficiency enabled by wide-band-gap copolymers. *IScience* 2019;**12**:1–12.
23. Jung HS, Park NG. Solar cells: perovskite solar cells: from materials to devices (small 1/2015). *Small* 2015;**11**(1). 2-2.
24. Li Y, Liu T, Liu H, Tian M-Z, Li Y. Self-assembly of intramolecular charge-transfer compounds into functional molecular systems. *Acc Chem Res* 2014;**47**(4):1186–98.
25. Choi H, Jo H, Paek S, Koh K, Ko HM, Lee JK, et al. Efficient hole-transporting materials with triazole core for high-efficiency perovskite solar cells. *Chem Asian J* 2016;**11**(4):548–54.
26. Klapötke TM, Stierstorfer J, Weyrauther M, Witkowski TG. Synthesis and investigation of 2, 6-Bis (picrylamino)-3, 5-dinitro-pyridine (PYX) and its salts. *Chem A Eur J* 2016;**22**(25):8619–26.
27. Li N, Xu J, Feng R, Hu T-L, Bu X-H. Governing metal–organic frameworks towards high stability. *Chem Commun* 2016;**52**(55):8501–13.
28. Wang S, Wang Q, Feng X, Wang B, Yang L. Explosives in the cage: metal–organic frameworks for high-energy materials sensing and desensitization. *Adv Mater* 2017;**29**(36):1701898.
29. Wang Q, Wang S, Feng X, Wu L, Zhang G, Zhou M, et al. A heat-resistant and energetic metal–organic framework assembled by chelating ligand. *ACS Appl Mater Interfaces* 2017;**9**(43):37542–7.
30. Yin P, Parrish DA, Shreeve JNM. Energetic multifunctionalized nitraminopyrazoles and their ionic derivatives: ternary hydrogen-bond induced high energy density materials. *J Am Chem Soc* 2015;**137**(14):4778–86.
31. Dong Y, Peng P, Hu B, Su H, Li S, Pang S. High-density energetic metal–organic frameworks based on the 5, 5′-dinitro-2H, 2′ H-3, 3′-bi-1, 2, 4-triazole. *Molecules* 2017;**22**(7):1068.
32. Boyd PG, Chidambaram A, García-Díez E, Ireland CP, Daff TD, Bounds R, et al. Data-driven design of metal–organic frameworks for wet flue gas CO_2 capture. *Nature* 2019;**576**(7786):253–6.

33. Shi Z, Tao Y, Wu J, Zhang C, He H, Long L, et al. Robust metal–triazolate frameworks for CO_2 capture from flue gas. *J Am Chem Soc* 2020;**142**(6):2750–4.
34. Li H, Zhao F-Q, Gao H-X, Tong J-F, Wang B-Z, Zhai L-J, et al. Three energetic salts based on oxy-bridged bis (gem-dinitro) furazan: syntheses, structures and thermal behaviors. *Inorg Chim Acta* 2014;**423**:256–62.
35. Gu H, Ma Q, Huang S, Zhang Z, Zhang Q, Cheng G, et al. Gem-dinitromethyl-substituted energetic metal–organic framework based on 1, 2, 3-triazole from in situ controllable synthesis. *Chem Asian J* 2018;**13**(19):2786–90.
36. Wang Y, Di Y, Antonietti M, Li H, Chen X, Wang X. Excellent visible-light photocatalysis of fluorinated polymeric carbon nitride solids. *Chem Mater* 2010;**22**(18):5119–21.
37. Chu S, Wang Y, Guo Y, Feng J, Wang C, Luo W, et al. Band structure engineering of carbon nitride: in search of a polymer photocatalyst with high photooxidation property. *ACS Catal* 2013;**3**(5):912–9.
38. Savateev A, Pronkin S, Epping JD, Willinger MG, Antonietti M, Dontsova D. Synthesis of an electronically modified carbon nitride from a processable semiconductor, 3-amino-1, 2, 4-triazole oligomer, via a topotactic-like phase transition. *J Mater Chem A* 2017;**5**(18):8394–401.
39. Ermakova T, Schaulina L, Kuznetsova N, Ratovsky G, Uzlova E, Prozorova G. Sorption concentration of palladium by a nitrogen-containing heterocyclic copolymer. *Russ Chem Bull* 2017;**66**(6):1104–8.
40. Pozdnyakov A, Sekretarev E, Emel'yanov A, Prozorova G. Hydrophilic functional copolymers of 1-vinyl-1, 2, 4-triazole with vinylsulfonic acid sodium salt. *Russ Chem Bull* 2017;**66**(12):2293–7.
41. Prozorova G, Kuznetsova N, Shaulina L, Bolgova Y, Trofimova O, Emel'yanov A, et al. Synthesis and sorption activity of novel cross-linked 1-vinyl-1, 2, 4-triazole–(trimethoxysilyl) methyl-2-methacrylate copolymers. *J Organomet Chem* 2020;**916**:121273.
42. Guzel M, Ak M. A solution-processable electrochromic polymer designed with reactive yellow 160 and 2-hydroxy carbazole. *Org Electron* 2019;**75**:105436.
43. Han Y, Dai L. Conducting polymers for flexible supercapacitors. *Macromol Chem Phys* 2019;**220**(3):1800355.
44. Gumusay O, Soganci T, Durur S, Soyleyici HC, Cetisli H, Ak M. Purpald containing poly (2, 5-dithienylpyrrole)-based multifunctional conducting polymer: synthesis, characterization, and electrochromic properties. *Ionics* 2020;**26**:1–11.
45. Lin WP, Liu SJ, Gong T, Zhao Q, Huang W. Polymer-based resistive memory materials and devices. *Adv Mater* 2014;**26**(4):570–606.
46. Song S, Ko Y-G, Lee H, Wi D, Ree BJ, Li Y, et al. High-performance triazole-containing brush polymers via azide–alkyne click chemistry: a new functional polymer platform for electrical memory devices. *NPG Asia Mater* 2015;**7**(12):e228.
47. Lau GP, Décoppet J-D, Moehl T, Zakeeruddin SM, Grätzel M, Dyson PJ. Robust high-performance dye-sensitized solar cells based on ionic liquid-sulfolane composite electrolytes. *Sci Rep* 2015;**5**(1):1–8.
48. Meyer D, Strassner T. 1, 2, 4-triazole-based tunable aryl/alkyl ionic liquids. *J Org Chem* 2011;**76**(1):305–8.
49. Zhou X, Weber J, Yuan J. Poly (ionic liquid) s: platform for CO_2 capture and catalysis. *Curr Opin Green Sustain Chem* 2019;**16**:39–46.
50. Yahia M, Mei S, Mathew AP, Yuan J. Linear main-chain 1, 2, 4-triazolium poly (ionic liquid) s: single-step synthesis and stabilization of cellulose nanocrystals. *ACS Macro Lett* 2019;**8**(10):1372–7.
51. Jeong Y, Ryu J-S. Synthesis of 1, 3-dialkyl-1, 2, 3-triazolium ionic liquids and their applications to the Baylis–Hillman reaction. *J Org Chem* 2010;**75**(12):4183–91.
52. Yue C, Fang D, Liu L, Yi T-F. Synthesis and application of task-specific ionic liquids used as catalysts and/or solvents in organic unit reactions. *J Mol Liq* 2011;**163**(3):99–121.

53. Savateev A, Liedel C, Tröger-Müller S, de León AS, Antonietti M, Dontsova D. Halogen free 1, 2, 3-and 1, 2, 4-triazolide based ionic liquids: synthesis and properties. *Chem Commun* 2017;**53**(73):10192–5.
54. Tang S, Edman L. Light-emitting electrochemical cells: a review on recent progress. In: *Photoluminescent materials and electroluminescent devices.* Springer; 2017. p. 375–95.
55. Sunesh CD, Ok S, Mathai G, Choe Y. Electroluminescent properties of yellow light-emitting electrochemical cells based on a cationic iridium complex and the effect of ionic liquids incorporation in an active layer. *Thin Solid Films* 2013;**531**:530–4.
56. Stroppa PHF, Martins JS, Dias RC, Salla CA, Bechtold IH, Legnani C, et al. High efficient light-emitting electrochemical cells based on ionic liquids 1, 2, 3-triazolium. *Org Electron* 2019;**73**:172–81.

Index

Note: Page numbers followed by *f* indicate figures and *s* indicate schemes.

O

P

Q

R

S

Kolibri 360 GmbH
Ferdinand-Jühlke-Straße 7
99095 Erfurt, Deutschland
produktsicherheit@kolibri360.de